METHODS EXPRESS

PCR

edited by **S. Hughes**

Centre for Tumour Biology,
Institute of Cancer and CR-UK Clinical Centre,
London, UK

and **A. Moody**

AstraZeneca, Macclesfield, UK

Scion

© Scion Publishing Ltd, 2007

First published 2007

A CIP catalogue record for this book is available from the British Library.

ISBN: 978 1 904842 28 6 (paperback)
ISBN: 978 1 904842 29 3 (hardback)

Scion Publishing Limited
Bloxham Mill, Barford Road, Bloxham, Oxfordshire OX15 4FF
www.scionpublishing.com

Important Note from the Publisher

The information contained within this book was obtained by Scion Publishing Limited from sources believed by us to be reliable. However, while every effort has been made to ensure its accuracy, no responsibility for loss or injury whatsoever occasioned to any person acting or refraining from action as a result of information contained herein can be accepted by the authors or publishers.

Typeset by Phoenix Photosetting, Chatham, Kent, UK
Printed by Ajanta Offset and Packagings Ltd, Delhi, India

Books are to be returned on or before
the last date below.

PCR

METHODS EXPRESS

The **METHODS EXPRESS** series

Series editor: B. David Hames

Faculty of Biological Sciences, University of Leeds, Leeds LS2 9JT, UK

METHODS EXPRESS

Bioinformatics

Biosensors

Cell Imaging

DNA Microarrays

Expression Systems

Genomics

Immunohistochemistry

PCR

Protein Arrays

Proteomics

Whole Genome Amplification

The editors would like to dedicate this book to the late Nat Bumstead,
a friend and mentor to us both.

*'Most of the fundamental ideas of science are essentially simple, and may, as
a rule, be expressed in language comprehensible to everyone.'* Albert Einstein

MX

Contents

Chapter 3.

A detailed guide to quantitative RT-PCR

Pete Kaiser

Chapter 4.

Use of quantitative PCR for the detection of genomic microdeletions or microduplications

Simon Hughes, Rosanna Weksberg, Laura Moldovan, and Jeremy A. Squire

Chapter 5.
Robust and unique PCR for single-nucleotide polymorphism genotyping applications
Xiangning Chen

Chapter 6.
Using PCR and linkage mapping to identify single genes and quantitative trait loci for livestock traits
Jillian F. Maddox, Imke Tammen, and Sonja Dominik

Chapter 7.
PCR restriction fragment length polymorphism analysis for genotyping of single-nucleotide polymorphisms
Simon Hughes

Chapter 11.

Rapid generation of gene-targeting constructs

Trevor J. Wilson, Dirk Truman, Antonietta Giudice, and Paul Hertzog

Chapter 12.

Construction of long DNA molecules from multiple fragments using PCR

Nikolai A. Shevchuk and Anton V. Bryksin

Chapter 13.

Efficient PCR-based mutagenesis method applicable to diverse mutagenesis strategies using type IIs restriction enzymes

Jae-Kyun Ko and Jianjie Ma

Chapter 14.

Inverse PCR-based restriction fragment length polymorphism for identifying low-level mutations in tumors

G. Mike Makrigiorgos

Chapter 15.

PCR methods for infectious disease diagnosis

Padmini Ramachandran, Andrew Hardick, Charlotte Gaydos, Samuel Yang, and Richard Rothman

Chapter 16.

Use of PCR for DNA methylation analyses

Mario F. Fraga and Manel Esteller

Chapter 17.

PCR–based methods to determine DNA methylation status at specific CpG sites using methylation-sensitive restriction enzymes

Helmtrud I. Roach and Ko Hashimoto

Chapter 18.
PCR-based whole genome amplification
Nona Arneson, Simon Hughes, Richard Houlston, and Susan Done

Chapter 19.
PCR sequencing of human genes for the discovery of DNA sequence variants
Abizar Lakdawalla

Appendix 1
List of suppliers

Index

Contributors

Nona Arneson, Division of Applied Molecular Oncology, Ontario Cancer Institute, Princess Margaret Hospital, 610 University Avenue, Toronto, Ontario, M5G 2M9, Canada. E-mail: arneson@uhnres.utoronto.ca

Simon Baker, ABgene, Blenheim Road, Epsom Surrey KT19 9AP, UK; and School of Biological & Molecular Sciences, Oxford Brookes University, Gipsy Lane, Oxford OX3 0BP, UK. E-mail: simon.baker@brookes.ac.uk

Claus Børsting, Department of Forensic Genetics, Institute of Forensic Medicine, University of Copenhagen, DK-2100 Copenhagen, Denmark. E-mail: claus.boersting@forensic.ku.dk

Anton V. Bryksin, Department of Microbiology and Immunology, New York Medical College, Valhalla, NY, USA; and Institute of Chemical Biology and Fundamental Medicine, 8 Lavrent'eva St, Novosibirsk, 630090, Russia. E-mail: anton_bryksin@gorodok.net

Xiangning Chen, Virginia Institute for Psychiatric and Behavioral Genetics and the Department of Psychiatry, Virginia Commonwealth University, Richmond, VA 23298, USA. E-mail: xchen@vcu.edu

Sonja Dominik, CSIRO Livestock Industries, Locked Bag 1, Armidale, New South Wales 2350, Australia. E-mail: sonja.dominik@csiro.au

Susan Done, Division of Applied Molecular Oncology, Ontario Cancer Institute, Princess Margaret Hospital, 610 University Avenue, Toronto, Ontario, M5G 2M9, Canada. E-mail: sdone@uhnres.utoronto.ca

Manel Esteller, Cancer Epigenetics Laboratory, Molecular Pathology Programme, Spanish National Cancer Centre (CNIO), Melchor Fernandez Almagro 3, 28029 Madrid, Spain. E-mail: mesteller@cnio.es

Mario F. Fraga, Cancer Epigenetics Laboratory, Molecular Pathology Programme, Spanish National Cancer Centre (CNIO), Melchor Fernandez Almagro 3, 28029 Madrid, Spain. E-mail: mffraga@cnio.es

Charlotte Gaydos, Johns Hopkins University, Division of Infectious Diseases, Baltimore, MD, USA. E-mail: cgaydos@jhmi.edu

Antonietta Giudice, Monash Immunology and Stem Cell Laboratories, STRIP Building 75, Monash University, Wellington Rd, Clayton 3800, Victoria, Australia. E-mail: antonietta.giudice@med.monash.edu.au

Andrew Hardick, Johns Hopkins University, Division of Infectious Diseases, Baltimore, MD, USA. E-mail: ahardic1@jhmi.edu

Ko Hashimoto, Department of Orthopaedic Surgery, Tohoku University School of Medicine, Sendai, Japan. E-mail: hasshie@mail.tains.tohoku.ac.jp

Paul Hertzog, CRC for Chronic Inflammatory Diseases and Centre for Functional Genomics and Human Disease, Monash Institute of Medical Research, Monash University, 27–31 Wright St, Clayton 3168, Victoria, Australia. E-mail: paul.hertzog@med.monash.edu.au

Richard Houlston, Section of Cancer Genetics, Institute of Cancer Research, 15 Cotswold Road, Surrey, SM2 5NG, UK.

Simon Hughes, Centre for Tumour Biology, Institute of Cancer and CR-UK Clinical Centre, Bart's and The London, Queen Mary's School of Medicine and Dentistry, Ground Floor, John Vane Science Centre, Charterhouse Square, London EC1M 6BQ, UK. E-mail: simon.hughes@cancer.org.uk

Pete Kaiser, Institute for Animal Health, Compton, Berkshire RG20 7NN, UK. E-mail: pete.kaiser@bbsrc.ac.uk

Ian Kavanagh, ABgene, Blenheim Road, Epsom, Surrey KT19 9AP, UK. E-mail: ian.kavanagh@thermofisher.com

Jae-Kyun Ko, Department of Physiology and Biophysics, UMDNJ-Robert Wood Johnson Medical School, Piscataway, NJ 08854, USA. E-mail: koja@umdnj.edu

Abizar Lakdawalla, Applied Biosystems, 45 West Gude Drive, Rockville, MD 20850, USA. E-mail: abizar.a.lakdawalla@appliedbiosystems.com

Jianjie Ma, Department of Physiology and Biophysics, UMDNJ-Robert Wood Johnson Medical School, Piscataway, NJ 08854, USA. E-mail: maj2@umdnj.edu

Jillian F. Maddox, Department of Veterinary Science, University of Melbourne, Victoria, Australia. E-mail: jillm@rubens.its.unimelb.edu.au

G. Mike Makrigiorgos, Dana Farber/Brigham and Women's Cancer Center, Harvard Medical School, Boston, MA 02115, USA. E-mail: mmakrigiorgos@lroc.harvard.edu

Meg Martel, ABgene, Blenheim Road, Epsom, Surrey KT19 9AP, UK.
E-mail: megm@abgene.com

Simon May, ABgene, Blenheim Road, Epsom, Surrey KT19 9AP, UK.
E-mail: simon.may@thermofisher.com

Laura Moldovan, Program in Genetics and Genomic Biology, The Research
Institute, The Hospital for Sick Children, Toronto, Canada.

Adrian Moody, AstraZeneca, 19B13, Mereside, Alderley Park, Macclesfield,
Cheshire, SK10 4TG, UK. E-mail: adrian.moody@astrazeneca.com

Niels Morling, Department of Forensic Genetics, Institute of Forensic Medicine,
University of Copenhagen, DK-2100 Copenhagen, Denmark.
E-mail: niels.morling@forensic.ku.dk

Padmini Ramachandran, Johns Hopkins University, Department of Emergency
Medicine, Baltimore, MD, USA. E-mail: pramach1@jhmi.edu

Helmtrud I. Roach, Bone and Joint Research Group, University of Southampton,
Southampton SO16 6YD, UK. E-mail: hr@soton.ac.uk

Richard Rothman, Johns Hopkins University, Division of Infectious Diseases and
Department of Emergency Medicine, Baltimore, MD, USA.
E-mail: rrothman@jhmi.edu

Juan J. Sanchez, Department of Forensic Genetics, Institute of Forensic
Medicine, University of Copenhagen, DK-2100 Copenhagen, Denmark.
E-mail: jjsanchz@ull.es

Nikolai A. Shevchuk, Center for Cancer and Immunology Research, Children's
Research Institute, Washington, DC, USA; and Institute for Biomedical
Sciences/Program in Molecular and Cellular Oncology, The George Washington
University, Washington, DC, USA. E-mail: shevchook@hotmail.com

Jeremy A. Squire, Ontario Cancer Institute and Department of Laboratory
Medicine, Pathology and Medical Biophysics, University of Toronto, Toronto,
Canada. E-mail: jeremy.squire@utoronto.ca

Imke Tammen, Reprogen: Centre for Advanced Technologies in Animal Genetics
and Reproduction, Faculty of Veterinary Science, The University of Sydney,
425 Werombi Rd, PMB 3, Camden, New South Wales 2570, Australia.
E-mail: itammen@camden.usyd.edu.au

Dirk Truman, Centre for Functional Genomics and Human Disease, Monash
Institute of Medical Research, Monash University, 27–31 Wright St, Clayton
3168, Victoria, Australia. E-mail: dirk.truman@med.monash.edu.au

Rosanna Weksberg, Program in Genetics and Genomic Biology, The Research Institute, The Hospital for Sick Children, Toronto, Canada; and Division of Clinical and Metabolic Genetics, Department of Paediatrics, The Hospital for Sick Children, Toronto, Canada. E-mail: rosanna.weksberg@sickkids.ca

Dagan Wells, Yale University Medical School, Department of Obstetrics and Gynecology, 333 Cedar Street, New Haven, Connecticut 06520, USA. E-mail: dagan.wells@yale.edu

Trevor J. Wilson, CRC for Chronic Inflammatory Diseases and Centre for Functional Genomics and Human Disease, Monash Institute of Medical Research, Monash University, 27–31 Wright St, Clayton 3168, Victoria, Australia. E-mail: trevor.wilson@med.monash.edu.au

Samuel Yang, Johns Hopkins University, Department of Emergency Medicine, Baltimore, MD, USA. E-mail: syang10@jhmi.edu

Foreword

Polymerase chain reaction (PCR) methods began to appear in the literature in the early 1980s, with a variety of adaptations of the basic technique emerging over the subsequent 20 years. These adaptations have relevance to many fields including basic research, clinical investigations, and forensic science. In this volume of the Scion book series *Methods Express*, we present a collection of some of the most interesting adaptations and applications of PCR. The chapters are supported by references covering the development, testing, and validation of each method, along with examples of research applications. References of particular interest are also indicated with asterisks to guide the reader to papers of importance in understanding the uses, capabilities, and expected performance of each method.

In any method-based book, it important to realise that information is not knowledge. We once assumed the role of books was the transfer of knowledge, but in actual fact, what is being transferred is information, which requires a context for the reader to maximize learning. In the laboratory, if an experienced researcher sees you struggle, they can offer a solution or a technique they learned after being in a similar situation. The work of Michael Polanyi on learning (Polanyi was a chemist in early life) established a two-way look at knowledge: explicit knowledge is written down, as in a book or manual, whilst tacit knowledge is the result of experience and typically resides in the expert's mind and is difficult to collect and distil. The key lies in the ability to transfer expert, or tacit, knowledge, and in finding ways of capturing the lessons learned by experts after years of working in a scientific field, or with a particular technique or tool, such as PCR, and convey this as a written document. As PCR continues to grow in use and value across scientific disciplines, this book will provide an approach to pass the knowledge of most value on to readers working in the field.

PCR: Methods Express is an up-to-date compendium of techniques and approaches. More than a reference manual for PCR methods, it is also a deliberate attempt to capture and share tricks-of-the-trade, lessons learned, and 'simple solutions to common problems'. Tell me how to do this, but also tell me which of the possible approaches offered is best for my circumstances, and help me troubleshoot when things happen differently than expected. This volume provides just that: it provides the tacit knowledge of PCR.

Jeff Witherly
Author of *An A to Z of DNA Science: What Scientists Mean When They Talk About Genes and Genomes*

Preface

This book is intended to supply fundamental practical information for basic, clinical, and student researchers interested in using PCR methods in their research. It is structured not only to impart protocols, but also to illustrate the great variety of applications in which PCR plays a fundamental role.

The motivation for preparing this book came from the realization that an up-to-date, affordable book covering a wide array of the practical aspects of PCR techniques does not exist. The offerings of information in the text are intended to cater to the broad range of abilities among students, clinicians, and technologists, and hopefully will permit more exploratory experiments using this amazingly versatile tool.

We wish to thank all those who kindly gave of their time and skill to prepare the exceptional chapters herein. We would also like to acknowledge all the rest of the people who have made this book possible. We hope you find this text valuable and we welcome comments and ideas for future editions.

Simon Hughes & Adrian Moody
February 2007

Abbreviations

22q11DS	chromosome 22q11 deletion syndrome
ADO	allele dropout
BAC	bacterial artificial chromosome
BS	bisulfite sequencing
BSA	bovine serum albumin
CCD	charge-coupled device
CE	capillary electrophoresis
CGH	comparative genomic hybridization
CODIS	Combined DNA Indexing System
CSF	cerebrospinal fluid
C_T	cycle threshold
DMSO	dimethyl sulphoxide
dNTP	deoxynucleoside triphosphate
ddNTP	dideoxynucleoside triphosphate
DOP-PCR	degenerate-oligonucleotide-primed PCR
EDNAP	European DNA Profiling Group
EDTA	ethylenediaminetetraacetic acid
EGFP	enhanced green fluorescent protein
EMPOP	EDNAP mitochondrial DNA population database
ePCR-RFLP	engineered PCR-RFLP
ES	embryonic stem
FISH	fluorescent *in situ* hybridization
FP-TDI	fluorescence polarization template-directed dye-terminator incorporation
FRET	fluorescence resonance energy transfer
FRT	Flp recognition target
GAPDH	glyceraldehyde 3-phosphate dehydrogenase
GBS	group B streptococci
HIV	human immunodeficiency virus
HPCE	high-performance capillary electrophoresis
HPLC	high-performance liquid chromatography
iFLP	inverse PCR-based amplified RFLP
I-PEP	improved PEP
LB	Luria–Bertani

LOD	logarithm of odds
MALDI-TOF	matrix-assisted laser desorption/ionization time-of-flight
MSP	methylation-specific PCR
MSRE	methylation-sensitive restriction enzyme
mtDNA	mitochondrial DNA
NTC	no-template control
PAGE	polyacrylamide gel electrophoresis
PBS	phosphate-buffered saline
PCR	polymerase chain reaction
PEG	polyethylene glycol
PEP–PCR	primer-extension pre-amplification PCR
PGD	pre-implantation genetic diagnosis
PRSG	PCR of randomly sheared genomic DNA
PVA	polyvinyl alcohol
qPCR	quantitative PCR
QTL	quantitative trait loci
RFLP	restriction fragment length polymorphism
RT	reverse transcriptase
RT-PCR	reverse transcriptase PCR
SAP	shrimp alkaline phosphatase
SBE	single-base extension
SCOMP	single-cell comparative genomic hybridization
SDS	sodium dodecyl sulfate
SNP	single-nucleotide polymorphism
STR	short tandem repeat
TEMED	tetramethylethylenediamine
VCFS	velocardiofacial syndrome
VNTRs	variable number of tandem repeats
WGA	whole genome amplification

Color Section

Chapter 5. Robust and unique PCR for SNP genotyping applications

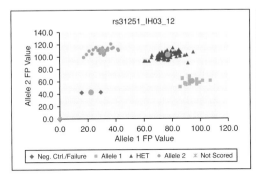
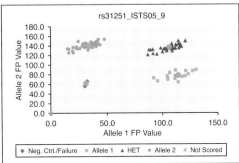
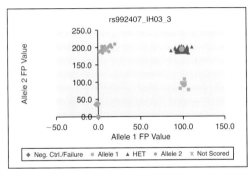
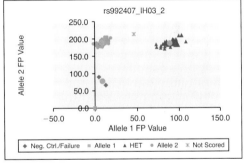

Figure 2. Some examples of SNPs typed by the FP-TDI protocol (see page 75).
Each panel contains 94 subjects and two negative controls. The top two panels show the results for the SNP rs31251 typed for two different samples. The results from a family sample are shown on the left and those from case–control samples on the right. Both demonstrate very similar results. The bottom two panels show the results for the SNP rs992407, typed for case–control samples. The panel on the right misses the minor allele homozygote group. Comparison with the left panel allows determination of the genotypes of the right panel.

Chapter 5. Robust and unique PCR for SNP genotyping applications

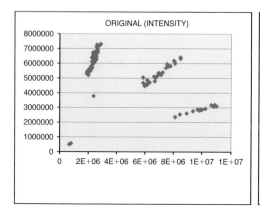

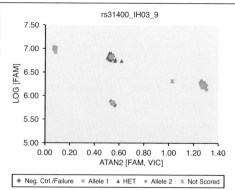

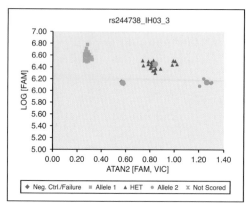

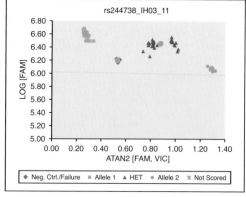

Figure 3. Examples of SNPs typed by the TaqMan protocol (see page 77).
The top two panels are results for rs31400 for the same sample plate scored by different algorithms. On the left is a scatter plot of fluorescence intensities of the two report dyes (FAM on the *x*-axis and VIC on the *y*-axis) as recorded by the fluorescence plate reader. On the right is a plot of the tangent-transformed value (the ratio of FAM/VIC) versus the intensity of FAM. Whilst the two methods produce identical results where the data quality is good, a tangent transformation can make the scoring more objective when the data are suboptimal. The bottom two panels show an unusual clustering pattern, where the heterozygotes are divided into two distinct subgroups, which is consistent across different sample plates. The results are a consequence of a second polymorphism, rs244737, located 8 bp upstream of the target SNP, rs244738. This second polymorphism creates a differential annealing affinity of the two allele-specific TaqMan probes between individuals, resulting in a consistent change in cleavage efficiencies of the allele-specific probes.

Chapter 8. Forensic genetic DNA typing with PCR-based methods

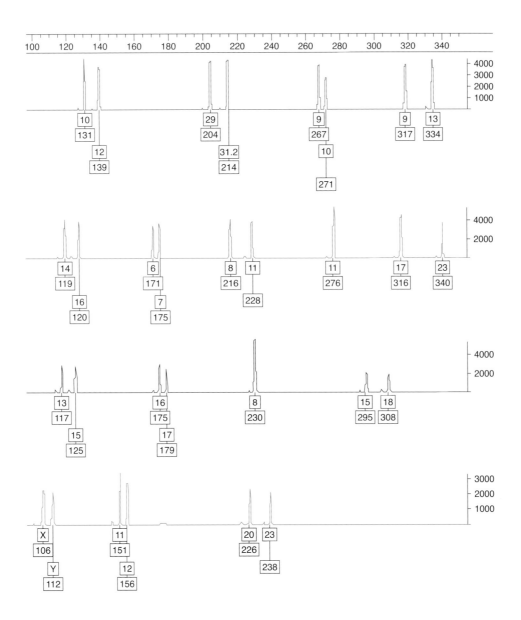

Figure 3. STR fragment analysis (see page 137).
A typical electropherogram obtained from one sample amplified with the AmpF/STR Identifiler PCR Amplification kit is shown. Each peak represents a STR allele. The number of STRs and the length of the PCR product in nucleotides is shown in boxes below the peak. Four, five, and four marker STR systems are detected in the results shown in blue, green, and black, respectively, and two marker STR systems and the sex marker amelogenin are shown in the results in red. The DNA profile identifies the person as a male with two alleles in 13 STR systems (heterozygous genotype) and one allele in two systems (homozygous genotype). The DNA profile of this person is shown in table format in *Table 1* (Suspect 1).

Chapter 8. Forensic genetic DNA typing with PCR-based methods

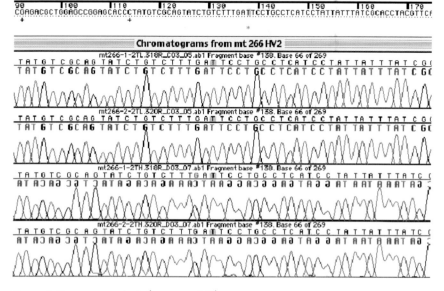

Figure 4. Sequence analysis (see page 140).
A typical alignment of two forward and two reverse DNA sequencing reactions is shown. At the top is shown the Cambridge reference sequence (29) and below are the results from the four sequencing reactions. A light blue background indicates that the PHRED quality score was higher than 20. A red letter indicates that manual correction of the sequence was performed. The proposed consensus sequence is shown above the electropherograms.

Chapter 16. Use of PCR for DNA methylation analyses

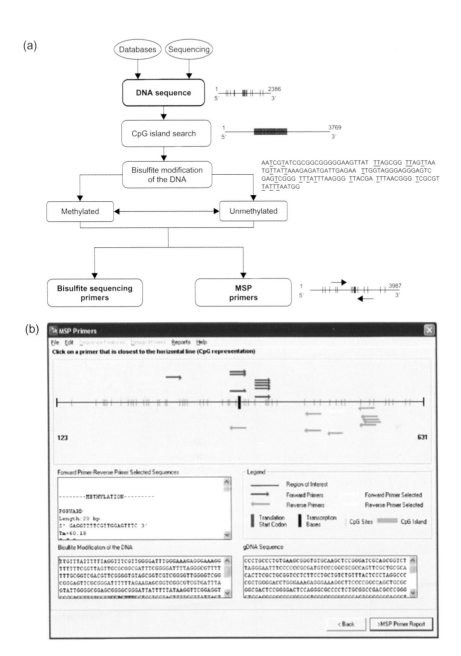

Figure 2. Designing experiments to study locus-specific DNA methylation status (see page 273).
(*a*) The first step is to identify a CpG island surrounding the transcription start point. Next, simulate the resulting DNA sequence after bisulfite modification of the DNA. Lastly, design oligonucleotide primers for MSP or BS. All of these steps can be achieved with the help of software such as METHYL PRIMER EXPRESS. (*b*) Representative output window in the design of MSP primers within the GSTP1 promoter obtained using METHYL PRIMER EXPRESS. The oligonucleotides are represented as red and orange arrows. Pink bars indicated CpG sites. The original and bisulfite-modified DNA sequences are shown in the lower boxes. The upper box shows the primer sequences.

Chapter 16. Use of PCR for DNA methylation analyses

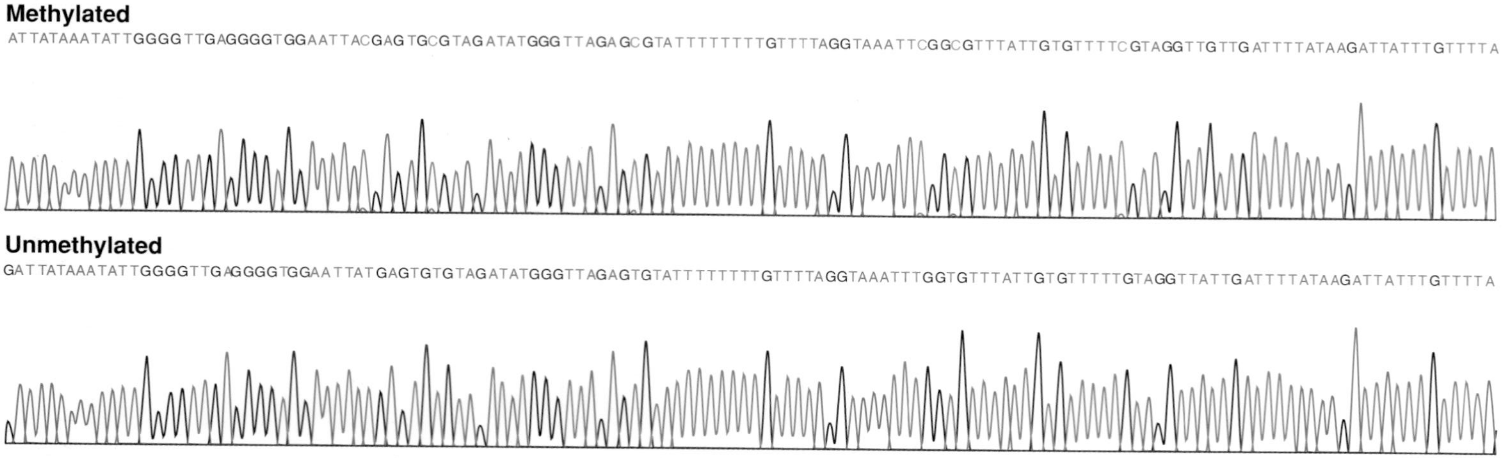

Methylated

ATTATAAATATTGGGGTTGAGGGGTGGAATTACGAGTGCGTAGATATGGGTTAGAGCGTATTTTTTTGTTTAGGTAAATTCGGCGTTATTGTGTTTTCGTAGGTTGTTGATTTTTATAAGATTATTTGTTTTA

Unmethylated

GATTATAAATATTGGGGTTGAGGGGTGGAATTATGAGTGTGAGATATGGGTTAGAGTGTATTTTTTTTGTTTAGGTAAATTGGTGTTTATTGTGTTTTGTAGGTTATTGATTTTTATAAGATTATTTGTTTTA

Figure 4. Bisulfite sequencing of methylated and unmethylated DNA (see page 276).
Example electropherograms obtained after BS of the promoter region of a gene in a methylated (protected cytosines, blue traces) and unmethylated sample (no cytosines because they have been converted to thymines).

Chapter 19. PCR sequencing of human genes for the discovery of DNA sequence variants

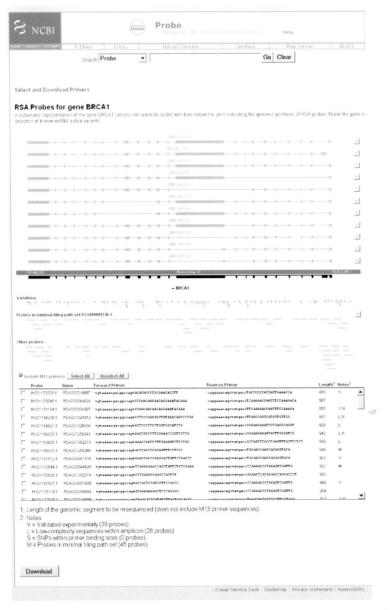

Figure 2. VariantSEQr resequencing primers for the breast cancer 1, early onset gene (BRCA1) (see page 324).

Resequencing amplicons (RSAs) are mapped to all known transcripts of BRCA1. Known SNPs are indicated by red slashes. Amplicons for a specific transcript can be selected by clicking * on the right. Individual amplicons can be selected by clicking on any individual amplicon in the map or in the table. The primer sequences (with or without an M13 sequencing tail) can be downloaded for the selected amplicon by clicking the 'Download' button.

Chapter 19. PCR sequencing of human genes for the discovery of DNA sequence variants

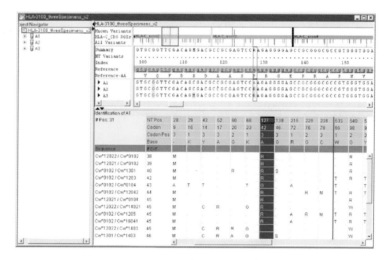

Figure 7. SEQSCAPE **Project Navigator (see page 337).**
In the project view, a summary of all of the specimen sequences is shown. The location of variants is displayed in the top panel for the regions (amplicons) that have been sequenced. In the second panel, a subsection of the sequenced region (blue square in the top panel) is shown in more detail. The assembled sequence is aligned with the reference sequence and variations are highlighted for the different specimens. The table in the lower panel summarizes the impact that the mutations would have on the encoded protein.

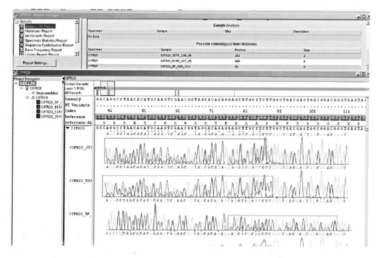

Figure 8. Indel mutations in the CIFR20 gene (see page 338).
Heterozygous insertions or deletions are detected and shown by the position and length of the indel (top panel) in the mutation report, which contains hyperlinks to the trace files (shown in the bottom panel). Three samples are shown with 2 bp deletions in the CIFR20 gene in one of the two chromosomes (region boxed in red).

CHAPTER 1

An introduction to the polymerase chain reaction

Adrian Moody

1. INTRODUCTION

The polymerase chain reaction (PCR) is ubiquitous in molecular biology laboratories the world over and is a 'workhorse' of modern molecular biology. The basic principal of PCR has changed little since its inception in the 1980s, with it being a three-stage cyclical reaction of DNA denaturation, primer annealing, and nucleotide extension resulting in the exponential increase of a targeted DNA sequence (see *Fig. 1*). Although PCR can be defined and applied as a distinct methodology, it can also be considered a set of principals that are extremely versatile and as such is the foundation for an ever-expanding range of PCR-based methods, many of which are described in this book.

1.1 History of PCR

As with many scientific breakthroughs and inventions, there is some controversy around who invented PCR as in essence it is the amalgam of many existing principals and ideas. In 1969, the Norwegian scientist Kjell Kleppe presented work on repair replication and described the doubling and quadrupling of small synthetic DNA molecules using two primers and DNA polymerase. Three years later, he released a 20-page research paper essentially describing the principals of PCR in the *Journal of Molecular Biology* (1). Despite this work, significant advances towards the PCR we know today were not made for a further 12 years, and Kary Mullis, working for the biotechnology company Cetus, is credited with the 'invention' of PCR in 1983. For his work he was awarded the Nobel Prize for Chemistry in 1993. However, accrediting the 'invention' of PCR to a single individual belies the complexity of making the process work, and other scientists including Henry Erlich, Norman Arnheim, Randall Saiki, Glen Horn, Corey Levenson, Steven Scharf, Fred Faloona, and Tom White, were also instrumental. As stated above, this area is contested and a detailed review is beyond the scope of

PCR: *Methods Express* (S. Hughes and A. Moody, eds.)
© Scion Publishing Limited, 2007

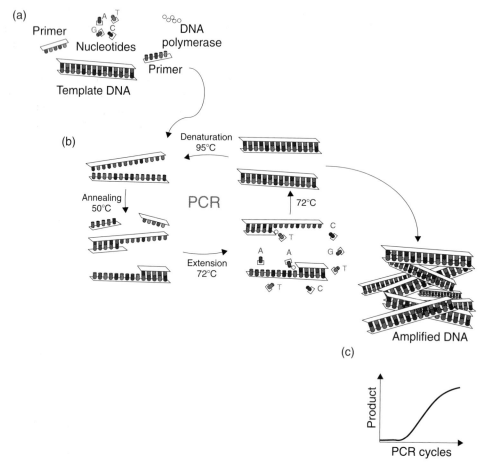

Figure 1. Schematic representation of basic PCR components and stages of a thermal cycling reaction.
The five basic components of a PCR are combined (*a*) and subjected to a three-stage thermal cycling reaction: denaturation (DNA strands separate), annealing (primers bind to complementary strands), and extension (polymerase enzyme adds nucleotides to complementary DNA strands starting at primer) (*b*), leading to an exponential increase of the targeted DNA sequence (*c*).

this chapter, but interested readers should refer to the book by Paul Rabinow (2) who concludes: 'Committees and science journalists like the idea of associating a unique idea with a unique person, the lone genius. PCR is, in fact, one of those classic examples of teamwork.'

The original publication describing PCR was in 1985 (3), but at this stage the amplification was achieved using heat-labile polymerases that needed to be added fresh after the denaturation step of each cycle. The first use of the thermally stable polymerase (*Taq*) from the bacteria *Thermus aquaticus* (for further details on DNA polymerases, see *Chapter 2*) was published 2 years later (4).

Kary Mullis received the Nobel Prize for PCR because he is credited with the idea of using *Taq* DNA polymerase in the reaction, and the importance of the *Taq* polymerase enzyme was recognized when it became the inaugural winner of *Science* magazines 'Molecule of the Year' in 1989. Following the invention, Cetus patented the technology and sold these patents to Roche Molecular Systems for US$300 million. Many of these patents have been contested over the years, and some are still being contested at the time of writing. However, despite this, PCR continues to be widely used (see *Fig. 2*) and the versatility of the technique continues to lead to ever more applications.

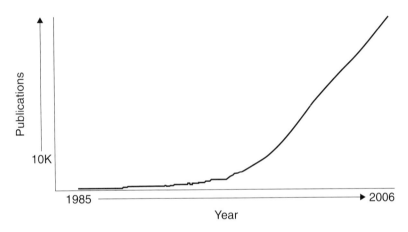

Figure 2. Illustration of the increase in the use of PCR following its invention in the mid-1980s.

1.2 Utility of PCR – modifications and applications

The PCR process has been extensively modified and adapted, which has allowed it to be applied to a range of novel applications. Many of the more common uses of PCR are covered in this book including: quantitative amplification of RNA (Chapter 3) and DNA (Chapter 4), introduction of specific mutations (Chapter 13), genotyping of polymorphisms (Chapters 5 and 7), identifying low-frequency somatic mutations (Chapter 14), generation of large molecules (Chapters 11 and 12), discriminating methylated and unmethylated DNA (Chapters 16 and 17), and amplification of whole genomes (Chapter 18).

The PCR process is used extensively to drive fundamental research, for example agricultural research (Chapter 6), but it is also being applied to applications that have a direct impact on the general population. Notable applications include: forensic genetics and paternity testing (Chapters 8 and 9), pre-implantation genetic diagnostics (Chapter 10), and diagnosis of infectious disease (Chapter 15).

2. METHODS AND APPROACHES

PCR is used in molecular biology laboratories throughout the world, and for experienced users its application is routine. However, for inexperienced users there are many basic processes that need to be learnt and potential pitfalls to be avoided (we recommend referring to section 2.8 before completing any of the methods described in this chapter). The aim of this chapter is to describe the components of PCR and provide context for their involvement in the reaction. The key protocols described will be a standard PCR using a commercially available PCR kit to which a researcher need only add DNA and primers (*Protocol 1*) and a standard reaction using individual PCR components (*Protocol 2*). In our experience, both of these methods give a very high PCR success rate; however, it is still necessary to perform rounds of PCR optimization for some products. Therefore, *Protocols 3* and *4* describe experimental approaches for optimizing primer annealing temperature, and *Protocol 5* describes a method for optimizing MgCl$_2$ concentration.

2.1 Components of PCR

There are five central components to a PCR: oligonucleotide primers, DNA polymerase, reaction buffer (including MgCl$_2$), dNTPs, and target DNA. With the exception of the target DNA, all of the others components are available from numerous commercial suppliers. Throughout this chapter, and the rest of the book, specific suppliers of reagents are named for guidance. However, it should be noted that in most cases equivalent reagents will be available from other suppliers.

1. *Oligonucleotide primers.* Each PCR requires a pair of primers that are directional and flank the region you wish to amplify. Oligonucleotide primers provide the specificity for the reaction and are the initiation point for the ssDNA molecules synthesized in each PCR cycle.
2. *DNA polymerase.* A wide variety of DNA polymerases is available from commercial suppliers and a detailed review of DNA polymerases is provided in Chapter 2.
3. *Reaction buffer (including MgCl$_2$).* The reaction buffer is supplied with the enzyme. For optimal results, use the reaction buffer as directed by the supplier (see also section 2.5 in Chapter 2 for further details). Usually, the only variable in the reaction buffer is MgCl$_2$, and this can be adjusted to optimize the specificity of the PCR. MgCl$_2$ can be purchased as a separate component to the PCR buffer and the optimum concentration empirically determined (see *Protocol 5*). Alternatively, some suppliers provide ready-to-use PCR master mixes, which contain MgCl$_2$ at differing concentrations, usually in increments of 0.5 mM (i.e. 1.5, 2.0 mM, etc.).

   It has been found that various organic additives can be added to the PCR to increase reaction specificity and/or to try and overcome secondary structure in the target DNA. These additives include:

- DMSO or glycerol (up to 10% final concentration). Both of these additives have been demonstrated to aid in DNA strand separation (in GC-rich difficult secondary structures) because they disrupt base pairing and have been shown to improve PCR efficiency.
- Formamide solution (up to 5% final concentration). Formamide has been demonstrated to decrease the temperature required for the reassociation of complementary nucleic acids and can improve PCR efficiency, particularly when templates contain difficult secondary structures.
- Betaine solution (0.8–1.6 M final concentration). Betaine has been demonstrated to improve the amplification of DNA by decreasing the formation of secondary structure in GC-rich regions. It is an isostabilizing agent, equalizing the contribution of GC and AT base pairing to the stability of the DNA duplex.

4. *dNTPs*. These provide the energy and building blocks for the extending molecule (see also section 2.5 in Chapter 2 for further details).

5. *Template DNA*. This provides the source of the DNA to be amplified. The template DNA can be the only component of the reaction supplied by the individual user and has been identified as a major cause of variability in many PCR applications. The quality, quantity, and purity of target DNA can all influence the success of the reaction (see section 3).

PCR components can either be purchased individually, i.e. separate dNTPs, DNA polymerase, $MgCl_2$, etc., or in ready-to-use PCR kits where the user need only add specific primers and DNA (although in certain cases companies now provide gene-specific amplification kits to which you need only add DNA). The major advantage of buying individual PCR components is that you can empirically optimize the conditions for each reaction, whereas with the ready-to-use kits you are restricted to the conditions provided. However, individually optimizing every PCR is often not necessary and the major disadvantages of using individual components are that setting up the reaction is more time-consuming and involves more steps, increasing the chances of manual error.

2.2 The PCR cycle

The PCR is a three-step process consisting of successive cycles (usually ~20–40) of DNA *denaturation*, oligonucleotide primer *annealing*, and DNA polymerase *extension* (see *Fig. 1*). In more detail:

1. *Denaturation*. A denaturation step of 94–95°C for ~30 s is used to separate the dsDNA to allow the oligonucleotide primers access to the template. The denaturation step in the initial cycle of the PCR can be much longer (~15 min) in order to activate the *Taq* polymerase (see Chapter 2, section 2.4, for more details).

2. *Annealing*. The annealing temperature is specific for each primer pair and is determined by the base composition of the oligonucleotide primers. The annealing step generally lasts for 60 s, and for basic PCR applications the

annealing temperature is fixed throughout the reaction. However, there are notable exceptions to this rule, one of which, touchdown PCR (5), will be described in more detail below.

3. *Extension.* The extension phase allows the enzyme to move along the template from the primer, synthesizing the complementary DNA strand. The temperature of the extension phase is generally 72°C. The length of the extension phase is determined by the length of the PCR product and the processivity of the enzyme (see Chapter 2 for more details). Standard *Taq* polymerase will add ~1000 bases in 60 s and as PCR products for routine applications are generally <1000 bp, a 60 s extension time is standard. Many PCR protocols also include a final 10–15 min extension phase after the final PCR cycle, which allows the enzyme to fill in the protruding ends of newly synthesized PCR products.

A standard PCR will consists of 20–40 cycles. Below 20 cycles, there will generally be insufficient product for downstream applications, whilst above 40 cycles, reaction components will become limiting, dramatically reducing the efficiency of amplification at each subsequent cycle. In addition, with a greater number of PCR cycles, you increase the possibility of detection of nonspecific background products.

Theoretically, the PCR process leads to a doubling of products at each cycle. However, reactions are rarely 100% efficient and the true extent of the amplification at each cycle will vary from reaction to reaction. For most applications, the efficiency of the PCR is not important, as long as you generate sufficient product for downstream analysis. However, in applications where PCR is used to quantify the absolute or relative amount of starting material, so-called quantitative PCR, the reaction efficiency is important and needs to be calculated using standard curves (further details of such applications are provided in Chapters 3 and 4).

2.3 Thermal cyclers

Thermal cyclers can be purchased from many different suppliers and in general they all have similar functionality. Most thermal cyclers now include heated lids so that the addition of mineral oil, to prevent evaporation of the reaction during the thermal cycling, is not required. However, it should be noted that, whilst thermal cyclers are comparable, they are not identical, and transferring a set of PCR conditions from one machine to another (especially when they are manufactured by different companies) is not always straightforward and can require additional rounds of reoptimization.

2.4 Oligonucleotide primers

Oligonucleotide primer design is critical in ensuring specific amplification of DNA. We recommend investing time in designing high-quality primers, as this will be rewarded with efficient and reproducible results in the laboratory.

Primers for routine PCR are short, usually 18–25 bases, and can be synthesized by commercial suppliers. Oligonucleotide primers can be ordered to different degrees of purity, from standard desalted primers to high-quality, high-performance liquid chromatography-purified primers, and with various modifications depending on the application, for example fluorescent labels (see Chapter 3) or generic sequence tags (see Chapter 19). Whilst high levels of purity and modified primers are required for some applications, for routine applications we recommend the most basic, and therefore cheapest, synthesis method. However, we would encourage users to monitor the quality of the oligonucleotides they purchase as, in our experience, not all suppliers consistently supply high-quality oligonucleotides.

There are numerous specialist primer design programs and tools available on the internet. However, four basic approaches are considered below:

1. Search the literature/internet. Whilst not strictly primer design, it is important to consider whether there are pre-designed primers for the region you wish to study, for example in the literature. In addition, in the field of human genetics, there are growing numbers of pre-designed primers available from both commercial suppliers (e.g. Applied Biosystems VariantSEQr available through the NCBI Probe Database; see also Chapter 19, section 2.1.4 for further details) and high-throughput academic groups (e.g. NHLBI Program for Genomic Applications, SeattleSNPs, http://pga.gs.washington.edu).

2. Manual primer design. This approach should only be considered by experienced users. Manual assessment of the DNA sequence flanking the target region and selecting primers based on basic considerations such as GC content, length of primer, length of PCR product, and the absence of known polymorphisms can be used for the successful design of primers. This method is relatively quick and has the advantage of not requiring primer design software or extensive manipulation of DNA sequence files. The major disadvantage of this method is that the quality of the chosen primers will be variable.

3. Design of primers using specific primer design applications, for example desktop programs such as PRIMERSELECT, part of the Lasergene suit of programs, or tools available on the internet (e.g. PRIMER3).

4. Design of primers for specific applications using proprietary software/algorithms. This will not be discussed in this chapter, but refers to specific applications that include proprietary PCR primer design processes such as ABI PRIMER EXPRESS (see Chapter 3) or Biotage Pyrosequencing primer design.

Regardless of the method used to design the primers, several variables must be taken into account to ensure high PCR success rates. Among the most critical are primer length/specificity, melting temperature (T_m), complementary primer sequences, GC content, repeat stretches of polypyrimidines (T and C) or polypurines (A and G), and the 3′-end sequence:

1. *Primer length.* The specificity and the annealing temperature of the primer are in part dependent on the length of the primer. In general, the specificity of the

primer will increase with the length, whilst the efficiency of primer binding decreases. Primers in the 18–25 nt range in general give the right balance of specificity and binding efficiency, although primers can be successfully designed outside of these parameters should this be required (e.g. in high-repeat-content regions, longer primers may be required in order to generate the required specificity). Primers should be designed to have an annealing temperature of at least 50°C, and the annealing temperature is generally considered to be 5°C lower than the T_m.

2. *Melting temperature (T_m)*. PCR requires two primers and it is important that both primers have similar T_m values (ideally within 1°C). There are numerous methodologies for calculating the T_m (users may witness this when they receive commercially supplied oligonucleotides where the recommended T_m from the supplier is not in agreement with the T_m calculated by the primer design tool), but the basic calculation: T_m (°C) = 2(A + T) + 4(G + C) provides a good approximation for primers in the 18–25 base range. In our laboratory, we aim to design primers with a T_m of 60–64°C.

3. *Complementary primer sequence*. A primer should not contain more than 3 bp of intraprimer homology. If regions of self-homology are present in the primer, partial double-stranded structures that will interfere with annealing to the template can occur. Primer pairs also should not contain homologous regions, as this can interfere with hybridization, and if the homology occurs at the 3′ end, primer dimers will form, reducing the efficiency of the PCR.

4. *GC content*. The base composition of the primers should be between 45 and 55% GC, and repeat stretches of nucleotides should be avoided (i.e. poly(G), poly(C), poly(A), or poly(T)). Repeat stretches of polypyrimidines and polypurines should also be avoided.

5. *3′-end sequence*. The 3′-end sequence is essential for controlling mispriming, and the inclusion of GC residues at the 3′ end helps to ensure correct binding due to the stronger hydrogen bonding of GC residues.

2.5 Standard PCR

A standard PCR should be quick, simple, reproducible, reliable, and able to generate products successfully in the absence of rounds of optimization. We recommend using a commercially available complete PCR kit (e.g. ReddyMix; Abgene) that only requires the addition of primers and DNA by the user. The advantages of such products are that they have been extensively optimized and are therefore reliable, and they reduce the number of steps and time required to set up a PCR, limiting the opportunity for user error. In our experience, testing all new products in a standard PCR often negates the need for rounds of optimization, and *Protocols 1, 2* and *3* presented below are routinely used in our laboratories where we achieve first-pass PCR success rates in excess of 75%.

Protocol 1

Standard PCR – ReddyMix

Equipment and Reagents
- Genomic DNA (25 ng/µl)
- ReddyMix (Abgene)[a,b]
- Nuclease-free water (Promega)
- Oligonucleotide primers (2.5 µM)
- Thermal cycler
- 1% Agarose gel containing 10 ng/ml ethidium bromide
- Equipment and reagents for agarose gel electrophoresis including 1× TBE agarose gel running buffer (10.8 g/l Tris base; 5.5 g/l boric acid; 4 ml/l 0.5 M EDTA, pH 8.0, diluted from a 10× stock; Sigma)
- DNA size marker (100 bp ladder; Invitrogen)
- UV light source

Method
1. Combine per reaction[c,d]:
   - 22 µl of ReddyMix
   - 2 µl of primer mix[e]
   - 1 µl of DNA

2. Seal the PCR vessel and mix briefly by vortexing. Centrifuge at 12 000 g for 5–10 s to consolidate sample.

3. Amplify the DNA using the following PCR conditions[f]:
   - Initial denaturation at 95°C for 2 min
   - 40 cycles of denaturation at 94°C for 30 s, annealing[g] at 58°C for 30 s, and extension at 72°C for 1 min
   - Final extension at 72°C for 10 min

4. Analyze the PCR products by running 3–5 µl[h] of the reaction mix on an agarose gel alongside a DNA size marker[i].

Notes
[a]Equivalent kits are available from other suppliers.

[b]Kits are available with different $MgCl_2$ concentrations. We routinely use 2 mM $MgCl_2$ for our standard PCR.

[c]We routinely use 96-well plates for PCR, but any appropriate-sized tube/plate can be used.

[d]We strongly recommend physical separation of pre- and post-PCR work areas to avoid contamination. If separate laboratories are not available, lamina flow cabinets can also be used for PCR set-up.

[e]To make a primer mix at 2.5 µM (assuming primer stocks are in suspension at 100 µM), combine 10 µl of forward primer, 10 µl of reverse primer, and 380 µl of nuclease-free water. Vortex and spin briefly to consolidate.

[f]In our laboratories, the thermal cyclers are in the post-PCR work area. However, as long as the sealed PCR vessels are not opened after completion of the PCR, the thermal cycling machines can be in either the pre- or post-PCR work area.

[g]The annealing temperature will vary depending on the primers used. Where possible, we recommend designing all primers to work at a standard annealing temperature.

[h]The addition of a gel loading dye, such as an orange loading dye solution, is not required as the ReddyMix already contains a gel loading dye.

[i]The addition of 5% DMSO can enhance the PCR success of ReddyMix.

The use of a pre-prepared PCR kit is quick, easy, relatively cheap, and in our hands generates strong products and is the method of choice for basic PCR. However, when using such an approach for applications such as DNA sequencing, we have found the sequence quality can be variable. Consequently we routinely use a 'hot-start' enzyme, e.g. AmpliTaq Gold (Applied Biosystems) (*Protocol 2*), for products that will be sequenced because we find they produce consistently higher-quality sequence data.

Protocol 2

Standard PCR – AmpliTaq Gold

Equipment and Reagents
- Genomic DNA (25 ng/μl)
- AmpliTaq Gold *Taq* DNA polymerase[a] (5 units/μl) and accompanying GeneAmp 10× PCR buffer II[b] and 25 mM MgCl$_2$ (Applied Biosystems)
- dNTP mix containing 10 mM of each dNTP (Invitrogen)
- Nuclease-free water (Promega)
- Oligonucleotide primers (2.5 μM)
- 6× Orange loading dye solution (Fermentas)
- Thermal cycler
- 1% Agarose gel containing 10 ng/ml ethidium bromide
- Equipment and reagents for agarose gel electrophoresis including 1× TBE agarose gel running buffer (10.8 g/l Tris base; 5.5 g/l boric acid; 4 ml/l 0.5 M EDTA, pH 8.0, diluted from a 10× stock; Sigma)
- DNA size marker (100 bp ladder; Invitrogen)
- UV light source

Method
1. Combine, in the order below, per reaction[c,d]:
   - 16.8 μl of nuclease-free water
   - 2.5 μl of GeneAmp 10× PCR buffer II
   - 0.5 μl of dNTP mix
   - 2 μl of primer mix[e]
   - 0.2 μl of AmpliTaq Gold
   - 2 μl of MgCl$_2$
   - 1 μl of DNA

2. Seal the PCR vessel and mix briefly by vortexing. Centrifuge at 12 000 *g* for 5–10 s to consolidate the sample.

3. Amplify the DNA using the following PCR conditions[f] :
   - Initial denaturation at 95°C for 10 min
   - 40 cycles of denaturation at 94°C for 30 s, annealing[g] at 55°C for 30 s, and extension at 72°C for 1 min
   - Final extension at 72°C for 10 min

4. Analyze the PCR products by mixing 5 μl of the reaction mix with 1 μl of 6× orange loading dye solution and resolve by agarose gel electrophoresis alongside a DNA size marker.

Notes

[a]AmpliTaq Gold *Taq* DNA polymerase is a hot-start enzyme and although we use this enzyme as standard, hot-start enzymes are available from other suppliers.

[b]AmpliTaq Gold can also be purchased with GeneAmp Buffer I, which contains 15 mM $MgCl_2$.

[c]We routinely use 96-well plates for PCR, but any appropriate-sized tube/plate can be used.

[d]We strongly recommend physical separation of pre- and post-PCR work areas to avoid contamination. If separate laboratories are not available, lamina flow cabinets can also be used for the PCR set-up.

[e]To make a primer mix at 2.5 µM (assuming primer stocks are in suspension at 100 µM), combine 10 µl of forward primer, 10 µl of reverse primer, and 380 µl of nuclease-free water. Vortex and spin briefly to consolidate.

[f]In our laboratories, the thermal cyclers are in the post-PCR work area. However, as long as the PCR vessels are not opened after completion of the PCR, the thermal cycling machines can be in either the pre- or post-PCR work area.

[g]The annealing temperature will vary depending on the primers used. Where possible, we recommend designing all primers to work at a standard annealing temperature.

2.6 Optimizing a PCR

If the standard PCR fails to generate the required product, optimization of the reaction may be required. However, before embarking on extensive rounds of optimization it is important to consider why your initial reaction failed. For example, if you have generated no PCR product at all under standards conditions with an appropriately calculated annealing temperature, then in our experience it is time-consuming and particularly difficult to optimize the PCR. Under these circumstances, we recommend repeating the PCR with appropriate positive controls for the DNA and master mix to eliminate the possibility of manual error in the original set-up. If the reaction still fails, we suggest redesigning your PCR primers, rather than embarking on rounds of optimization, as these are the mostly likely cause of gross PCR failure if all of your positive controls work as expected. Alternatively, if you have a PCR product but it is weak or smeared, then rounds of optimization may help to generate a stronger, more specific product (see also section 3 for a broader discussion on reasons for PCR failure).

2.6.1 Optimizing primer concentration

We would not recommend routinely optimizing primer concentration because, whilst primer concentration can effect the PCR (i.e. too low a concentration and they become rate-limiting; too high a concentration and they can lead to nonspecific amplification), most reactions are sufficiently robust to work with primers in a 0.2–0.5 µM range (final concentration). Consequently this chapter will not include a protocol for optimizing primer concentrations. However, should this be necessary, primer concentration can be optimized empirically by modifying *Protocol 1* to include test primer concentrations in the range of 0.1–1 µM (final concentration). We would suggest avoiding primer concentrations higher than this as it can cause primer-dimer formation. Primer dimers are 'nontarget' amplification products, caused by mispriming events or primer self-annealing. Primer pairs that demonstrate complementarity at their 3′ ends can anneal to each other and primer extension creates a small double-stranded primer dimer. This process is favored at high primer concentrations, typically >1 µM of each primer in a 50 µl reaction.

2.6.2 Optimizing primer annealing temperature

Primer design programs will calculate the annealing temperature of primers, and many laboratories will design primers to anneal at specific temperatures so that they can run PCRs under standard conditions. However, there is no universally agreed standard for calculating annealing temperatures and as such some oligonucleotides may not work optimally at the first annealing temperature tested.

Described below are two different methods that can be used to address the issue of accurately predicting the annealing temperature of oligonucleotide primers. *Protocol 3*, Touchdown PCR, is not an optimization technique but a simple modification to the thermal cycling conditions that can be applied to *Protocol 1* or *2*. Touchdown PCR was first described by Don *et al.* (1991) (5) as a PCR technique to circumvent spurious priming during PCR amplification. The method was designed to by-pass nonspecific PCR amplification without the need for lengthy optimization procedures. The principal behind Touchdown PCR is that the annealing temperature at the start of the reaction is higher than the anticipated oligonucleotide primer annealing temperature, and during the first few cycles the annealing temperature is progressively lowered, reducing the stringency of the reaction. The exponential nature of PCR means that those products generated during the stringent phase of the reaction (i.e. the desired specific products) will predominate during the less-stringent cycles, resulting in preferential amplification of the desired product (5). Touchdown PCR uses a cycling program where the annealing temperature decreases by 0.2–0.5°C every cycle. The first cycle annealing temperature should be approximately 2–5°C above the maximum melting temperature (T_m) of the primers, whilst the final cycle annealing temperature should be 2–5°C below the T_m of the primers.

Protocol 4 describes empirical optimization of the annealing temperature for each new primer pair. This can be done one annealing temperature at a time for each new product, i.e. perform *Protocol 1* or *2* at multiple different annealing temperatures, i.e. 55, 56, 57°C, etc., although this is time- and labor-intensive. An alternative approach, and the one described below, is to use the gradient function on the thermal cycler, which enables multiple different annealing temperatures to be tested in a single experiment.

Protocol 3

Standard PCR – Touchdown

Equipment and Reagents
- As for *Protocol 1*[a]

Method
As for *Protocol 1* but with the following cycling conditions to replace those in step 3:
- Initial denaturation at 95°C for 2 min
- 14 cycles of 94°C for 20 s, 63°C[b] – reducing 0.5°C per cycle – for 1 min, and 72°C for 1 min
- 20 cycles of 94°C for 20 s, 56°C for 1 min, and 72°C for 1 min
- Final extension at 72°C for 10 min

Notes

[a]The Touchdown cycling conditions can also be applied to the hot-start PCR protocol (*Protocol 2*), but ensure that the initial denaturation phase is extended to 10 min.
[b]The starting temperature, and hence the final annealing temperature used for 20 cycles, can be altered to reflect the predicted T_m of your primers.

2.7 Optimizing MgCl$_2$ concentration

The correct concentration of magnesium ions is essential for optimal polymerase activity in PCR and magnesium ions are usually added in the form of MgCl$_2$. If the concentration of magnesium ions is too high or too low, the amplification efficiency and specificity of the reaction can be compromised (further details on magnesium ion concentration can be found in Chapter 2). It is important to remember that the magnesium ion concentration in the PCR is affected not only by the amount of MgCl$_2$ added to the reaction, but also by the presence of magnesium ion-collating agents within the reaction, e.g. proteins, dNTPs, or DNA. Contaminants in the DNA sample can be a source of magnesium ion-collating agents, and it is therefore important to optimize MgCl$_2$ concentration empirically for each specific set of primers and template (see *Protocol 5*). However, in our experience this is not always necessary if standard DNA templates and primer design parameters are used.

Protocol 4

Gradient PCR

Equipment and Reagents
- As for *Protocol 1*[a,b]
- Thermal cycler with gradient PCR functionality (e.g. MJ Tetrad)

Method
As for *Protocol 1*, but with the following cycling conditions to replace those in step 3:
- Initial denaturation at 95°C for 2 min
- 40 cycles of denaturation at 94°C for 30 s, annealing at 50–70°C[c] for 30 s, and extension at 72°C for 1 min
- Final extension at 72°C for 10 min

Typical results of a gradient PCR are shown in *Fig. 3*.

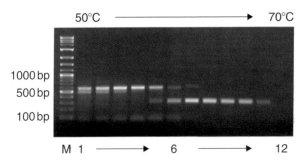

Figure 3. Representative gel demonstrating the outcome of a gradient PCR experiment. The PCR products are resolved in lanes 1–12. The gradient nature of the PCR program used means that the reaction in lane 1 was at 50°C and the reaction at lane 12 was at 70°C and the lanes in between were at varying temperatures (see notes section of *Protocol 5*). The optimal temperature for the specific product of interest (300 bp) is 66.8°C. This figure demonstrates the necessity of annealing temperature optimization, as at a low temperature several nonspecific products are generated. M, DNA size standard. See notes in Protocol 4.

Notes
[a]The principal of gradient PCR is to run the same PCR mix at multiple temperatures across the PCR block. A typical gradient reaction would be run at 12 different temperatures (see note b).
[b]The Touchdown cycling conditions can also be applied to the hot-start PCR protocol (*Protocol 2*), but ensure that the initial denaturation phase is extended to 10 min.
[c]The temperature gradient will typically be calculated across the PCR block from left to right. An example of the temperature in individual wells generated by an MJ Tetrad 50–70°C gradient is shown below:

						Column						
	1	2	3	4	5	6	7	8	9	10	11	12
Temperature (°C)	50	50.5	51.5	53.2	55.5	58.4	61.8	64.6	66.8	68.4	69.6	70

Protocol 5

Optimizing MgCl$_2$ concentration[a]

Equipment and Reagents

- Genomic DNA (25 ng/μl)
- AmpliTaq Gold *Taq* DNA polymerase[b] (5 units/μl) and accompanying GeneAmp 10× PCR buffer II and 25 mM MgCl$_2$ (Applied Biosystems)
- dNTP mix containing 10 mM of each dNTP (Invitrogen)
- Nuclease-free water (Promega)
- Oligonucleotide primers (2.5 μM)
- 6× Orange loading dye solution (Fermentas)
- Thermal cycler
- 1% Agarose gel containing 10 ng/ml ethidium bromide
- Equipment and reagents for agarose gel electrophoresis including 1× TBE agarose gel running buffer (10.8 g/l Tris base; 5.5 g/l boric acid; 4 ml/l 0.5 M EDTA, pH 8.0, diluted from a 10× stock; Sigma)
- DNA size marker (100 bp ladder; Invitrogen)
- UV light source

Method

1. For each reaction to be optimized, prepare and label four reaction vessels (i.e. 1, 2, 3, and 4 mM MgCl$_2$).

2. Combine, in the order below, per reaction[c,d]:
   - Nuclease-free water (see table[e] below for amount)
   - 2.5 μl GeneAmp 10× PCR buffer II
   - 0.5 μl dNTP mix
   - 2 μl primer mix[f]
   - 0.2 μl AmpliTaq Gold
   - MgCl$_2$ (see table below for amount)
   - 1 μl DNA

	Final concentration of MgCl$_2$ (mM)			
	1	2	3	4
MgCl$_2$ (25 mM)	1 μl	2 μl	3 μl	4 μl
H$_2$O	17.8 μl	16.8 μl	15.8 μl	14.8 μl

3. Seal the PCR vessel and mix briefly by vortexing. Centrifuge at 12 000 ***g*** for 5–10 s to consolidate the sample.

4. Amplify the DNA using the following PCR conditions[g]:
   - Initial denaturation at 95°C for 10 min
   - 40 cycles of denaturation at 94°C for 30 s, annealing[h] at 58°C for 30 s, and extension at 72°C for 1 min
   - Final extension at 72°C for 10 min

5. Analyze the PCR products by mixing 5 μl of the reaction mix with 1 μl of 6× orange loading dye solution and resolving the sample by agarose gel electrophoresis alongside a DNA size marker. Typical results of a MgCl$_2$ optimization experiment are shown in *Fig. 4*.

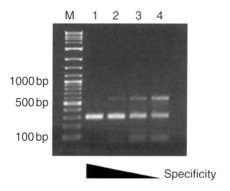

Figure 4. Representative gel demonstrating the outcome of a MgCl$_2$ optimization experiment.
The PCR products resolved in lanes 1–4 were amplified with MgCl$_2$ concentrations of 1, 2, 3, and 4 mM, respectively. For this reaction, it can be seen that, as the MgCl$_2$ concentration increases, the specificity of the reaction decreases. M, DNA size standard.

Notes

[a]This protocol describes manual alteration of the MgCl$_2$ concentration by varying the amount of 25 mM MgCl$_2$ added to the reaction mix. Many suppliers provide ready-to-use PCR kits/buffers (e.g. ReddyMix), which are supplied at different specific MgCl$_2$ concentrations, and these can also be used to find the optimal magnesium concentration.

[b]AmpliTaq Gold *Taq* DNA polymerase is a hot-start enzyme and although we use this enzyme as standard, hot-start enzymes are available from other suppliers.

[c]We routinely use 96-well plates for PCR, but any appropriate-sized tube/plate can be used.

[d]We strongly recommend physical separation of pre- and post-PCR work areas to avoid contamination. If separate laboratories are not available, flow cabinets can also be used for the PCR set-up.

[e]In this protocol, four MgCl$_2$ concentrations are tested, but any number of concentrations can be used. However, it is widely recognized that MgCl$_2$ concentrations in the range of 1–4 mM work well in most circumstances.

[f]To prepare a primer mix at 2.5 μM (assuming primer stocks are in suspension at 100 μM), combine 10 μl of forward primer, 10 μl of reverse primer, and 380 μl of nuclease-free water. Vortex and spin briefly to consolidate.

[g]In our laboratories, the thermal cyclers are in the post-PCR work area. However, as long as the PCR vessels are not opened immediately after completion of the PCR, the thermal cycling machines can be in either the pre- or post-PCR work area.

[h]Annealing temperature will vary depending on the primers used. However, where possible we recommend designing all primers to work at a standard annealing temperature.

2.8 General hints and tips

2.8.1 Reaction set-up

If not using an enzyme with hot-start activity, it is recommended to set up reactions on ice and keep everything on ice until immediately before placing the reactions into the thermal cycler. Keeping samples on ice will suppress misannealing of primers to the template and will also prevent *Taq* polymerase from extending misannealed primers prior to thermal cycling. When using a hot-start enzyme, this is not necessary, as the enzyme is inactive prior to thermal cycling (see Chapter 2) and the denaturation step will separate any primers misannealed to the template.

2.8.2. Positive controls

Positive controls should always be included. High-quality control genomic DNA can be purchased commercially and should be used for optimizing reactions to eliminate poor-quality DNA as a variable. Once a reaction is optimized, high-quality DNA should be run as a positive control alongside test samples.

2.8.3 Negative controls

A negative control should always be included. Contamination is a significant issue when performing PCR. Each experiment should include at least one no-DNA-template negative control to ensure there is no contamination. Amplification in the no-DNA-template negative control indicates contamination (see section 3).

2.8.4 Reduce the risk of contamination

The probability of contamination is increased if a laboratory is routinely analyzing the same set of products on multiple samples for a prolonged period. Steps to reduce the risk of contamination include:

• Routine use of dedicated laboratory coats and gloves for pre- and post-PCR
• Aliquotting of reagents into small volumes for personal use to prevent contamination of stock solutions
• Decontamination of the PCR work area, PCR pipettes, tube racks, etc. after use with a appropriate solution (e.g. DNAaway)
• The use of aerosol-resistant pipette tips
• The use of designated pipettes for PCR set-up, DNA/RNA handling and post-PCR manipulation
• Physical separation of the pre- and post-PCR work areas (ideally in separate rooms/laboratories)
• Ensuring stock DNA tubes are not used in the post-PCR work area to eliminate the chance of contaminating DNA sample with PCR product

2.8.5 Monitor changes of reagent batch

Commercially supplied reagents and oligonucleotides have simplified PCR and reduced variability. However, poor-quality batches of products are occasionally produced and it is important to note when new reagents are first used in case there is a reduction in the quality/reliability of your reaction.

2.8.6 Standardize input DNA

Where possible, ensure that the sample DNA used is of high quality and the concentration is consistent across all samples within an experiment. We recommend quantifying and normalizing DNA to a pre-defined optimal concentration prior to PCR analysis. Significant variability in the quality/quantity of the DNA across an experiment can lead to variable PCR success.

2.8.7 Ensure purity of DNA

The test DNA sample is a major source of contaminants that can potentially inhibit the PCR (e.g. proteases, PCR inhibitors, magnesium ion-collating agents). If test DNA samples give significantly poorer PCR success rates than controls, consider additional rounds of DNA purification (e.g. through commercially available spin columns) to increase the purity of the sample.

3. TROUBLESHOOTING

- **No amplification using new primers under standard PCR conditions**
  Ensure that a high-quality source of DNA is being used. For new products where a positive control reaction is not available, repeat the reaction to ensure that no mistakes were made in the original set-up and thermal cycling. If no product is generated the second time around, we suggest redesigning the PCR primers. Alternatively, rounds of annealing temperature and $MgCl_2$ optimization can be performed, although the success of this approach may be limited.
- **No amplification using previously optimized primers**
  Has the positive control reaction worked? If not, there may be a problem with the PCR master mix and the reaction should be repeated ensuring successful amplification in the positive control. Alternatively, if the primers have not been used for some time, PCR failure may indicate primer degradation, and fresh primer aliquots should be prepared from the stock or a new batch ordered. If the positive control reaction has worked, it may indicate PCR inhibitors in the test samples used. In this case, the DNA can be put through an additional round of purification (e.g. spin-column purification) to increase the sample purity and eliminate PCR inhibitors. Alternatively, a twofold serial dilution series of input DNA can be performed, as the PCR inhibitor will often be diluted out before the concentration of DNA becomes limiting for the reaction.

- **Amplification in negative controls**
  This indicates contamination of reagents and/or equipments and all possible sources of contamination should be eliminated from the PCR set-up area:
  - Wear a clean laboratory coat.
  - Discard all potentially contaminated consumables (PCR tubes and pipette tips) and reagents (primers, dNTPs, PCR mix, and water).
  - Clean and decontaminate the work area and pipettes.
  - Prepare fresh aliquots of all reagents, ensuring that none have been accessed previously.
  - Prepare a fresh aliquot of test DNA.

  If low-level nonspecific amplification is consistently present in the no-DNA-template control, reducing the number of PCR cycles can eliminate the problem.

- **Unexpected variability in PCR success rate across a collection of DNA samples**
  - This may reflect sample-to-sample variability in DNA quantity. Quantify the DNA in each sample and normalize the collection to a standard concentration.
  - This may reflect sample-to-sample variability in DNA quality, for example the presence of DNA inhibitors, which requires repurification of the sample. However, if the DNA is from a low-quality source, e.g. paraffin-embedded tumor samples, the variability may be due to DNA fragmentation. If this is suspected, redesign the PCR products to give the smallest possible amplicon for analysis.
  - This may reflect well-to-well variability in the performance of the thermal cycler. Over time, thermal cycling blocks will become less efficient and more variable. If this is suspected, perform an experiment using a positive control reaction in each well and compare the amplification efficiency across the block.
  - This may reflect polymorphisms in the primer-binding sites, which will affect the efficiency of primer binding. Ensure that there are no polymorphisms in the primer-binding sites

- **Weak PCR products**
  - Increase the number of PCR cycles.
  - Optimize the $MgCl_2$ and annealing temperature for your assay as described.
  - Include a PCR enhancing agent such as:
    - DMSO (up to 10% final concentration)
    - Glycerol (up to 10% final concentration)
    - Formamide solution (up to 5% final concentration)
    - Betaine solution (0.8–1.6 M final concentration)

- **Smeared/nonspecific PCR products**
  - Optimize the $MgCl_2$ and annealing temperature for your assay as described.
  - Ensure that the primers are not in known DNA repeat regions.
  - Redesign the PCR primers avoiding repeat regions.

- Excess primer dimers
  - Switch to using a hot-start PCR enzyme.
  - Redesign the primers avoiding 3′ complementarity.
- Product of an unexpected size
  - Optimize the $MgCl_2$ and annealing temperature for your assay as described (see *Fig. 3*).
  - This may be due to contamination – refer to the section above on amplification in negative control samples for actions.
  - Repeat the experiment to ensure that no manual error occurred during PCR set-up, for example the use of the wrong primer pair.
  - This may be a genuine result and reflect a lack of knowledge about the DNA region being amplified.
- Nonspecific secondary products
  - Optimize the $MgCl_2$ and annealing temperature for your assay as described.

4. REFERENCES

1. Kleppe K, Ohtsuka E, Kleppe R, Molineux I & Khorana HG (1971). *J. Mol. Biol.* **56**, 341–361.
2. Rabinow P (1996). *Making PCR: a Story of Biotechnology.* University of Chicago Press, Chicago, IL, USA.
★ 3. Saiki RK, Scharf S, Faloona F, *et al.* (1985). *Science,* **230**, 1350–1354. – *Original publication describing PCR.*
★ 4. Saiki RK, Gelfand DH, Stoffel S, *et al.* (1988). *Science,* **239**, 487–491. – *Original publication describing the use of Taq DNA polymerases in PCR.*
★ 5. Don RH, Cox PT, Wainwright BJ, Baker K & Mattick JS (1991) *Nucleic Acids Res.* **19**, 4008. – *Original publication describing Touchdown PCR.*

CHAPTER 2

Polymerases for PCR

Meg Martel, Simon Baker, Ian Kavanagh, and Simon May

1. INTRODUCTION

1.1 Function of DNA polymerases

DNA polymerases catalyze the template-directed synthesis of long polynucleotide chains by the incorporation of monomeric deoxynucleoside triphosphates, using one of the parental strands as a template for the synthesis of a new complementary strand. DNA polymerases require a short segment of DNA or RNA (the primer) to anneal to a complementary sequence to initiate synthesis. Deoxynucleoside triphosphates (dNTPs) used in DNA synthesis comprise deoxyadenosine triphosphate (dATP), deoxycytidine triphosphate (dCTP), deoxyguanosine triphosphate (dGTP) and deoxythymidine triphosphate (dTTP). These dNTPs are covalently joined to the free hydroxyl group of the primer and form a newly synthesized strand complementary to the template.

1.2 Origins of DNA polymerases

The majority of thermostable DNA polymerases for PCR originated from two classes of organisms, the thermophilic bacteria and the hyperthermophilic archaea. *Thermus aquaticus*, a thermophilic bacterium, is the source of the best understood of the thermostable DNA polymerases, known as *Taq* polymerase. It remains the workhorse of PCR in most laboratories. The most commonly used archeal enzyme was originally purified from *Pyrococcus furiosus* and is known as *Pfu* polymerase. Whilst the action of the polymerases in PCR is identical, the cellular function of the enzymes is different. The bacterial DNA polymerases used in PCR are DNA polymerase type I enzymes, whose cellular role is in the repair of DNA. The archeal polymerases belong to the DNA polymerase A or B families and are the main replicative DNA polymerases in this phylum.

PCR: *Methods Express* (S. Hughes and A. Moody, eds.)
© Scion Publishing Limited, 2007

2. METHODS AND APPROACHES

2.1 Unit definition of DNA polymerase

The unit definition of DNA polymerase is subject to considerable reinterpretation in commercial laboratories. Ultimately, this means that preparations of *Taq* and *Pfu* sold by different manufacturers are not identical. Variations due to *Taq* in standard PCR alone have been reported to affect the yield, the size of product achievable, and the fidelity of the amplified product generated (1). It is therefore important to optimize the polymerase concentration of each new enzyme that a laboratory obtains or to assay for DNA polymerase activity (see *Protocol 1*).

2.2 DNA polymerase activity

As well as polymerization functions, DNA polymerases may also have associated 3′→5′ and/or 5′→3′ exonuclease activities (see *Protocol 2*). The 5′→3′ activity allows the enzyme to excise bases ahead of it as it moves along the double-stranded DNA (a crucial function for applications involving probes annealing mid-amplicon), whilst the 3′→5′ activity checks base pairing (see below). For any given application, both of these exonuclease activities should be considered. This is particularly important for polymerases possessing a 3′→5′ activity, as this can cause degradation of the primers before PCR commences. This effect can be reduced by preparing reactions on ice, which will decrease enzyme activity; alternatively, a 'hot-start' enzyme can be used (discussed below). The presence of the 3′→5′ exonuclease activity is associated with 'proofreading'; therefore, the fidelity of these enzymes is greater than that of polymerases without this activity.

Commercially available thermostable enzymes can normally be divided into those with and those without a 3′→5′ exonuclease activity (see *Table 1*). The latter group adds a nontemplate-specified dATP to the 3′ termini of both strands of an amplicon. This can be exploited as a method of cloning by ligation of the PCR product into a linearized vector with overhanging 'T' ends (TA cloning). The lack of proofreading function leads to increased error rates (see *Table 1*) compared with an enzyme with 3′→5′ exonuclease activities. *Taq* DNA polymerase (both native and recombinant forms, with the latter being prepared from the *T. aquaticus* gene expressed in *Escherichia coli*) lacks 3′→5′ exonuclease activity, whilst *Pfu* (in native and recombinant forms) has a good proofreading ability.

2.3 Choice of enzyme

Ultimately, the choice of enzyme should be determined by the purpose of the experiment, with the presence or absence of a proofreading function being intrinsic to this decision. Enzymes without 3′→5′ exonuclease proofreading activity, such as *Taq*, are suitable for most applications including conventional PCR, quantitative PCR, primer extension, microarray analysis, and high-throughput PCR, whereas proofreading enzymes, such as *Pfu*, are the enzyme of

Table 1. Activities and source of DNA polymerase enzymes

DNA polymerase	Source	5'→3' exonuclease activity	3'→5' exonuclease activity	Resulting ends of amplicons	Error rate	Extension rate at 72°C	Applications
Taq	Thermus aquaticus	Yes	No	3'A	8×10^{-6}	2–4 kb/min	PCR, QPCR, DHPLC, microarray analysis
Taq preparation (N-terminal deleted)	T. aquaticus	No	No	3'A	8×10^{-6}	2–4 kb/min	PCR, QPCR, DHPLC, microarray analysis
Pfu	Pyrococcus furiosus	No	Yes	Blunt	1.3×10^{-6}	0.5 kb/min	Long PCR, high-fidelity PCR, PCR of GC-rich templates, PCR of templates with secondary structure
Psp	Pyrococcus species GB-D gene expressed in E.coli	No	Yes	95% Blunt	2.8×10^{-6}	0.5 kb/min	Long PCR, high-fidelity PCR, PCR of GC-rich templates, PCR of templates with secondary structure
Pwo	Pyrococcus woesei	No	Yes	Blunt	1.3×10^{-6}	0.5 kb/min	Long PCR, high-fidelity PCR, PCR of GC-rich templates, PCR of templates with secondary structure
Tbr	Thermus brockianus	Yes	No	3'A after 15 min at 72°C	9.5×10^{-6}	1–2 kb/min	PCR, QPCR, RT-PCR
Tfl	Thermus flavus	Yes	No	3'A	8×10^{-6}	2–4 kb/min	PCR
Tli	Thermococcus litoralis	No	No	>95% Blunt		2–4 kb/min	PCR
Tma	Thermotoga maritima	No	Yes	Blunt	5×10^{-5}	–	Long PCR, high-fidelity PCR
Tth	Thermus thermophilus	Yes	No	3'A	2.8×10^{-6}	2–4 kb/min	PCR, RT-PCR

DHPLC, denaturing high-performance liquid chromatography; QPCR, quantitative PCR; RT-PCR, reverse transcriptase PCR.

choice when carrying out PCR for cloning, where fidelity is essential, or amplification of templates that are GC-rich or contain secondary structure. However, proofreading enzymes tend to have a lower processivity (extension times are twice that of *Taq*). A mixture of both proofreading and nonproofreading enzymes, which are available from many commercial companies, can utilize the advantages of both enzyme systems and significantly increase the yield, length, and fidelity of the PCR product (2–4). This enhanced activity is thought to be due to the capacity of one enzyme to complement the inability of another to extend through obstructions on the template strand. These obstructions include regions of high secondary structure (5), abasic gaps that cannot be bridged by polymerases lacking terminal transferase activity (6), and mispaired bases that cause nonproofreading polymerases to pause and dissociate from the primer–template complex (2).

Once an enzyme has been chosen, it should be recognized that several factors including mutagenesis, chemical modification, or even changes in buffer composition can have a large effect on an enzyme's efficacy in PCR. Measurement of extension rate (see *Protocol 3*) gives an estimation of the number of bases that can be added in a given time. Whilst we are accustomed to *Taq* adding 1000 bp per min for up to 10 min, many polymerases cease to add bases after only 2–3 min.

The processivity (see *Protocol 4* and above) and fidelity (see *Protocol 5*) of a particular polymerase can be measured. The fidelity of the reaction measures the number of mispaired bases that the polymerase adds per 100 or 1000 template bases. The error rate in PCR is highly dependent on the type of *Taq* used and the buffer in which the reaction is performed. Indicative error rates are given in *Table 1*. For both processivity and fidelity measurements, it is important that the enzyme is not incubated at suboptimal temperatures. Periods spent above or below the optimum (e.g. if the reaction tube is left at room temperature for a few minutes prior to incubation in a water bath) can have a significant and misleading effect on the results. Whilst it may seem that an enzyme that introduces no errors is the desired result, there are applications for error-prone enzymes, such as PCR-mediated random mutagenesis.

2.4 Hot-start PCR

In the majority of cases in which PCR is used, DNA amplification of specific sequences takes place in a milieu of background DNA, and so the possibility of nonspecific amplification must be eliminated. This is particularly important when trying to amplify low-copy-number target sequences (7). Primer dimerization and nonspecific primer binding can cause erroneous product formation that is then amplified during subsequent cycles. Nonspecific amplification can result from incubation of reaction mixtures at room temperature during preparation, which can allow initiation of the reaction (7–10). Hot-start PCR is a method that can reduce these problems by removing at least one critical component from the reaction mix until a higher temperature is reached, when the primer-binding specificity is at its greatest. The traditional approach to hot-start PCR was to withhold an essential reaction component, usually the polymerase, until the temperature exceeded 55°C, at which point the tubes were opened and the critical component was added to permit polymerization to commence. However, this method increases the possibility of introducing contaminants and so other methods have been developed whereby a reagent is included that only functions when an optimal temperature is reached (7–10).

The current range of methods aimed at achieving hot-start PCR is listed below.

2.4.1 Chemical modification of *Taq* DNA polymerase

Taq DNA polymerase can be chemically modified using citraconylation (ThermoStart DNA polymerase, ABgene; AmpliTaq Gold, ABI). Dicarboxylic acid anhydrides introduce a negative charge in place of a positive charge on the amino group of lysine residues in the polymerase. This reaction is reversible by pre-incubation at 94–95°C for 9–12 min.

2.4.2 Oligonucleotide ligands

Screening of large collections of chemical compounds (combinatorial library screening) has been used to identify oligonucleotide ligands that bind *Taq* DNA polymerase in a temperature-dependent and reversible manner below 55°C (NeXstar Pharmaceuticals). The binding of the oligonucleotide ligand to the polymerase prevents enzyme activity at temperatures below 55°C, but acts in a reversible manner, and so, unlike some hot-start methods, the ligand is not denatured during the first heating (denaturation) step. This means that the enzyme is inactivated whenever the temperature drops below a specified temperature. Pre-activation of the inhibited enzyme is not required by this method (7).

2.4.3 Temperature-dependent *Taq* DNA polymerase inhibition using an antibody

An antibody specific to the *Taq* DNA polymerase catalytic site can be used to block *Taq* DNA polymerase activity (Platinum *Taq*, Invitrogen; JumpStart *Taq*, Sigma). At elevated temperatures, the antibody is denatured and the polymerase enzyme activity is regained, allowing the PCR to commence.

2.4.4 Wax/agarose beads or barriers

Heat-labile materials such as wax and jelly can separate reaction components until the reaction temperature exceeds a specified temperature. *Taq* DNA polymerase can be separated from the rest of the reaction mix through the use of a temperature-dependent physical barrier. Examples include TaqBead (Promega) in which a bead of paraffin wax melts when the temperature reaches 60°C, releasing the enzyme. AmpliGrease (ABI) uses petroleum jelly that acts as a barrier between the reagents; it melts at 55°C allowing the reagents to mix. Alternatively, in magnesium hot beads, the $MgCl_2$ is contained within wax pellets and when the temperature reaches 68°C, the $MgCl_2$ is released, enabling *Taq* DNA polymerase activity to commence.

2.5 Buffer composition

In PCR, a Tris buffer (10–75 mM) is required to maintain the reaction mix at the correct pH so that polymerization will occur. Other common reaction buffer constituents are potassium chloride (10–100 mM) or ammonium sulfate (10–30 mM). The pH of a PCR buffer should be in the range of 8–9 for the DNA polymerase enzyme to work efficiently. When carrying out PCR, it is best to use the buffer supplied by the manufacturer, as this has been optimized specifically for use with their enzyme. Most commercial buffers are supplied at 10× or 5× concentration and need to be thoroughly defrosted and mixed before use.

The reaction mix contains DNA, DNA polymerase, dNTPs, primers, and magnesium chloride. By varying the concentrations of one of these components, it is possible to alter the specificity of the reaction:

- High concentrations of DNA polymerase can result in the production of nonspecific PCR products.
- High concentrations of dNTPs can result in misincorporation by DNA polymerase.
- dNTPs provide energy and nucleotides for synthesis of DNA and it is important that equal amounts of each dNTP (dATP, dCTP, dGTP, dTTP) are added to the reaction mix to prevent mismatches of bases.
- Primers are short pieces of DNA that bind to the template allowing the DNA polymerase to initiate incorporation of the dNTPs.
  - Primers should be approximately 20 bases in length, and should be an appropriate distance apart to produce a product length of between 100 and 5000 bp.
  - It is highly advantageous for the primers to have a similar melting point (T_m) to each other, which is dependent on their length and GC content.
  - Inter- and intracomplementarity of bases within the primer(s) should be avoided to prevent the formation of primer dimers, as should stretches of polypurines and polypyrimidines.
- Magnesium ions are essential for enzyme activity. Proteins, dNTPs, and DNA chelate Mg^{2+} and it is therefore important to ensure that the absolute concentration of Mg^{2+} is sufficient to allow enzyme activity. If too much magnesium is present, the fidelity of the enzyme is compromised. Chelating agents such as ethylenediaminetetraacetic acid (EDTA) may reduce this problem; however, optimal concentrations of such compounds need to be determined empirically, as they can also inhibit the PCR (for a protocol on optimizing $MgCl_2$ in PCR, see Chapter 1).
- Enhancing agents, including dimethyl sulphoxide (DMSO), polyethylene glycol (PEG 6000), glycerol, nonionic detergents, and formamide (11), can be added to increase reaction specificity.

2.6 Inhibitors

Several compounds can inhibit PCR if present in excess. Common culprits include:

- Proteinase K (which can degrade thermostable DNA polymerase)
- Phenol
- EDTA, a magnesium chelator
- Ionic detergents (12)
- Heparin (13)
- Polyanions, such as spermidine (14)
- Hemoglobin
- Gel loading dyes such as bromophenol blue and xylene cyanol (15)

In many cases, the chief causes of low or erratic yields are contaminants in the template DNA, which is often the only component of the reaction supplied by the investigator. Many problems with PCR can be cured simply by cleaning up the template by dialysis, ethanol precipitation, extraction with chloroform, and/or chromatography through a suitable resin.

2.7 Detection methods for assessing enzyme activity

All methods either use a ^{32}P or biotin label as the primary means of detection. Results using ^{32}P (both the intensity of the bands and the absolute figures that may be obtained from scintillation counting) depend on:

- The specific activity of the radiolabel supplied by Amersham Biosciences (which varies from batch to batch);
- The age of the radiolabel (due to the decay of the radiolabel once manufactured).

This means that proper controls must be carried out in each case with every iteration to ensure that the figures obtained are a valid representation of the activity measured.

Protocol 1

Assay for DNA polymerase activity

Equipment and Reagents

- 0.5–1.5 µg DNA polymerase
- [α-^{32}P]ATP (Amersham Biosciences)
- DNA polymerase reaction buffer (20 mM Tris/HCl (pH 8.0), 10 mM KCl, 2 mM MgSO$_4$, 10 mM (NH$_4$)$_2$SO$_4$, 0.1% Triton X-100, 100 µg/ml bovine serum albumin (BSA), 66.67 µM dNTPs (dATP, dCTP, dGTP, dTTP; ABgene), 200 µg/ml activated calf thymus DNA (Sigma), 1 µCi [α-^{32}P]ATP (3000 Ci/mmol))
- 500 mM EDTA (Sigma)
- Carrier solution (2 mM EDTA, 50 µg/ml activated calf thymus DNA)
- 20% (w/v) Trichloroacetic acid (Sigma)
- 2% (w/v) Sodium pyrophosphate (Sigma)
- GF/B glass fiber filters (Whatman)
- Block heater or thermal cycler
- Liquid scintillation counter (Packard Tri-Carb 2200 CA; Packard BioScience)

Method

1. Add 0.5–1.5 µg of DNA polymerase enzyme in a volume of 15–135 µl of pre-warmed (to 75°C) reaction buffer (containing 200 µg/ml of activated calf thymus DNA)[a].

2. Incubate at 75°C.

3. Remove a 15 µl sample every minute for up to 10 min.

4. Terminate the reaction by adding EDTA to a final concentration of 30 mM.

5. Dilute samples with 1 ml of carrier solution.

6. Precipitate the DNA by the addition of 1 ml each of 20% (w/v) trichloroacetic acid and 2% (w/v) sodium pyrophosphate.

7. Incubate on ice for 15 min.

8. Collect the DNA on GF/B glass fiber filters

9. Measure the amount of radioactivity incorporated by scintillation counting. Each sample is counted for 2 min and radioactivity is expressed as disintegrations per min. Determine the specific activity from plots of counts versus time[b].

Notes

[a]Quantify the DNA polymerase using the Bradford method for protein quantification. The assay is based on the observation that the absorbance maximum for an acidic solution of Coomassie Brilliant Blue G-250 shifts from 465 to 595 nm when binding to protein occurs, causing a visible color change.

[b]One unit of enzyme activity is defined as the amount of enzyme required to incorporate 10 nmol of dNTP into an acid-insoluble form at 75°C in 30 min.

Protocol 2

Exonuclease activity assays of the DNA polymerase

Equipment and Reagents

- 50 units of DNA polymerase
- [γ-^{32}P]ATP (Amersham Biosciences)
- [^{3}H]dTTP (Amersham Biosciences)
- 5–10 units of *Hind*III (New England Biolabs)
- 100% Ethanol
- 3 M Sodium acetate (Sigma)
- 10 µg of λ DNA (ABgene)
- 5–10 units of T4 polynucleotide kinase (ABgene)
- 5–10 units of the Klenow fragment of DNA polymerase I (ABgene)
- Exonuclease reaction buffer (10 mM Tris/HCl (pH 8.0), 1.5 mM MgCl$_2$, 50 mM KCl, 0.1% Triton X-100)
- 10% Trichloroacetic acid (Sigma)
- 1% (w/v) BSA (Sigma)
- EDTA (Sigma)
- GF/B glass fiber filters (Whatman)
- Block heater or thermal cycler
- Liquid scintillation counter
- Centrifuge

Method

1. Digest 10 µg of λ DNA using *Hind*III in 50 µl of the appropriate buffer at 37°C for 1 h.

2. Stop the reaction by heating at 65°C for 15 min.

3. Label *Hind*III-digested λ DNA fragments with [γ-^{32}P]ATP at its 5′ end using T4 polynucleotide kinase, according to the manufacturer's instructions.

4. Precipitate the DNA by adding 0.1 vols of 3 M sodium acetate and 2.5 vols of 100% ethanol[a]. Mix the solution by vortexing, incubate at –20°C for 20 min, and centrifuge at 12 000 r.p.m. for 20 min. Remove supernatant and air dry the pellet at room temperature for 10 min.

5. Label the 5′-labeled DNA fragment at the 3′ end with [^{3}H]dTTP and the Klenow fragment of DNA polymerase I, according to the manufacturer's instructions.

6. Add 0.2 µg of the double-labeled DNA fragments to 50 µl of the exonuclease reaction buffer.

7. Pre-incubate for 5 min at 75°C.

8. Add 50 units of DNA polymerase.

9. Incubate for 30 min at 75°C.

10. Terminate the reaction by adding 50 µl of 1% BSA and 100 µl of 10% trichloroacetic acid.

11. Place the reaction on ice for 10 min.

12. Centrifuge for 10 min at 12 000 r.p.m.

13. Collect the DNA on GF/B glass fiber filters.

14. Measure the amount of radioactivity incorporated by scintillation counting. Determine the 3′→5′ and 5′→3′ exonuclease activity of the DNA polymerase from the release of ^{3}H and ^{32}P from the 3′ and 5′ ends, respectively, of the labeled *Hind*III-digested λ DNA fragments[b].

Notes

[a]Cool the ethanol in a −20°C freezer to aid precipitation.

[b]When 3′→5′ exonuclease activity of DNA polymerase is present, ^{3}H-labeled nucleotides from the 3′ end of the DNA will be released into an acid-soluble form and detected by the scintillation counter. When 5′→3′ exonuclease activity of DNA polymerase is present, ^{32}P-labeled nucleotides from the 5′ end of the DNA will be released into an acid-soluble form and detected by the scintillation counter.

Protocol 3

Extension rate measurements

Equipment and Reagents
- 5 units of DNA polymerase
- 50 µM P7 primer (5′-CGCCAGGGTTTTCCCAGTCACGAC-3′)
- [α-^{32}P]dCTP (Amersham Biosciences)
- 1 µg/µl M13 ssDNA (Amersham Biosciences)
- 100 mM dNTPs (dATP, dCTP, dGTP, dTTP; ABgene)
- Reaction buffer (20 mM Tris/HCl (pH 8.8), 10 mM KCl, 10 mM $(NH_4)_2SO_4$, 2 mM $MgSO_4$, 0.1% Triton X-100)
- NaOH (Sigma)
- EDTA (Sigma)
- Block heater or thermal cycler
- 1% Agarose gel containing 10 ng/ml ethidium bromide (Sigma)
- 6× Gel loading buffer (ABgene)
- UV light source and gel documentation system

Method
1. Combine the following:
   - 1.6 µl (1.6 µg) of M13 ssDNA
   - 0.32 µl (16 pmol) of P7 primer
   - 0.16 µl (0.2 mM) each of dATP, dCTP, dGTP, and dTTP
   - 0.1 µM [α-^{32}P]dCTP (1.11 MBq)
   - 5 units of DNA polymerase
   - Reaction buffer to a final volume of 80 µl[a]

2. Incubate at 75°C.

3. Remove a 15 µl sample every minute for up to 10 min.

4. Terminate the reaction by adding stop solution containing EDTA and NaOH to final concentrations of 60 mM each.

5. Analyze a 10 µl aliquot of each sample by agarose gel electrophoresis.

Notes

[a]Ensure that the reaction buffer pH is 8.8, as this is critical for the reaction.

Protocol 4

Processivity measurements[a]

Equipment and Reagents

- DNA polymerase
- 50 μM 5′-biotin-labeled M13 primer (5′-CGCCAGGGTTTTCCCAGTCACGAC-3′)
- 1 μg/μl M13 ssDNA (Amersham Biosciences)
- Reaction buffer (20 mM Tris/HCl (pH 8.0), 2 mM MgCl$_2$, 6 mM (NH$_4$)$_2$SO$_4$, 10 mM KCl, 0.1% Triton X-100, 0.01% BSA)
- Stop solution (95% formamide, 20 mM EDTA, 0.05% bromophenol blue, 0.05% xylene cyanol)
- 6% Polyacrylamide gel
- Chemical luminescence method (ECL kit; Amersham)

Method

1. Anneal 200 fmol of 5′-biotin-labeled M13 primer to 100 fmol of the M13 ssDNA template using two different ratios of DNA polymerase (10:1 and 100:1 ratios of primer–template to DNA polymerase) at 75°C in reaction buffer.

2. Take 3.5 μl aliquots after 20 s, 1 min, and 3 min, and place in 1.5 μl of stop solution.

3. Analyze the products by denaturing 6% polyacrylamide gel electrophoresis (PAGE) and detect using a chemical luminescence method.

Notes

[a]This method was first described by Tabor *et al.* (1987) (16). Processivity is one of the most important parameters for DNA polymerases and is defined as the number of nucleotides that can be extended in one catalytic reaction by one DNA polymerase molecule.

Protocol 5

Fidelity of DNA polymerase

Equipment and Reagents
- *PvuI* and *PvuII* and Reaction Buffer 3 (New England Biolabs)
- 5 units of DNA polymerase
- 1 µg/µl M13mp18 dsDNA (Fermentas)
- 100 µg/µl M13 ssDNA (Amersham Bioscience)
- 1% Low-melting-point agarose gel containing 10 ng/ml ethidium bromide
- 100 mM dNTPs (dATP, dCTP, dGTP, dTTP; ABgene)
- Block heater or thermal cycler

Method
1. Perform a double digest of 50 µg of M13mp18 dsDNA with *PvuI* (5 units) and *PvuII* (5 units) in Reaction Buffer 3 in a final reaction volume of 20 µl, according to the manufacturer's instructions.

2. Resolve the digestion products on an agarose gel and purify the 6.5 kb band by phenol extraction[a,b].

3. Heat the digested dsDNA with 100 µg of M13 ssDNA at 95°C for 10 min.

4. Cool in an ice bath and then incubate at 65°C for 30 min.

5. Purify the M13mp18 molecule (which is now incompletely double stranded, with gaps in one of the strands) by agarose gel electrophoresis (1% low-melting-point agarose gel), phenol extraction, and ethanol precipitation[a].

6. Add 100 ng of the purified M13mp18 molecule into a DNA polymerase reaction mixture containing 200 µM dNTPs and 5 units of DNA polymerase and incubate at 75°C for 30 min to fill in the gaps.

7. Use 2 µl (approximately 4 ng) of the remaining product for transfection.

8. Calculate the mutation frequencies by counting the light blue and white plaques[c].

Notes

[a]Melt the slice by incubating the tube at 65°C. Add 1 vol. of Tris/EDTA-saturated phenol and vortex for 30 s. Centrifuge in a microcentrifuge at 12 000 r.p.m. for 5 min at room temperature to separate the phases. Transfer the aqueous phase to a clean tube, extract twice with an equal volume of chloroform, and precipitate the DNA by adding 0.1 vols of 3 M sodium acetate and 2.5 vols of ethanol. Mix by vortexing, incubate at –20°C for 20 min and centrifuge at 12 000 r.p.m. for 20 min. Air dry the pellet at room temperature for 10 min.

[b]The DNA could also be purified using a QIAquick PCR Purification kit (Qiagen).

[c]A mutation, as detected as a blue colony, shows that misincorporation of a dNTP has occurred during the extension of the complementary M13 ssDNA.

3. TROUBLESHOOTING

- **Which type of DNA polymerase should be chosen for PCR?**
  The size and end use of the PCR product needs to be considered when choosing which DNA polymerase enzyme to use.

- **Can one set of reaction conditions be used for all DNA polymerases in PCR?**
  When changing the DNA polymerase enzyme, for example from *Taq* to *Pfu*, it is important to remember that the PCR conditions, such as magnesium concentration, annealing temperature, and extension times, may need to be reoptimized.

- **Will a single buffer work for all types of *Taq?***
  In order to obtain accurate results, it is important to use the appropriate buffer with the corresponding DNA polymerase enzyme, as buffer pH is critical for the reaction to work optimally.

- **Is there any carry-over of bacterial DNA from the DNA polymerase enzyme preparation?**
  Yes. It is important to include a negative (no template) control when using DNA polymerase enzymes for PCR.

- **Can the purity of the genomic DNA influence the outcome of the PCR?**
  Yes. The best results are obtained when starting with good-quality genomic DNA.

4. REFERENCES

★★★ 1. Linz U, Delling U & Rübsamen-Waigmann H (1990) *J. Clin. Chem. Clin. Biochem.* **28**, 5–13. – *Description of the various factors that can influence the efficiency and specificity of PCR.*

2. Barnes WM (1994) *Proc. Natl. Acad. Sci. U. S. A.* **91**, 2216–2220.

3. Cheng S, Fockler C, Barnes WM & Higuchi R (1994) *Proc. Natl. Acad. Sci. U. S. A.* **91**, 5695–5699.

4. Cohen J (1994) *Science*, **263**, 1564–1565.

5. Eckert KA & Kunkel TA (1993) *Nucleic Acids Res.* **21**, 5212–5220.

6. Hu G (1993) *DNA Cell Biol.* **12**, 763–770.

7. Dang C & Jayasena SD (1996) *J. Mol. Biol.* **264**, 268–278.

8. Birch DE, Kolmodin L, Wong J, *et al.* (1996) *Nature*, **381**, 445–446.

9. Nilsson J, Bosnes M, Larsen F, Nygren PA, Uhlen M & Lundeberg J (1997) *BioTechniques*, **22**, 744–751.

★★ 10. Kellogg DE, Rybalkin I, Chen S, *et al.* (1994) *BioTechniques*, **16**, 1134–1137. – *Description of the basic 'hot-start' technique for PCR.*

★★ 11. Roux KH (1995) *PCR Methods Appl.* **4**, S185–S194. – *Provides useful optimization and troubleshooting tips for PCR.*

12. Weyant RS, Edmonds P & Swaminathan B (1990) *BioTechniques*, **9**, 308–309.

13. Beutler E, Gelbart T & Kuhl W (1990) *BioTechniques*, **9**, 166.

14. Ahokas H & Erkkila MJ (1993) *PCR Methods Appl.* **3**, 65–68.

15. Hoppe BL, Conti-Tronconi BM & Horton RM (1992) *BioTechniques*, **12**, 679–680.

16. Tabor S, Huber HE & Richardson CC (1987) *J. Biol. Chem.* **262**, 16212–16223.

CHAPTER 3

A detailed guide to quantitative RT-PCR

Pete Kaiser

1. INTRODUCTION

Quantitative reverse transcriptase PCR (qRT-PCR), also known as real-time qRT-PCR (as the technique is designed to detect and quantify sequence-specific PCR products as they accumulate in 'real-time' during the PCR amplification process), is characterized by high sensitivity and specificity, and has become the method of choice for quantitative gene expression measurements. In particular, it has revolutionized molecular diagnostics. Real-time PCR was first described in 1993 (1). Its logical successor, qRT-PCR, was first described 3 years later in 1996 (2) and remains the most sensitive technique available for detection and quantification of mRNA. It has a dynamic range of $>10^7$-fold, compared with 2–3 logs for normal end-point RT-PCR techniques, such as relative RT-PCR and competitive RT-PCR.

Relative RT-PCR uses primers for an internal control gene that are multiplexed in the same RT-PCR as the target gene-specific primers. The primers must be compatible, i.e. they must not produce additional bands or hybridize to each other. The use of internal controls in relative RT-PCR requires substantial optimization. For relative RT-PCR data to be meaningful, the PCR must be terminated when the products from both the internal control gene and the target gene of interest are detectable and are being amplified within the exponential phase. As internal control RNAs are typically constitutively expressed reference genes of high abundance, their amplification surpasses the exponential phase with very few PCR cycles. It is therefore difficult to identify compatible exponential-phase conditions where the PCR product from a rare message is detectable.

Competitive RT-PCR quantitates a message by comparing RT-PCR product signal intensity with a concentration curve generated by a synthetic competitor RNA sequence. This competitor RNA transcript is designed to be amplified by the same primers and with the same efficiency as the endogenous target, but it produces a different-sized product so that it can be distinguished from the endogenous target product by gel analysis. The competitor must be carefully

PCR: *Methods Express* (S. Hughes and A. Moody, eds.)
© Scion Publishing Limited, 2007

quantified and titrated into replicate RNA samples to find the range of competitor concentrations where the experimental signal is most similar. Finally, the mass of product in the experimental samples is compared with the curve to determine the amount of a specific RNA present in the sample.

Real-time PCR/RT-PCR has been applied to mRNA expression studies in tissues, cell types, paraffin-embedded tissues (3–5), and even laser-capture micro-dissected cells (6–9); for copy number measurements, both RNA and DNA, of viruses, bacteria, and protozoan pathogens (10–15); for allelic discrimination assays (16, 17), and for analyzing expression of different splice variants of the same gene (18, 19).

For accurate gene expression profiling by qRT-PCR, several parameters must be considered and carefully optimized. These include careful primer and probe selection for the genes under investigation and the use of reference genes for data normalization in order to compensate for variables in the PCR process. Accurate normalization is an absolute prerequisite for correct measurement of gene expression. We will provide a detailed guide for the planning, preparation, performance, and analysis of a qRT-PCR experiment using Applied Biosystems TaqMan chemistry. This will include pointers towards primer/probe design, correct selection of reference genes for normalization, and discussion of modes of data normalization.

2. METHODS AND APPROACHES

2.1 Technologies available

There are two general chemistries available for use on the various platform technologies. Both detect PCR products via the generation of a fluorescent signal. The first, and most common, uses fluorescent probes specific for the target mRNA species of interest. These rely on Forster resonance energy transfer, or FRET, to quench a fluorescence signal by coupling a fluorogenic dye molecule and a quencher moiety to either the same or different oligonucleotides. *Fig. 1* shows the principles of, and differences between, the various techniques.

The original chemistry was developed by Applied Biosystems (ABI) using TaqMan probes. These probes are template specific, and also require two template-specific primers which anneal either side of, and in close proximity to, the probe (see *Fig. 1a*). TaqMan probes have a fluorescent reporter dye attached to the 5′ end and a quencher moiety attached to the 3′ end. Normally, the proximity of the fluor and the quencher prevents the detection of any fluorescence from the probe. In addition to polymerase activity, *Taq* polymerase also has 5′-exonuclease activity. During each PCR step, as the polymerase replicates the template to which the probe is bound, the 5′ end of the probe is displaced from the template and the 5′-exonuclease activity can cleave the probe. Once the fluor and the quencher are decoupled, FRET no longer occurs and detectable fluorescence thus increases with each PCR cycle in proportion to the amount of probe cleaved and therefore the amount of specific template originally present.

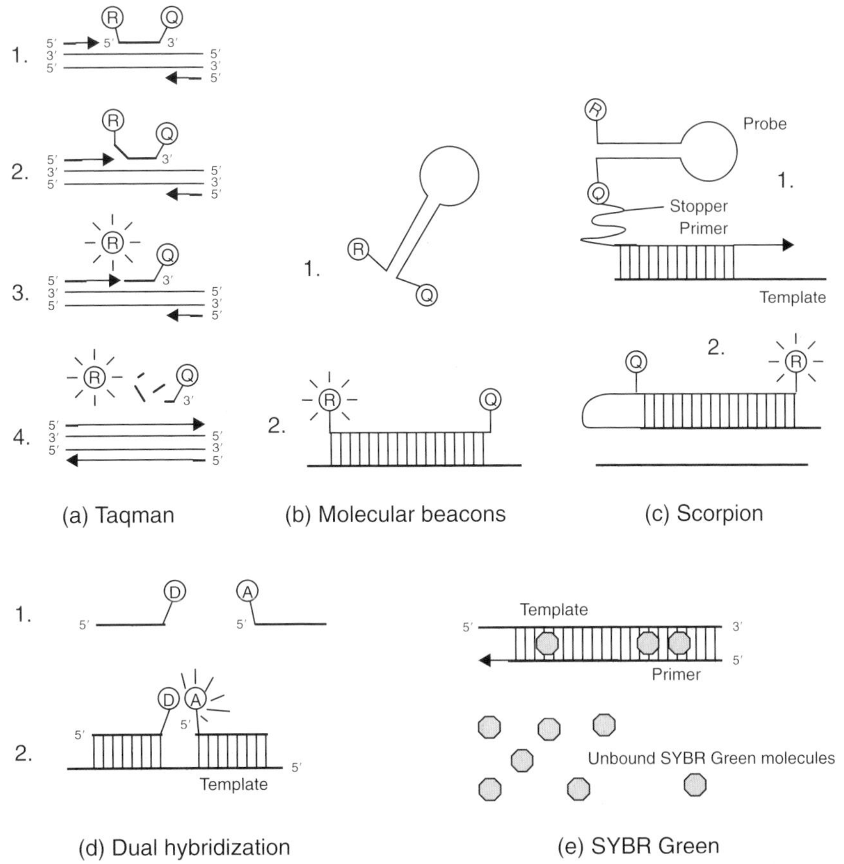

(a) Taqman   (b) Molecular beacons   (c) Scorpion

(d) Dual hybridization   (e) SYBR Green

Figure 1. Schematic of the different technologies available for quantitative RT-PCR.
(a) TaqMan probes. Two primers (forward and reverse) are designed to the target template, with a target-specific probe with two fluorescent dyes attached designed to anneal between the two primers. (1) The primers are extended by Taq polymerase as in a normal PCR. (2) As Taq extends the forward primer, the 5′ end of the probe is displaced. (3) In addition to polymerase activity, Taq also has 5′ exonuclease activity, so the reporter dye is cleaved from the probe. (4) As the reporter dye is no longer held in close proximity to the quencher dye, fluorescence emitted by the reporter dye can be detected. (b) Molecular beacons. (1) These probes form a hairpin loop, keeping the reporter and quencher in close proximity and blocking fluorescence. (2) Upon annealing to their target sequence, the probes change conformation, the two fluorophores are separated, and fluorescence can be detected. (c) Scorpion probes. (1) With this technology, a template-specific primer and probe are linked by a nonamplifiable 'stopper'. The probe is a hairpin loop structure, bringing the reporter and quencher into close proximity, as with the molecular beacons probes, and therefore quenching any fluorescence. The 'stopper' prevents readthrough of the hairpin loop. During PCR, the Scorpion primer is extended to become part of the amplicon and generates a target sequence complementary to the probe. (2) During the next annealing step of the PCR, the hairpin loop probe binds to the newly formed complementary target sequence in the PCR product. The two dyes are therefore separated and fluorescence emitted by the reporter dye can be detected. (d) Dual hybridization. (1) This technology uses two target-specific probes, one carrying a donor fluorophore at its 3′ end and the other an acceptor fluorophore at its 5′ end. (2) The two probes are designed to anneal in a head-to-tail conformation, bringing the two dyes together, resulting in emission of fluorescence at a different wavelength. (e) SYBR Green. When unbound and in solution, SYBR Green gives off little fluorescence. During PCR, SYBR Green molecules are intercalated into the dsDNA amplification product, which results in an increase in the amount of fluorescence detected.

Molecular beacons are similar to TaqMan probes, but in this case the probe forms a stem–loop structure when in solution, putting the fluor and quencher in close proximity (see *Fig. 1b*). When the molecular beacon anneals to its specific template, the fluor and quencher are separated, FRET no longer occurs, and fluorescence can be detected. The probe, however, remains intact.

Scorpion probes, like molecular beacons, form a stem–loop structure putting the fluor (5′) and quencher (3′) into close proximity (see *Fig. 1c*). The specific probe sequence is within the hairpin loop and is attached to the 5′ end of another template-specific PCR primer by a nonamplifiable monomer, the PCR stopper. After extension of the Scorpion primer, the specific probe sequence can now bind to its template, thus opening up the hairpin loop and decoupling the fluor and quencher. Once again, FRET no longer occurs and fluorescence can be detected.

With dual hybridization probes, one carries a donor fluor at the 3′ end and the other an acceptor fluor at the 5′ end (see *Fig. 1d*). The probes are designed so that, during annealing, they both hybridize close enough to each other for FRET to occur. However, in this case FRET causes the acceptor fluor to emit energy at a different wavelength, which is then detected.

The second chemistry uses dsDNA-intercalating agents, or DNA-binding dyes, such as SYBR Green I or II (see *Fig. 1e*). There is therefore no need for an amplicon-specific probe, and this approach is therefore simpler and cheaper. In solution, SYBR Green emits little fluorescence. Fluorescence is therefore only detected when SYBR Green is intercalated into the minor groove of dsDNA, and the intensity of fluorescence is proportional to the amount of dsDNA present. However, SYBR Green will bind to all dsDNA present, such as primer dimers and nonspecific PCR products.

2.2 Choice of reference gene

The first thing to state is that there is no 'perfect' reference gene, but some are in general better than others. The most common genes used include β-actin, glyceraldehyde 3-phosphate dehydrogenase (GAPDH), and 18S and 28S rRNA. β-Actin expression levels are linked strongly to cell-cycle stage, whereas GAPDH expression can also be modified by cell proliferation, as well as by glucose, heat shock, and other treatments (20–22). In general, rRNA expression levels are less variable (23, 24), although some consider it a poor indicator of the total cellular mRNA population, as it is expressed at a much higher level than mRNA. An alternative approach is to use multiple reference genes, and to use the mean expression of these multiple genes for normalization. We routinely use 18S or 28S rRNA as reference genes (10), but others working in the avian system disagree (25).

2.3 Analysis of real-time qRT-PCR data

There are two main methods of quantifying the results obtained from real-time qRT-PCR: the absolute quantitation method (using a standard curve) and the relative quantitation method (or the $2^{-\Delta\Delta CT}$ method).

2.3.1 Absolute quantitation method

The absolute quantitation method generates copy numbers for each test sample by relating test sample cycle threshold (C_T) values back to a standard curve of C_T values based on a dilution series of plasmid containing the target species of interest. There are two problems with this approach. Firstly, the method does not provide a control for the RT step, using plasmid DNA as a template for the PCR amplification steps of the reaction only. More importantly, mRNA for any target is not a single species, as poly(A) tails will be of different lengths, each mRNA molecule may be degraded to a different degree, etc. Therefore, generating copy numbers of mRNA from a standard curve based on a DNA dilution series is, at best, an estimate.

2.3.2 Relative quantitation method

The $2^{-\Delta\Delta Ct}$ method compares target species mRNA levels between different samples by correcting for a reference gene and relating expression levels to a reference control (e.g. an untreated sample). This assumes, however, that the relative amplification efficiencies of the reference and target genes are approximately equal. In practice, this is a dangerous assumption, as the relative amplification efficiencies are often quite different. It is therefore best to correct for the relative amplification efficiencies of the reference and target genes using the slopes of the respective standard curves, as detailed below in section 2.3.3.

2.3.3 Quantification

Using the ABI system, quantification is based on the increased fluorescence detected due to hydrolysis of the target-specific probes by the 5′-exonuclease activity of the recombinant *Tth* (*rTth*) DNA polymerase during PCR amplification. The passive reference dye 6-carboxy-c-rhodamine, or ROX, which is not involved in amplification, is used for normalization of the reporter signal necessary to correct for fluorescent fluctuations due to changes in concentration or volume in the wells. Normalization is carried out automatically by the software provided by ABI, which divides the emission intensity of the reporter dye by that of the passive reference to obtain a ratio defined as Rn (normalized reporter) for a given reaction well. Results are expressed in terms of C_T value, the cycle at which the change in the reporter dye passes a significance threshold (ΔRn).

We use the following calculations, first described by Withers *et al.* (26), to calculate corrected C_T values. To account for variation in sampling and RNA preparation, the C_T values for target-specific product for each sample are standardized using the C_T value of the reference product for the same sample. To normalize RNA levels among samples within an experiment, the mean C_T value for the reference gene-specific product is calculated by pooling values from all samples in that experiment. Tube-to-tube variations in reference gene C_T values about the experimental mean are calculated. The slope of the reference gene log_{10} dilution series regression line is used to calculate differences in input total RNA. Using the slopes of the respective target gene or reference gene log_{10} dilution series regression lines, the difference in input total RNA, as represented by the reference gene, is then used to adjust target gene-specific C_T values, as follows:

$$\text{Corrected } C_T \text{ value} = C_T + (Nt - C_T') * S/S'$$

where C_T is the mean sample target gene C_T, Nt is the experimental reference gene mean, C_T' is the mean reference gene value of the sample, S is the target gene slope, and S' is the reference gene slope. Results are then expressed as $40 - C_T$ values, or fold-change differences from a control sample.

Fig. 2 shows examples of standard curve raw data, similar raw data transformed to give reference and target gene slopes, actual raw data samples for

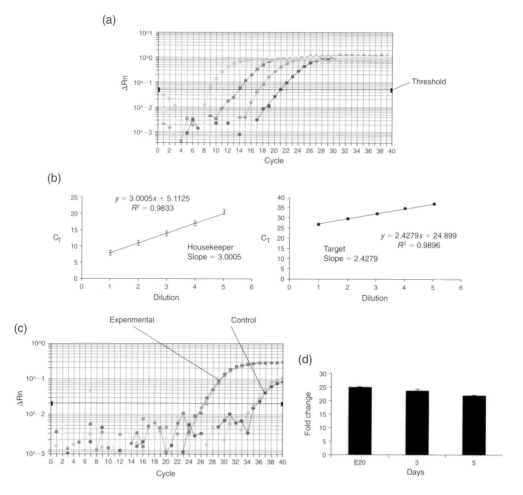

Figure 2. Typical examples of data generated from TaqMan experiments.
(*a*) Raw standard curve data. The curves represent fluorescence detected (ΔRn) per cycle of the PCR for tenfold dilutions of template RNA. C_T values are calculated as the number of cycles necessary for ΔRn to pass the threshold. (*b*) Standard curves for a reference and a target gene. The slope is taken from the equation of the line, $y = mx + c$, where m is the slope. (*c*) Raw data for experimental and control samples. (*d*) Fold-change graphs comparing expression levels in experimental samples with those in age-matched controls.

a control and an experimental sample for the target gene, and fold-change graphs for expression levels of the experimental samples compared with the controls.

2.4 Designing TaqMan primers and probes

As standard, we use the PRIMER EXPRESS program (Applied Biosystems) for primer and probe design. The program will find a series of possible primer/probe sets and the 'best' set should be selected on the basis of their compatibility with the following rules (those marked * are default parameters already in the program).

1. Either one of the primers or the probe must cross the intron/exon boundary.
2. The primers:
   - Should be between 9 and 40 nt, with a 20–80% GC content*;
   - Should have a T_m of 58–60°C and there should be less than 2°C difference in T_m between the two primers*;
   - Must have a maximum of two C or G bases in the last 5 nt at the 3′ end.
3. The probe:
   - Should be between 9 and 40 nt, with a 20–80% GC content*;
   - The T_m should be 69°C (i.e. 10°C higher than the T_m of the primers)*;
   - Cannot start with a G at the 5′ end;
   - Must have C≥G;
   - Must have fewer than four contiguous Gs* (the probe sequence can always be reverse-complemented to make sure it obeys these rules).
4. The amplicon:
   - Should be between 50 and 150 nt*;
   - Should have the 3′ end of the primer as close to the probe as possible without overlapping*.

Although there are alternative programs and parameters that are also suitable for primer and probe design, we have always obtained good results when following these guidelines.

2.5 Recommended protocols

These protocols are based on almost a decade of experience of using TaqMan real-time qRT-PCR assays in both avian and murine models (for examples, see 27–40).

Protocol 1

Homogenization of animal tissues for subsequent RNA extraction

Equipment and Reagents
- Bead mill (Retsch MM 300)
- 5 mm Stainless steel beads, one per tube (RNase-free)
- RLT buffer (Qiagen) (10 μl of β-mercaptoethanol (Sigma) per 1 ml RLT is added before use)
- RNAlater (Ambion)

Method
1. Weigh out 30 mg of tissue stabilized in RNAlater.

2. Place the tissue into a 2 ml safe-lock Eppendorf tube[a] with 600 μl of RLT lysis buffer (10 μl of β-mercaptoethanol per 1 ml of RLT buffer is added before use).

3. Add one 5 mm stainless steel bead to each tube.

4. Insert the tubes into the bead mill and set the program for 4 min, 20 Hz.

5. Carry out normal RNA extraction using an RNeasy mini kit (Qiagen), following the manufacturer's instructions[b].

Notes
[a]Take care when labeling tubes, as writing on lids tends to rub off during homogenization; it is better to write on the side of the tube.
[b]We usually pass the homogenate through a QIAshredder (Qiagen); however, this step may be omitted if the tissue is thoroughly disrupted.

Protocol 2

Optimizing primer concentrations for real-time qRT-PCR assays

Equipment and Reagents
- ABI Prism 7700 Sequence Detector or ABI 7500 FAST (Applied Biosystems)
- RNase-free 0.5 ml tubes
- RNase-free H_2O
- Microfuge and plate centrifuge
- DNase- and RNase-free Thermofast 96-well PCR plates and domed lids (ABgene), plate holder, and lid applicator
- Reverse transcriptase qPCR master mix (Eurogentec)
- Gene-specific primers[a] (100 μM in DEPC-treated H_2O) and fluorescent-labeled probe[a] (5' FAM, 3' TAMRA; 5 μM in DEPC-treated H_2O)
- 'Standard' RNA, i.e. positive-control RNA used for the standard curves for the gene of interest

Method

1. Prepare sufficient master mix (see table below). This needs to be sufficient for six replicates of the standard curve (with a minimum of five tenfold dilutions), three no-template-control (NTC) wells, and extra volume to allow for pipetting errors.

	Basic mix (µl)	Master mix: total for 38 wells (µl)
2× PCR master mix (in kit)	12.5	475.0
Primer mix[b]	–	–
Probe (5 µM)	0.50	19.0
'Euroscript' enzyme (in kit)	0.125	4.75
DEPC-treated H_2O	5.875	223.25

2. Thaw all reaction components (including primers and probe, *but excluding the RNA*) on ice.

3. Prepare the primer dilutions first[b] (see *Table 1*) and then prepare the master mix, vortex to mix, pulse spin, and place on ice.

4. Place a Thermofast 96-well plate in a rack on ice, and add 19 µl of master mix to each well and 1.0 µl of each of the various primer dilutions to the appropriate wells. Then add 5.0 µl of DEPC-treated H_2O to the NTC wells and cap these wells off to prevent contamination.

5. Thaw the standard RNA samples on ice and prepare the serial tenfold dilutions for these[c].

6. Pipette 5.0 µl of each RNA standard into each well, as appropriate[d].

7. Cap off all wells properly (use the lid applicator) and spin the plate down briefly using the plate centrifuge.

8. Place the plate in the real-time RT-PCR machine and use the following program:
   - 2 min at 50°C
   - 30 min at 60°C
   - 5 min at 95°C
   - 40 cycles of 20 s at 94°C, 1 min at 59°C

Notes

[a]Designed using PRIMER EXPRESS software and produced by any reputable oligonucleotide manufacturer. It is important to ensure that the primer and probes are desalted, as salts can interfere with enzyme activity and significantly reduce the intensity of the signal.

[b]A range of primer dilutions needs to be prepared for optimization, ranging from 1.0 to 0.1 µM (final concentration); details are given in *Table 1*.

[c]You will need a total of 30 µl of each dilution (5 µl/well, and there are six replicates), so dilute 3.4 µl of RNA in 30.6 µl of DEPC-treated H_2O, vortex to mix thoroughly (*this step is very important*), and continue with further tenfold dilutions by transferring 3.4 µl of the first dilution to the next tube containing 30.6 µl of DEPC-treated H_2O, and so on. Place all tubes on ice.

[d]Vortex the RNA before pipetting to ensure that it is mixed thoroughly.

2.6 Analyzing primer optimization results

The amplification plots should show even increments between each dilution. Select the first two rows of standards (i.e. two replicates, although the second row

Table 1. Primer dilutions

Primer	Dilution	[μM][a]	[Final μM][b]	Forward (100 μM stock)	Reverse (100 μM stock)	DEPC-treated H₂O
I	1:2	50	1.0	2.5 μl	2.5 μl	5.0 μl
II	1:2.5	40	0.8	2.0 μl	2.0 μl	6.0 μl
III	1:3.3	33	0.6	1.5 μl	1.5 μl	7.0 μl
IV	1:5	20	0.4	1.0 μl	1.0 μl	8.0 μl
V	1:10	10	0.2	0.5 μl	0.5 μl	9.0 μl
VI	1:20	5	0.1	0.25 μl	0.25 μl	9.5 μl

[a]Primer concentration after dilution of 100 μM stocks.
[b]Final concentration of primer per well.

(a)

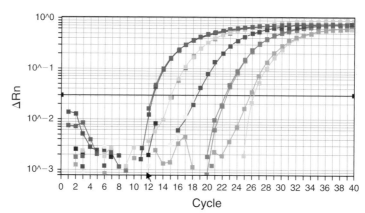

(b)

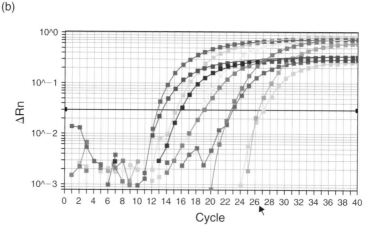

Figure 3. Raw standard curve data for optimization of primer concentrations.
Until primer concentrations become limiting, different sets (i.e. different concentrations of primers) of standard dilutions of template RNA should give identical amplification curves (*a*). Once the primer concentrations begin to become limiting, there will be deviation in the amplification plots compared with those for the previous dilution (*b*). The optimal concentration for the primers is therefore the lowest dilution that shows no deviation in amplification plot.

has a lower concentration of primer than the first row) and again check the amplification plot. The two sets of standard dilutions should give identical amplification curves (see *Fig. 3a*) until the primer concentrations begin to become limiting, at which stage there should be some deviation in the amplification plots compared with those for the previous dilution (see *Fig. 3b*). The optimal concentration for the primers is therefore the lowest dilution that shows no deviation in amplification plot.

Protocol 3

Running a TaqMan assay

Equipment and Reagents
- ABI Prism 7700 Sequence Detector or ABI 7500 FAST (Applied Biosystems)
- 0.5 ml RNase-free tubes
- RNase-free H_2O
- Microfuge and plate centrifuge
- DNase- and RNase-free Thermofast 96-well PCR plates and domed lids (ABgene), plate holder, and lid applicator
- Reverse Transcriptase qPCR Master Mix (Eurogentec)
- Gene-specific primers (100 µM in DEPC-treated H_2O) and fluorescent-labeled probe (5′ FAM, 3′ TAMRA; 5 µM in DEPC-treated H_2O)
- 'Standard' RNA, i.e. positive-control RNA used for the standard curves for (a) a reference gene and (b) the gene of interest
- Test RNA

Method
1. Prepare sufficient master mix. This needs to be sufficient for all samples in triplicate (maximum of 29 test samples per plate), a standard curve (with a minimum of five tenfold dilutions), three no-template-control (NTC) wells, and extra volume to allow for pipetting errors. Prepare the master mix by multiplying the number of wells by the basic mix volumes shown below:

	Basic mix (µl)
2× PCR master mix (in kit)	12.5
Primer mix (at optimal conc.)	1.00
Probe	0.50
'Euroscript' enzyme (in kit)	0.125
DEPC-treated H_2O	5.875

2. Thaw all reaction components (including primers and probe, *but excluding the RNA*) on ice.

3. Prepare the primer mix first (using the optimal concentration as determined in *Protocol 2*) and then prepare the master mix, vortex to mix, and keep on ice.

4. Place a Thermofast 96-well plate in a rack on ice and add 5.0 µl DEPC-treated H_2O to the NTC wells, followed by 20 µl master mix and cap these wells off to prevent contamination.

5. Thaw the RNA samples on ice and prepare the appropriate dilutions (usually 1:10, but often 1:100 for the reference gene). You will require 15 µl of diluted RNA for each sample

(5.0 µl/well, triplicate wells). Therefore, for a 1:10 dilution, pipette 1.6 µl of RNA into 14.4 µl DEPC-treated H_2O. Vortex to mix thoroughly and place on ice.

6. Thaw the positive control 'standard' RNA on ice and prepare serial dilutions on ice.

7. Add 20 µl of master mix to all wells being used in the plate.

8. Vortex the RNA before pipetting to ensure that it is thoroughly mixed and then pipette 5.0 µl of RNA sample per well.

9. Cap off all wells properly (use the lid applicator) and spin the plate down briefly using the plate centrifuge.

10. Place the plate in the real-time RT-PCR machine and use the following program:

Number of cycles	Program
1	2 min at 50°C, 30 min at 60°C, 5 min at 95°C
40	20 s at 94°C, 1 min at 59°C

3. TROUBLESHOOTING

- **The results obtained are not as expected**
  To control for this possibility, it is important to include both positive controls (the standard curve for each primer–probe set run on a plate acts as a good positive control and provides data for analysis of the results) and NTCs on each plate of samples assayed. Ideally, a minimum of five standard curves should be run for each specific primer–probe set and the data from these combined to generate an average standard curve to use in data analysis. If controls give expected results, then reaction constituents can be ruled out, suggesting a problem with the test samples.

- **The control samples do not give the expected results**
  TaqMan probes do occasionally degrade at higher cycle numbers (typically >35), giving false-positive readings. One simple way to check this is to look at the multi-component plot for any suspect sample. With 'real' amplification, the fluorescence emitted by the reporter dye (e.g. FAM) and that emitted by the quencher dye (TAMRA) should diverge once the threshold value is reached, with the FAM fluorescence increasing exponentially and the TAMRA fluorescence decreasing exponentially but with a shallower gradient. With 'false' amplification, usually due to probe degradation, FAM fluorescence will increase, but TAMRA fluorescence will remain static.

  It is recommended that probes are stored at working concentrations in small aliquots and that they should not be freeze-thawed more than once.

- **Has my RNA been degraded?**
  Most RNA samples will be diluted 1:10 or more before being assayed. Repeated freeze-thawing of diluted samples should be avoided to prevent degradation of the mRNA template. However, in our experience, repeated freeze-thawing of the concentrated RNA samples has little effect on the mRNA samples, with

little or no increase in C_T value for any particular target with repeated freeze-thawing.

- **My primers/probe do not span intron/exon boundaries**

  It is not always possible to design primer–probe sets across exon boundaries, either because no set fits the TaqMan parameters, or because you wish to quantify a gene without introns. In these cases, on-column DNase treatment (available with Qiagen RNeasy kits) successfully removes any contaminating genomic DNA. In addition, use of the QuantiTect Reverse Transcription kit (Qiagen) can also help to eliminate contaminating genomic DNA.

- **The NTC C_T values are less than 40 cycles**

  NTC C_T values should normally be 40. However, environmental contamination, despite relevant precautions being taken, can lead to contamination of NTCs. In these cases, the best approach is often to set up the reactions in a laboratory that has never seen the target species nucleic acid. Alternatively, instead of expressing results as 40 – C_T values, they can be expressed as NTC value – C_T values.

- **Is there an absolute requirement for triplicates?**

  Triplicate samples, in our opinion, is the minimum number that can be used to identify, and therefore avoid, bias based on pipetting errors. Others recommend duplicate samples, and a few use single samples, but both of these approaches would not enable one to distinguish a genuine value from a pipetting error.

- **Are there alternative optimization protocols?**

  Applied Biosystems recommend a slightly different primer optimization protocol to that given above. They also recommend optimizing the concentration of the probe. We have never felt the need to do this. These protocols are available from ABI if required (http://docs.appliedbiosystems.com/search.taf).

- **Is the specific design of primers and probe together truly necessary?**

  For the TaqMan probe technology, it is essential to use PRIMER EXPRESS to design your primer–probe sets. In our experience, using existing primers designed for RT-PCR of your target species of interest and bolting on a probe that hybridizes in between simply does not work.

4. REFERENCES

★ 1. Higuchi R, Fockler C, Dollinger G & Watson R (1993) *Biotechnology*, **11**, 1026–1030. – *The original publication describing real-time PCR.*
2. Gibson UE, Heid CA & Williams PM (1996) *Genome Res.* **6**, 995–1001.
3. Godfrey TE, Kim SH, Chavira M, *et al.* (2000) *J. Mol. Diagn.* **2**, 84–91.
4. Andreassen CN, Sorensen FB, Overgaard J & Alsner J (2004) *Radiother. Oncol.* **72**, 351–356.
5. Hillemann D, Galle J, Vollmer E & Richter E (2006) *Int. J. Tuberc. Lung Dis.* **10**, 340–342.
6. Fink L, Seeger W, Ermert L, *et al.* (1998) *Nat. Med.* **4**, 1329–1333.
7. Glockner S, Lehmann U, Wilke N, Kleeberger W, Langer F & Kreipe H (2000) *Pathobiology*, **68**, 173–179.
8. Elkahloun AG, Gaudet J, Robinson GS & Sgroi DC (2002) *Cancer Biol. Ther.* **1**, 354–358.
9. Shieh DB, Chou WP, Wei YH, Wong TY & Jin YT (2004) *Ann. N.Y. Acad. Sci.* **1011**, 154–167.

★ 10. Moody A, Sellers S & Bumstead N (2000) *J. Virol. Methods*, **85**, 55–64. – *First application of real-time qRT-PCR to an avian virus.*

11. Desire N, Dehee A, Schneider V, *et al.* (2001) *J. Clin. Microbiol.* **39**, 1303–1310.

12. Zhang Z, Wilson F, Read R, Pace L & Zhang S (2006) *J. Vet. Diagn. Invest.* **18**, 204–208.

13. Lee CW & Suarez DL (2004) *J. Virol. Methods.* **119**, 151–158.

14. Balaji B, O'Connor K, Lucas JR, Anderson JM & Csonka LN (2005) *Appl. Environ. Microbiol.* **71**, 8273–8283.

15. Wargo AR, Randle N, Chan BH, Thompson J, Read AF & Babiker HA (2006) *Exp. Parasitol.* **112**, 13–20.

16. Johnson VJ, Yucesoy B & Luster MI (2004) *Cytokine*, **27**, 135–141.

17. Petersen K, Vogel U, Rockenbauer E *et al* (2004) *Mol. Cell. Probes*, **18**, 117–122.

18. Medhurst AD, Lezoualc'h F, Fischmeister R, Middlemiss DN & Sanger GJ (2001) *Brain Res. Mol. Brain Res.* **90**, 125–134.

19. Richard V, Luchin A, Brena RM, Plass C & Rosol TJ (2003) *Clin. Chem.* **49**, 1398–1402.

20. Yperman J, de Visscher G, Holvoet P & Flameng W (2004) *J. Heart Valve Dis.* **13**, 848–853.

21. Steele BK, Meyers C & Ozbun MA (2002) *Anal. Biochem.* **307**, 341–347.

22. Rhoads RP, McManaman C, Ingvartsen KL & Boisclair YR (2003) *J. Dairy Sci.* **86**, 3423–3429.

23. Bas A, Forsberg G, Hammarström S & Hammarström ML (2004) *Scand. J. Immunol.* **59**, 566–573.

24. Dheda K, Huggett JF, Bustin SA, Johnson MA, Rook G & Zumla A (2004) *BioTechniques*, **37**, 112–119.

25. Li YP, Bang DD, Handberg KJ, Jorgensen PH & Zhang MF (2005) *Vet. Microbiol.* **110**, 155–165.

★★ 26. Withers DR, Davison TF & Young JR (2005) *Dev. Comp. Immunol.* **29**, 651–662. – *Description of the* $2^{-\Delta\Delta CT}$ *method with allowance for different amplification efficiencies.*

27. Kaiser P, Rothwell L, Galyov EE, Barrow PA, Burnside J & Wigley P (2000) *Microbiology*, **146**, 3217–3226.

28. Kaiser P, Rothwell L, Vasícek D & Hala K (2002) *J. Immunol.* **168**, 4216–4220.

29. Kaiser P, Underwood G & Davison F (2003) *J. Virol.* **77**, 762–768.

30. Kogut MH, Rothwell L & Kaiser P (2003) *J. Interferon Cytokine Res.* **23**, 319–327.

31. Kogut MH, Rothwell L & Kaiser P (2003) *Mol. Immunol.* **40**, 603–610.

32. Peters MA, Browning GF, Washington EA, Crabb BS & Kaiser P (2003) *Immunology*, **110**, 358–367.

33. Tötemeyer S, Foster N, Kaiser P, Maskell DJ & Bryant CE (2003) *Infect. Immun.* **71**, 6653–6657.

34. Ferro PJ, Swaggerty CL, Kaiser P, Pevzner IY & Kogut MH (2004) *Epidemiol. Infect.* **132**, 1029–1037.

35. Swaggerty CL, Kogut MH, Ferro PJ, Rothwell L, Pevzner IY & Kaiser P (2004) *Immunology*, **113**, 139–148.

36. Withanage GSK, Kaiser P, Wigley P, *et al.* (2004) *Infect. Immun.* **72**, 2152–2159.

37. Rothwell L, Young J, Zoorob R, *et al.* (2004) *J. Immunol.* **173**, 2675–2682.

38. Smith CK, Kaiser P, Rothwell L, Humphrey T, Barrow PA & Jones MA (2005) *Infect. Immun.* **73**, 2094–2100.

39. Tötemeyer S, Kaiser P, Maskell DJ & Bryant CE (2005) *Infect. Immun.* **73**, 1873–1878.

40. Eldaghayes I, Rothwell L, Williams A, *et al.* (2006) *Viral Immunol.* **19**, 83–91.

CHAPTER 4

Use of quantitative PCR for the detection of genomic microdeletions or microduplications

Simon Hughes, Rosanna Weksberg, Laura Moldovan, and Jeremy A. Squire

1. INTRODUCTION

Screening for DNA copy number alterations has been achieved traditionally by the use of either array comparative genomic hybridization (1) or classical cytogenetic techniques such as metaphase comparative genomic hybridization and fluorescent *in situ* hybridization (FISH) (2). However, each of these approaches has their own specific limitations (3). DNA arrays providing sufficient coverage of the human genome to detect small copy number alterations are not widely available and are too expensive to screen large numbers of patients diagnostically, whilst cytogenetic methods are time-consuming, require a high level of technical expertise, and do not provide a high enough resolution to detect microalterations (microdeletions or microduplications). As a consequence, we have developed a robust, quick, and sensitive technique, based on the quantitative analysis of PCR, which requires relatively little labor and can be conducted on a large number of samples in parallel.

Quantitative PCR (qPCR) has been used commonly for quantifying levels of gene expression (4), but has not been applied routinely to the screening of copy number alterations in human genomic DNA. qPCR exceeds the limitations of traditional end-point PCR methods by allowing the user to measure the accumulation of a specific product/amplicon in 'real-time' as the reaction proceeds. This collection of data allows fast, precise, and accurate DNA (or RNA) quantitation.

In this chapter, we will use chromosome 22q11 deletion syndrome as a test model to demonstrate the utility of qPCR for detecting microalterations in genomic DNA.

PCR: *Methods Express* (S. Hughes and A. Moody, eds.)
© Scion Publishing Limited, 2007

1.1 Test model

The most frequent interstitial deletion found in humans (approximately 1 in every 4000 live births) (5) involves a microdeletion at chromosome 22q11. This chromosome abnormality is inherited from an affected parent in 5–10% of cases, thus suggesting recessive inheritance, and occurs *de novo* in the remainder (6).

The clinical phenotype is complex, consisting of many different birth defects and malformation, as well as behavioral and psychiatric components. It is also variable, with different combinations of defects occurring, sometimes even within the same family. The complexity and variability of features associated with 22q11 deletion has resulted historically in the definition of a number of overlapping clinical syndromes including DiGeorge syndrome, Velocardiofacial syndrome (VCFS), Shprintzen syndrome, and conotruncal anomaly face syndrome. As these diagnostic categories probably reflect variable outcomes of a common cytogenetic anomaly, the term chromosome 22q11 deletion syndrome (22q11DS) is now used to refer to all individuals with these clinical features.

The most common clinical features include congenital heart disease, palatal abnormalities, immunodeficiency, hypocalcemia, renal anomalies, feeding problems, learning difficulties, characteristic facial features, and psychiatric illness. Many other less common features have been reported as well, including hearing loss, autoimmune disorders, skeletal anomalies, and growth hormone deficiency, to name a few.

Clinical testing for 22q11DS only became widely available in ~1995 and the most common associated deletions are identifiable using FISH with D22S75 and TUPLE1 probes. This screening technique is now available in clinical laboratories (6, 7). Although this assay can detect the majority of affected patients (85–90%), clinical FISH testing has its limitations, as it does not determine the extent of the deletion, nor can it detect the atypical deletions or duplications that have been

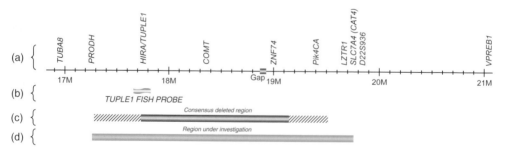

Figure 1. Schematic of the genomic region on chromosome 22 containing the causative deletion for VCFS.
(*a*) A selection of the markers present on the affected region (~4 megabases) on 22q11.2. (*b*) The location of the hybridization site for the TUPLE1 FISH probe that is used clinically for the diagnosis of VCFS patients. (*c*) A review of the VCFS literature identified a consensus deleted region (black) common to many of the reports. The hatched extensions to this consensus region correspond to the extreme 5′ and 3′ deleted regions identified in these reports and are atypical deletions. (*d*) The gray region corresponds to the region that was screened for deletion in the analysis of VCFS patients.

reported. As a consequence, these patients will go undiagnosed due to the presence of atypical deletions that map outside the area covered by the D22S75 and TUPLE1 probes (see *Fig. 1*).

2. METHODS AND APPROACHES

2.1 SYBR Green

The assay described in this chapter is based on the SYBR Green I dye (available from Applied Biosystems, Qiagen, and Stratagene) approach to qPCR. SYBR Green binds to the minor groove of dsDNA in a sequence-independent manner. When bound to DNA, it emits green fluorescence (521 nm) on excitation with blue light (495 nm). When not bound to DNA, SYBR Green cannot fluoresce. During the course of the PCR, after each cycle of replication, the dye is incorporated into the dsDNA and fluoresces. When the DNA is denatured in the next amplification cycle, the dye is released back into solution, where it loses its capacity to fluoresce. With subsequent cycles, the levels of PCR product increase, which in turn means that there is an increase in the target to which SYBR Green can bind, thus resulting in a higher level of fluorescence (8). When using SYBR Green, it is important to perform a melting-curve analysis for each primer set. The positive control (with DNA) should yield a sharp peak at the melting temperature of the specific product; conversely, the negative/no-template control should not generate a significant fluorescent signal. If this trend is observed, the PCR is specific and fluorescence is a direct measure of product accumulation.

An alternative approach for qPCR is the use of sequence-specific probes (i.e. molecular beacons or TaqMan probes). Probe-based qPCR utilizes a fluorescently labeled target-specific probe that can provide increased specificity and sensitivity. However, when assaying a large number of primer sets, the cost of probes can become prohibitive. Furthermore, SYBR Green can be purchased as a 'ready mix', which is ideal for high-throughput applications (only primers and template are required) and can easily be automated.

The DNA that can be analyzed using SYBR Green can come from virtually any source, e.g. fresh tissue, cultured cells, fixed tissue, and DNA obtained from macro- or microdissected tissue sections. However, it is important to ensure that the DNA is free of any contaminants that might affect the PCR adversely (9). Therefore, it is recommended that the DNA is isolated using a kit or protocol that will generate DNA of high purity (see *Protocol 1*).

2.2 Measuring amplification

qPCR results are displayed as an amplification plot (see *Fig. 2a*), which reflects the change in fluorescence during cycling. qPCR data is expressed in terms of the threshold cycle value, C_T, the cycle at which fluorescence of the reporter dye (i.e. SYBR green) is determined to have passed above a significance threshold. This value is inversely proportional to the initial copy number (i.e. the lower the C_T value, the more template was present at the start of the reaction), and so C_T can be used to quantitate initial copy number. The more template that is present initially, the fewer cycles it takes to reach the point where the fluorescence signal

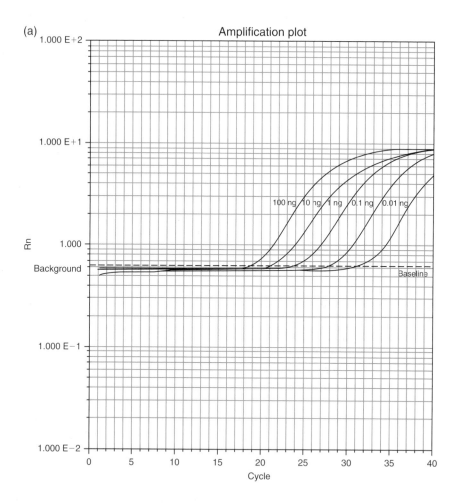

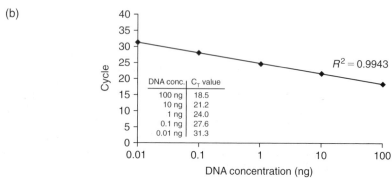

is detectable above the threshold. The threshold cycle is based on measurements taken during the exponential phase of PCR amplification.

To enable accurate C_T determination, it important to set a precise baseline, which then must be kept consistent for all subsequent experiments for that primer pair. The baseline should be high enough to remove the background amplification present in the early cycles of PCR, but should not be above the level at which the amplification signal begins to rise above background (see *Fig. 2a*). To allow accurate comparison between different primer pairs, all baselines must be set in the same way.

2.3 Primer design

Optimal design of the PCR primers is crucial for precise and specific quantification using real-time PCR. The primers described in this chapter were designed using the PRIMER EXPRESS version 2.0 (Applied Biosystems) program, but alternative internet-based programs such as PRIMER3 (http://frodo.wi.mit.edu/cgi-bin/primer3/primer3.cgi) can also be used. When designing primers, is it important to adhere to specific parameters to ensure consistency. Although these parameters may vary slightly among different laboratories and investigators, we have found that those described in Chapter 1 work well in our assays.

Once identified, primer sequences should be compared with the human genome. There are a number of free tools available for users to perform such comparisons including BLAT (http://www.genome.ucsc.edu/cgi-bin/hgBlat), BLAST (http://www.ncbi.nlm.nih.gov/BLAST) and LALIGN (http://www.ch.embnet.org/software/LALIGN_form.html). This step ensures that the sequences show 100% homology to only the sequence from which they were designed, and also guarantees that the forward and reverse primers are free of single-nucleotide polymorphisms.

Prior to carrying out qPCR, it is advisable to ensure that the primers work in a standard PCR and produce just a single band when a PCR profile similar to that used for qPCR is used. In certain situations (when facing consistent qPCR

Figure 2. Example of standard curves produced for primer optimization.
(*a*) Once the optimal concentration for each primer pair has been determined, it is necessary to determine the efficiency of the qPCR assay. This is accomplished by generating a set of standard curves for each primer pair. A $\log_{10}$ dilution series of genomic DNA template is prepared at concentrations ranging from 0.01 to 100 ng. When used in a qPCR, the different dilutions will produce a set of amplification curves that have different C_T values due to differing amounts of starting DNA. The C_T values for each dilution can then be used to plot a standard curve of C_T against template DNA concentration, the slope of which can be calculated using a software package such as EXCEL or with the aid of the Applied Biosystems SDS software or similar programs. (*b*) The slope of the standard curve is used to determine reaction efficiency. When C_T is plotted against template DNA concentration, as in *Fig. 2(b)*, regression analysis of the data should give R^2 values close to 1. Values >0.98 are generally acceptable. Rn is the value obtained by dividing the emission intensity of reporter dye by the emission intensity of the passive reference.

problems), it might be necessary to sequence the PCR products. If this is the case, it is advisable to sequence from a standard PCR product, as the SYBR Green dye will interfere optically with the sequencing process.

2.4 Housekeeping/reference genes and data normalization

Data normalization in qPCR is essential for accurate quantification analysis. To this end, it is essential to include primers for one or more invariant endogenous controls, also referred to as 'housekeeping' or reference genes/controls, to correct for sample-to-sample and run-to-run variations in qPCR efficiency and the range of factors that can adversely affect PCR (DNA quality, pipetting errors, etc.). Care must be taken in the selection of appropriate controls. The reference control must not be present within a region of copy number variation. Reference primers must then be designed from the unaffected chromosomes/regions. qPCR must be performed using the reference primers on all experimental samples and the C_T values obtained. Some commonly used reference genes include β-actin, glyceraldehyde 3-phosphate dehydrogenase, glucose 6-phosphate dehydrogenase, β-2 microglobulin, ubiquitin C, and proline dehydrogenase.

2.5 Primer optimization

When developing a qRT-PCR assay, it is strongly recommended that the primer concentrations be optimized first, using a source of well-characterized, easily replenishable DNA obtained from a commercial source (e.g. human genomic DNA from Promega). Primer optimization is essential to reduce the formation of nonspecific amplification products, which is of particular importance when the amplification target is at low concentration. To maximize reaction sensitivity, it is necessary to use the lowest primer concentration that results in the lowest C_T and a sufficient fluorescent signal (ΔRn), with little or no primer-dimer formation. However, it is also important not to compromise the efficiency of the PCR. It is recommended that the concentration of each primer be titrated, starting at 100 nM and increasing in 100 nM intervals up to 800 nM (final concentration in a PCR). The primer grid shown in *Table 1* is a good guide for primer optimization. Although in our assays we have found that the best concentrations of the upstream and downstream primers are equal, this might not always be the case, and optimal results can be obtained with forward and reverse primers at different molarities. When assaying several different primer sets, each primer pair can work optimally at different nanomolar concentrations.

2.6 Generating a standard curve

Once the optimal concentration for each primer pair has been determined, it is then necessary to determine the quality/efficiency of the qPCR assay. This is accomplished by generating a set of standard curves for each primer pair. For the

Table 1. Optimization grid for forward (sense) and reverse (antisense) primer concentrations

	Forward (sense) primer (nM)							
	100/100	100/200	100/300	100/400	100/500	100/600	100/700	100/800
	200/100	200/200	200/300	200/400	200/500	200/600	200/700	200/800
	300/100	300/200	300/300	300/400	300/500	300/600	300/700	300/800
	400/100	400/200	400/300	400/400	400/500	400/600	400/700	400/800
	500/100	500/200	500/300	500/400	500/500	500/600	500/700	500/800
	600/100	600/200	600/300	600/400	600/500	600/600	600/700	600/800
	700/100	700/200	700/300	700/400	700/500	700/600	700/700	700/800
	800/100	800/200	800/300	800/400	800/500	800/600	800/700	800/800

Reverse (antisense) primer (nM)

primers under investigation, a $\log_{10}$ dilution series of genomic DNA should be prepared at concentrations ranging from 0.01 to 100 ng. These dilutions can then be used as template in a qPCR (see *Protocol 2*).

It is advisable to have two separate areas in the laboratory for the qPCR set-up: one area to set up the reaction mixture and pipette this into the PCR plates and a second area to add the DNA. When carrying out qPCR, it is essential to perform reactions in triplicate, with replicates being performed on different days. This permits an average to be obtained, as any one reaction can either fail or give an incorrect reading. However, it is highly unlikely that several identical reactions would suffer the same problems.

The dilution series should produce a set of amplification curves that have different C_T values due to the differing amounts of starting DNA (see *Fig. 2a*). The C_T value for each dilution can then be used to calculate the slope of the standard curve (the slope is equal to the average C_T change across the dilution series) using a software package such as EXCEL (Microsoft; see *Fig 2b*) or with the aid of the Applied Biosystems SDS software or similar programs. The slope of the standard curve is used to determine reaction efficiency. When studying a $\log_{10}$ dilution series, it will take approximately 3.32 cycles for tenfold amplification of product, a value that is equal to the slope of the standard curve. If all conditions are optimal and the reactions are 100% efficient, the amount of template will exponentially increase (2^n) with each cycle (where n = cycle number). Thus, one cycle copies one molecule into two; the second cycle copies two molecules into four; the third cycle copies four into eight and 0.32 cycles copies eight into ten. Regression analysis of the C_T values generated by the $\log_{10}$ dilution series should give R^2 values close to 1. As R^2 decreases, the accuracy of quantification will decrease. R^2 values >0.98 are generally acceptable, indicating good reaction efficiency.

2.7 Recommended protocols

Protocol 1

DNA extraction and quantification

Equipment and Reagents
- QIAamp DNA extraction kit (Qiagen)
- 1% Agarose gel containing 10 ng/ml ethidium bromide
- 6× Orange loading dye solution (Fermentas)
- Equipment and reagents for agarose gel electrophoresis including 1× TBE agarose gel running buffer (10.8 g/l Tris base; 5.5 g/l boric acid; 4 ml/l 0.5 M EDTA, pH 8.0; diluted from a 10× stock; Sigma)
- UV light source
- RediPlate 96 PicoGreen dsDNA quantitation kit (Invitrogen)
- Sterile water (Sigma)
- Fluorescence-based microplate reader or fluorometer

Method
1. Extract the DNA using a QIAamp DNA extraction kit[a].

2. Determine the quality of the DNA by analyzing a 5 µl aliquot by agarose gel electrophoresis (using a 1% agarose gel containing 10 ng/ml ethidium bromide)[b]. Detect the DNA under UV light.

3. For quantification, dilute the DNA to at least 1:1000 in sterile water[c].

4. Quantify the DNA concentration using the RediPlate 96 PicoGreen dsDNA quantitation kit (or a similar kit) in conjunction with a fluorescence-based microplate reader, following the manufacturer's instructions[d].

5. Prepare a DNA solution at a concentration of 25 ng/µl[e].

Notes

[a]These kits are suitable for extracting DNA from a number of sources, including buffy coats, buccal swabs, cultured cells, blood spots, and mouthwashes. They are also suitable for extracting DNA from fixed tissue; however, the fixation process can introduce sequence variations and reduce overall DNA quality.

[b]The agarose gel electrophoresis analysis will also assess the degree of DNA degradation. When using microdissected cells, it will not be possible to analyze the DNA using agarose gel electrophoresis due to the low DNA concentration.

[c]It may be necessary to dilute the DNA further, but a 1:1000 dilution is a good starting point.

[d]DNA concentration (ng/µl) can also be quantified using a standard spectrophotometer by taking the absorbance reading at 260 nm and multiplying it by 50 and then by the dilution factor. When quantifying using this method, it is not necessary to do a 1:1000 dilution. It is advisable to do a 1:50 or 1:100 dilution.

[e]If the DNA is to be used immediately, store it at 4°C. Alternatively, keep it at –20°C for long-term storage (more than 1 year). Avoid repeated freeze-thawing.

Protocol 2

qPCR

Equipment and Reagents

- 2× SYBR Green PCR I Master Mix (Applied Biosystems)[a]
- Primers (test and reference[b])
- Sterile water (Sigma)
- PCR plates and caps/seals (Applied Biosystems)[c]
- ABI Prism 7900 high-throughput sequence detection system and accompanying SDS analysis software (Applied Biosystems)

Method

1. For each reaction, combine 12.5 µl of 2× SYBR Green PCR Master Mix, 1 µl each of forward and reverse primer at optimized concentrations[d], and sterile water up to a final volume of 24 µl[e].

2. Pipette the reaction mixture into separate wells of a PCR plate and then add 1 µl of DNA (25 ng/µl) to each well (excluding the negative controls). Seal the plate with either caps or a plastic seal.

3. Mix the sample by brief vortexing.

4. Consolidate the sample by centrifugation at 5000 **g** for 5–10 s.

5. Incubate the samples in an ABI Prism 7900 high-throughput sequence detection system using the recommended conditions. The following is an example of the typical conditions utilized with ABIs SYBR Green: initial step, 50°C for 2 min, denaturation at 95°C for 10 min, followed by 40 cycles of denaturing at 95°C for 15 s and combined annealing and extension at 60°C for 60 s[f,g].

6. Analyze the results using Applied Biosystems SDS software.

Notes

[a]The methods described in this chapter were performed using SYBR Green I PCR Master Mix (Applied Biosystems), which includes the internal reference ROX. ROX is a passive reference dye that does not take part in the PCR and its fluorescence remains constant during the PCR. ROX is used to normalize for nonPCR-related fluorescence signal variation.

[b]The sequences of the primers used will be assay specific and are not provided here.

[c]The plates and caps/seals used for qPCR are nonstandard and are required to be of high optical clarity. We used plates obtained from Applied Biosystems, but other manufacturers produce equally suitable plates.

[d]The concentration of the primers used has to be determined experimentally; the process for this is described in section 2.5.

[e]We recommend the use of filter tips for all pipetting steps to avoid the introduction of contamination by aerosols.

[f]We recommend including three negative (i.e. water instead of DNA) controls. We suggest that these be the first, middle, and last reactions that are pipetted, as this will ensure that there are no contaminants in the starting reaction mixture and will check that no contaminants have been introduced during pipetting of the test samples.

[g]The exact incubation conditions could change depending on the reaction mix supplier. Check with the supplier for the correct reaction conditions.

2.8 DNA quantification and data analysis

Test primers and reference gene primers should always be included on the same reaction plate. The output of the Applied Biosystems sds software is expressed in terms of the C_T values. For further analysis, the results should be exported in tab-delimited text file format into a program such as Microsoft EXCEL.

The scientific literature describes several methods for qPCR data normalization:

- Absolute standard curve method (10);
- Comparative C_T method; also called the $2^{-\Delta\Delta CT}$ method (reviewed in 11);
- Relative standard or comparative quantification method (12).

We employ elements of all of these methods to achieve the sensitivity required for genomic copy number determination.

2.8.1 Data normalization

qPCR data for each of the primer pairs from the 22q11.2 region can be normalized using the following formula. A worked example is displayed in *Fig. 3*.

$$KC_{Tt} = \left(\frac{AC_{Tr} - C_{Tr}}{S_r} \right) \times S_t + C_{Tt}$$

where KC_{Tt} is the corrected C_T of the test primer set (t); AC_{Tr} is the average C_T value for the reference primer set (r) for all of the samples included in one qPCR run (control and patient); C_{Tr} is the average C_T value for the reference primer set for the sample to be corrected; S_r is the slope value (from the standard curve) for the reference primer set; S_t is the slope value (from the standard curve) for the test primer set; and C_{Tt} is the C_T value for the test primer set.

2.8.2 Copy number calculation

The fold copy number (ΔKC_T) change for each of the markers, here from the 22q11.2 region, is obtained using the following formula. A worked example is displayed in *Fig. 4*.

$$\Delta KC_T = KC_{Tt\ control} - KC_{Tt\ affected}$$

where ΔKC_T is copy number gain or loss; $KC_{Tt\ control}$ is corrected C_T of the test primer for the control samples; and $KC_{Tt\ affected}$ is corrected C_T of the test primer for the affected sample.

2.9 Case study

As a case study, ten test primer sets and two reference genes were used to analyze 12 patients with clinical symptoms of 22q11DS, one patient trisomic for 22q11, and four normal controls. In the context of 22q11DS patients, we were trying to discriminate between two copies of a product (normal) versus one copy of the

(a) Raw C_T values for samples 1–14

A1 Sample 1 18.31	A2 Sample 1 17.87	A3 Sample 1 17.97	A4 Sample 2 18.03	A5 Sample 2 17.89	A6 Sample 2 17.91	A7 Sample 3 17.49	A8 Sample 3 17.59	A9 Sample 3 17.02	A10 Sample 4 18.66	A11 Sample 4 18.09	A12 Sample 4 17.8
B1 Sample 5 18.17	B2 Sample 5 18.31	B3 Sample 5 18.22	B4 Sample 6 19.88	B5 Sample 6 20.17	B6 Sample 6 21.47	B7 Sample 7 17.87	B8 Sample 7 17.91	B9 Sample 7 18.01	B10 Sample 8 17.4	B11 Sample 8 17.45	B12 Sample 8 16.96
C1 Sample 9 18.03	C2 Sample 9 18.17	C3 Sample 9 17.79	C4 Sample 10 17.23	C5 Sample 10 17.29	C6 Sample 10 16.88	C7 Sample 11 17.31	C8 Sample 11 17.65	C9 Sample 11 17.26	C10 Sample 12 21.1	C11 Sample 12 19.82	C12 Sample 12 19.88
D1 Sample 13 18.23	D2 Sample 13 18.18	D3 Sample 13 18.54	D4 Sample 14 18.06	D5 Sample 14 18.02	D6 Sample 14 17.86	D7 NTC 0	D8 NTC 0	D9 NTC 0	D10 Blank	D11 Blank	D12 Blank
E1 Sample 1 20.71	E2 Sample 1 20.98	E3 Sample 1 20.94	E4 Sample 2 20.55	E5 Sample 2 20.86	E6 Sample 2 20.48	E7 Sample 3 21.16	E8 Sample 3 21.15	E9 Sample 3 20.96	E10 Sample 4 20.98	E11 Sample 4 20.82	E12 Sample 4 20.52
F1 Sample 5 21.79	F2 Sample 5 21.77	F3 Sample 5 21.95	F4 Sample 6 21.86	F5 Sample 6 22.14	F6 Sample 6 22.09	F7 Sample 7 21.07	F8 Sample 7 21.05	F9 Sample 7 19.87	F10 Sample 8 20.16	F11 Sample 8 20.46	F12 Sample 8 20.14
G1 Sample 9 21.07	G2 Sample 9 20.94	G3 Sample 9 21.06	G4 Sample 10 21.01	G5 Sample 10 21.05	G6 Sample 10 20.94	G7 Sample 11 20.55	G8 Sample 11 20.32	G9 Sample 11 20.09	G10 Sample 12 23.04	G11 Sample 12 23.15	G12 Sample 12 23.39
H1 Sample 13 21.23	H2 Sample 13 21.29	H3 Sample 13 20.36	H4 Sample 14 20.85	H5 Sample 14 21.03	H6 Sample 14 21.25	H7 NTC 0	H8 NTC 0	H9 NTC 0	H10 Blank	H11 Blank	H12 Blank

Rows A–D: Reference (slope 3.29)
Rows E–H: Test (slope 3.19)

(b) Average C_T values for samples 1–14

Reference

Sample 1	Sample 2	Sample 3	Sample 4	Sample 5	Sample 6	Sample 7	Sample 8	Sample 9	Sample 10	Sample 11	Sample 12	Sample 13	Sample 14
18.05	17.94	17.37	18.18	18.23	20.51	17.93	17.27	18.00	17.13	17.41	20.27	18.32	17.98

Test

Sample 1	Sample 2	Sample 3	Sample 4	Sample 5	Sample 6	Sample 7	Sample 8	Sample 9	Sample 10	Sample 11	Sample 12	Sample 13	Sample 14
20.88	20.63	21.09	20.77	21.84	23.03	20.66	20.25	21.02	21.00	20.32	23.19	21.13	21.04

(c) Normalization (example for sample 1)

Equation:

$$KC_{Tt} = \left(\frac{AC_{Tr} - C_{Tr}}{S_r}\right) \times S_t + C_{Tt}$$

$$KC_{Tt} = \left(\frac{\text{Average of A1 to D6} - \text{Average of A1 to A3}}{\text{Slope for reference primer}}\right) \times \text{Slope for test primer} + \text{Average of E1 to E3}$$

$$= \left(\frac{18.18 - 18.05}{3.29}\right) \times 3.19 + 20.88$$

$$= 0.0395 \times 3.19 + 20.88$$

$$= 21.01$$

(d) Normalized C_T values for samples 1–14

Sample 1	Sample 2	Sample 3	Sample 4	Sample 5	Sample 6	Sample 7	Sample 8	Sample 9	Sample 10	Sample 11	Sample 12	Sample 13	Sample 14
21.01	20.86	21.88	20.77	21.79	19.78	20.91	21.14	21.21	22.02	21.07	21.17	21.00	21.24

Figure 3. Data normalization using a housekeeping/reference gene.
To control for differences in sampling, DNA preparation, reaction efficiency, accurate pipetting, and other variables that can adversely affect the PCR, C_T values for each primer pair are corrected/normalized using a reference gene. We use a derivation of the comparative quantification method for data normalization (12). (a) Each experiment is performed in triplicate, with replicates being performed on different days. An example of one set of triplicates is shown here. (b) An average C_T value is determined for each triplicate (i.e. A1 to A3 for sample 1). (c) Data are normalized using the equation described in section 2.8.1 and repeated here. (d) Normalization generates a set of corrected C_T values, which can then be used for copy number calculation.

same product (deleted) and three copies (duplicated) for the trisomic sample. In order to do this, we used data normalization, using reference genes, to control for PCR variations among samples (13), as the copy number of the reference genes will be the same in all of the samples under investigation. Any variation in copy number for the reference among samples will be the result of differences in PCR efficiency and initial template concentration, as long as the same DNA is sampled

Copy number calculation

Equation: $\Delta KC_T = KC_{Tt\ control} - KC_{Tt\ affected}$

$$\Delta KC_T = KC_{Tt}\ for\ sample\ 1 - KC_{Tt}\ for\ sample\ 2$$

$$= 21.01 - 20.86$$

$$= 0.15$$

Copy number changes for samples 1–14

Sample 1	Sample 2	Sample 3	Sample 4	Sample 5	Sample 6	Sample 7	Sample 8	Sample 9	Sample 10	Sample 11	Sample 12	Sample 13	Sample 14
0	0.15	−0.88	0.23	−0.78	1.23	0.10	−0.13	−0.20	−1.01	−0.07	0.83	0.01	−0.24
	No change	Loss	No change	Loss	Gain	No change	No change	No change	Loss	No change	No change	No change	No change

Figure 4. Copy number calculation.
We compared the test (patient sample) and control (normal controls) results to obtain the changes in the patient samples compared with the normal DNA. ΔKC_T represents copy number gain or loss per sample (fold changes). Values of 0 (± 0.35) indicate an equal ratio of the test and reference, corresponding to no change in ploidy and therefore no genetic abnormality. Values of −1 (± 0.35) indicate loss of one copy (microdeletion) for the affected samples and values of +1 (± 0.35) indicate copy gain consistent with microduplication.

for both the control and target primers. Once the C_T values for the reference are determined, these are used to correct the values of the test markers for PCR variation.

For the VCFS analyzed here, this correction permitted determination of copy number differences among the samples under investigation. Six of the 22q11DS patients (group 1) had known hemizygous deletions (loss of one copy of 22q11.2), as detected by a standard diagnostic TUPLE1 FISH assay. Screening of the group 1 patients by qPCR confirmed the TUPLE1 FISH assay results with 100% concordance and was used to determine the sensitivity of the qPCR assay and to identify ΔKC_T values corresponding to loss (see *Table 2*). We obtained ΔKC_T values of 0 ± 0.35 corresponding to no loss, and −1 ± 0.35 indicating loss (microdeletion).

The remaining six patients (group 2) were classified initially as negative by standard diagnostic TUPLE1 FISH assay. qPCR analysis of group 2 patients determined that they were also deletion negative by qPCR, showing ΔKC_T values similar to the normal controls (14; see *Table 2*). Finally, the trisomy 22q11.2 patient, which has been shown to have symptoms complementary to 22q11DS (15), gave values of +1 ± 0.35, indicating gain of one copy (microduplication).

2.10 Concluding remarks

The advantages of the approach described in this chapter are the ease with which one can investigate a specific genomic region and increase the resolution by increasing the number of primers in the region under investigation. This facilitates

Table 2. Copy number change (ΔKC_T) for the VCFS patients, 22q11 duplication, and controls

ΔKC_T values of 0 ± 0.35 indicate an equal ratio of the target and reference and are represented as NC (no change). Values of -1 ± 0.35 indicate loss of one copy (microdeletion) and are represented as – (loss). Values of 1 ± 0.35 indicate gain of one copy (microduplication) and are represented as + (gain).

Sample type	qPCR primers of the VCFS deleted region									
	Primer 1	Primer 2	Primer 3	Primer 4	Primer 5	Primer 6	Primer 7	Primer 8	Primer 9	Primer 10
Normal controls	NC	NC	NC	NC	NC	NC	NC	NC	NC	NC
	NC	NC	NC	NC	NC	NC	NC	NC	NC	NC
	NC	NC	NC	NC	NC	NC	NC	NC	NC	NC
	NC	NC	NC	NC	NC	NC	NC	NC	NC	NC
Group 1	NC	–	–	–	–	–	–	–	–	NC
	NC	–	–	–	–	–	–	–	–	NC
	NC	–	–	–	–	–	–	–	–	NC
	NC	–	–	–	–	–	–	–	–	NC
	NC	–	–	–	–	–	–	–	–	NC
	NC	–	–	–	–	–	–	–	–	NC
Group 2	NC	NC	NC	NC	NC	NC	NC	NC	NC	NC
	NC	NC	NC	NC	NC	NC	NC	NC	NC	NC
	NC	NC	NC	NC	NC	NC	NC	NC	NC	NC
	NC	NC	NC	NC	NC	NC	NC	NC	NC	NC
	NC	NC	NC	NC	NC	NC	NC	NC	NC	NC
	NC	NC	NC	NC	NC	NC	NC	NC	NC	NC
Trisomy sample	NC	+	+	+	+	+	+	+	+	NC

accurate mapping of genomic microalterations. This approach has the potential of being a fast and robust validation tool for other high-throughput methodologies such as microarray analysis. It has the advantage of providing greater flexibility and adaptability than currently available cytogenetic methods and will be beneficial in molecular classification and diagnosis.

3. TROUBLESHOOTING

- **The sequence in the region of interest will not permit primers to be designed that conform to the guidelines**
  Unfortunately the DNA sequence in some regions of interest will not lend itself to optimal primer design. If this is the case, it will be necessary to deviate from suggested guidelines and design 'best fit' primers. We have found that when nucleotide composition and complementarity guidelines are followed, it is possible to select primers that will work well in qPCR.
- **There are multiple bands on the test agarose gel or extra peaks in the melting curve**
  If multiple bands or extra peaks are observed, this reflects the generation of nonspecific PCR products; therefore, optimization of the reaction conditions is

necessary. This may be as simple as incrementally increasing the annealing/extension temperature from 60°C up to 65°C or re-optimizing primer concentrations. A more expensive high-fidelity enzyme reaction mix pre-treated with uracil-N-glycosylase could potentially solve the problem of nonspecific priming. However, if problems persist, it may be necessary to redesign the primers to solve the problem.

- **The negative controls produce positive results**
  - This result indicates DNA contamination; therefore, the reactions need to be repeated, ideally using fresh reagents. When a new kit arrives, it is always a good idea to aliquot it into separate tubes so that if one tube becomes contaminated, only one aliquot needs to be discarded and not the whole kit. It is also advisable always to use barrier pipette tips to avoid aerosols when pipetting and to use dedicated pipettes for all set-up steps apart from the addition of DNA.
  - Primer dimers may be present. Recheck the optimization of primer concentrations to eliminate this.
- **There is no amplification product on data analysis**
  The most likely reason is that there is no PCR product. This can be verified by resolving the PCR product on a standard agarose gel. If there is still no product seen on the agarose gel, there is probably a problem with the PCR itself. Use the DNA and primers used for qPCR in a standard PCR to check that they are working as expected. If they are, repeat the qPCR. If the 'no product' problem persists, check the cycling parameters and the machine and computer settings. If this does not solve the problem, contact the manufacturers.

4. REFERENCES

1. Shaw-Smith C, Redon R, Rickman L, *et al.* (2004) *J. Med. Genet.* **41**, 241–248.
2. Babovic-Vuksanovic D, Jenkins SC, Ensenauer R, Neuman DC & Jalal SM (2004) *Am. J. Med. Genet. A*, **124**, 318–322.
3. Hoffman JD, Zhang Y, Greshock J, *et al.* (2005) *J. Med. Genet.* **42**, 49–53.
4. Bustin SA, Benes V, Nolan T & Pfaffl MW (2005) *J. Mol. Endocrinol.* **34**, 597–601.
5. Oskarsdottir S, Vujic M & Fasth A (2004) *Arch. Dis. Child.* **89**, 148–151.
6. Driscoll DA, Spinner NB, Budarf ML, *et al.* (1992) *Am. J. Med. Genet.* **44**, 261–268.
7. Lindsay EA, Goldberg R, Jurecic V, *et al.* (1995) *Am. J. Med. Genet.* **57**, 514–522.
8. Bustin SA (2000) *J. Mol. Endocrinol.* **25**, 169–193.
9. Burgos JS, Ramirez C, Tenorio R, Sastre I & Bullido MJ (2002) *Mol. Cell. Probes*, **16**, 257–260.
10. Giulietti A, Overbergh L, Valckx D, Decallonne B, Bouillon R & Mathieu C (2001) *Methods*, **25**, 386–401.
★★★ 11. Livak KJ & Schmittgen TD (2001) *Methods*, **25**, 402–408. – *First description of the analysis of relative gene expression data using real-time quantitative PCR and the $2^{-\Delta\Delta CT}$ method.*
★★★ 12. Pfaffl MW (2001) *Nucleic Acids Res.* **29**, e45. – *Description of a new mathematical model for the relative quantification real-time RT-PCR data.*
13. Meijerink J, Mandigers C, van de Locht L, Tonnissen E, Goodsaid F & Raemaekers J (2001) *J. Mol. Diagn.* **3**, 55–61.
14. Weksberg R, Hughes S, Moldovan L, Bassett AS, Chow EW & Squire JA (2005) *BMC Genomics*, **6**, 180.
15. Hassed SJ, Hopcus-Niccum D, Zhang L, Li S & Mulvihill JJ (2004) *Clin. Genet.* **65**, 400–404.

CHAPTER 5

Robust and unique PCR for single-nucleotide polymorphism genotyping applications

Xiangning Chen

1. INTRODUCTION

Since the invention of DNA sequencing techniques, it has been possible to identify naturally occurring variation in DNA at the sequence level, and the completion of numerous genome sequencing projects in recent years has greatly accelerated the identification and categorization of such variation. Naturally occurring variations in DNA sequences can be classified broadly as insertions, deletions, duplications, translocations/rearrangements, or single-nucleotide polymorphisms (SNPs). As many of these variations are relatively stable, heritable, and easy to interrogate experimentally, geneticists use them as tools for numerous applications including evolutionary studies, disease gene mapping, and forensic applications. Of the many sequence variations, two forms are widely used today for genetics studies: variable number of tandem repeats (VNTRs) and SNPs.

1.1 Variable number of tandem repeats

VNTRs, also referred to as either short tandem repeat polymorphisms or microsatellite markers, are simple repeats of short sequence motifs. These sequence motifs can be two, three, four, five, or more nucleotides that can be repeated several times. VNTR markers were used to construct human genetic linkage maps in the early phase of the human genome project and most of these markers are now mapped to specific positions in the human genome. The alleles of VNTR markers are characterized by the number of tandem repeats. Readers interested in VNTRs are referred to (1) and (2).

PCR: *Methods Express* (S. Hughes and A. Moody, eds.)
© Scion Publishing Limited, 2007

1.2 Single-nucleotide polymorphisms

SNPs constitute a substitution or change of a single nucleotide. As there are only four possible nucleotides at any given position, the maximum number of alleles for any SNP is four, although the majority of SNPs have only two alleles. In the last few years, the human genome project and the HapMap project have identified millions of SNPs in the human genome. These SNPs are deposited in the dbSNP database at the National Center for Biological Information (http://www.ncbi.nlm.nih.gov/SNP/index.html) and they have become a popular resource for genetics studies.

There are numerous methods for performing SNP genotyping (3–5), and most current protocols require PCR amplification. A typical protocol for SNP typing has several steps, which include (see *Fig. 1*):

- Target DNA amplification. PCR is the most popular method; however, other amplification methods include isothermal rolling-circle amplification and ligation-mediated amplification.
- Purification of amplified products (not necessary for some techniques, e.g. *Protocol 3*). This can be achieved via physical separation (gel filtration, precipitation, electrophoresis, etc.) or enzymatic digestion (alkaline phosphatase and endonuclease I).

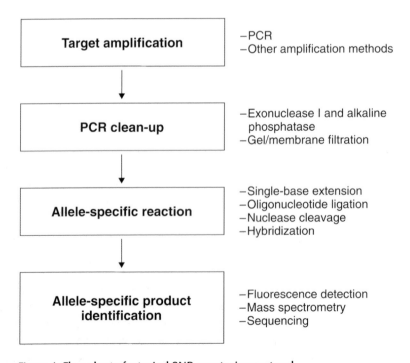

Figure 1. Flow chart of a typical SNP genotyping protocol.
The major steps of the protocol are shown on the left. The popular methods for each step are listed on the right. For more details, see (3).

- Allele-specific reaction and allele-specific product identification. Allele-specific reactions are the core of genotyping procedures by which allele-specific products are produced. Many biochemical reactions (DNA synthesis, DNA ligation, endonuclease digestion, etc.) and biophysical processes (DNA hybridization, melting temperature) have been explored to generate target template-dependent products, i.e. if the target template contains an A allele, then product A will be produced; if the target template contains a B allele, then product B will be produced. The accuracy and efficiency of allele-specific reactions largely determine the overall accuracy, throughput, and cost of genotyping protocols. The efficiency, reliability, and reproducibility of the PCR are critical for the success of SNP genotyping.

More detailed descriptions of SNP typing are available elsewhere (3–5). This chapter will focus on methods used in research laboratories that use SNP genotyping for applications such as population studies, evolutionary studies, and genetic disease mapping. Genetic disease testing and forensic testing require different protocols and they will not be covered here. In research laboratories, the requirements for genotyping are high accuracy, efficiency, and throughput, but low cost. Therefore, protocols are designed and optimized to meet these requirements.

2. METHODS AND APPROACHES

2.1 General guidelines for genotyping applications

2.1.1 Essential requirements for PCR in genotyping applications

Research-oriented genotyping is normally conducted for many subjects and many markers at the same time. Under these conditions, in addition to accuracy, factors such as cost, success rate and reproducibility are important parameters. To accomplish these goals, each step in the protocols needs to be designed and optimized accordingly, and, where possible, made amenable to automation. This section will outline the parameters for PCR design and illustrate how they can influence the overall quality and cost of the genotyping procedure. In addition to conventional guidelines for PCR design and operation, special attention should be extended to the issues of:

1. *Uniqueness.* To be used as a genetic marker, the location of a SNP has to be unique, occurring at only one location in the genome. However, this is not necessarily sufficient to guarantee that the genotypes obtained are unique. This is because many genotyping techniques cannot identify the alleles without first amplifying the surrounding sequence by PCR. As there are many repetitive sequences in the genomes of higher organisms, there is a realistic probability that the PCR may amplify other unintended sequences along with the target sequence, which can cause genotyping errors. Typically, when more than one DNA sequence is amplified, they cause skewed genotype/allele distribution, because those unintended sequences may interfere with the

allele-specific reaction, or they may contain extra alleles. These errors may be detected by the Hardy–Weinberg equilibrium test (which states that, in a random mating population, genotype frequencies of a single locus would become fixed at a particular value after one generation, i.e. $a^2 + 2ab + b^2 = 1$, where a and b are the frequencies of the two alleles). However, this alone is not sufficient to exclude PCR errors. In fact, several high-profile studies have arrived at wrong conclusions because the authors were unaware of these unintended amplifications when they performed their genotyping (6).

2. *Even amplification of both alleles.* In higher organisms, there are two copies of each chromosome and two alleles for each locus. For genotyping applications, we want to identify both alleles. However, most genotyping techniques, although in principle capable of quantification, do not normally take quantitative measurements into consideration when they score the genotypes. This means that these techniques do not discriminate hemizygotes from homozygotes. Because of this weakness, a biased PCR that preferentially amplifies one allele over the other will go unnoticed. This will lead to artificial loss of heterozygosity and spurious allele and genotype distribution. If no preventive measurements are taken, this kind of error can go undetected.

3. *Robustness.* To be efficient and cost-effective, a PCR protocol has to be reliable and reproducible. The quality and quantity of DNA can vary considerably among different samples. Different markers are located in genomic regions with different GC content and genomic structure. These factors can affect PCR amplification significantly and impose challenges on PCR design and performance. Under many circumstances, elevated failure rate not only increases genotyping costs but also reduces the statistical power.

2.1.2 PCR design and performance

To ensure that the PCR is specific and amplifies the two alleles evenly, several factors need to be considered when selecting parameters for designing PCR primers. As the PCR is conducted for many subjects and markers, the parameters selected should, wherever possible, be kept constant, so that the same operational procedures and reagents can be used for many different individuals and markers (see *Protocol 1*). This is of particular importance for large-scale applications, as frequent changes in procedures and supplies increase the difficulty of quality control and cause human errors.

The key issues important for consistency, reliability, and reproducibility in PCR primer design include masking out repetitive sequences, using a constant set of parameters, and verifying potential polymorphism sites. There are many software packages that can handle this kind of job, including some commercial packages such as OLIGO and PRIMER EXPRESS. However, many of the commercial software programs do not disclose details of their algorithms (T_m calculation, GC content and surrounding sequence weighing, self-priming, etc.) or other primer design parameters. This lack of details makes it hard to incorporate them into an automatic protocol.

We recommend the open-sourced PRIMER 3 program for PCR and extension primer design for fluorescence polarization template-directed dye-terminator incorporation (FP-TDI) genotyping protocols. The general protocol is as follows:

1. Obtain sequences for SNPs from the dbSNP site (http://www.ncbi.nlm.nih.gov/SNP/index.html).
2. Mask repetitive sequences using the REPEATMASKER program (7).
3. Design primer sequences using the PRIMER 3 program (http://frodo.wi.mit.edu/cgi-bin/primer3/primer3_www.cgi) (8).

The parameters used for primer design are as follows (for other parameters use the default parameters suggested on the PRIMER 3 website):

Product size range = 100–350 bp

- Primer Size
  - ○ Minimum (Min) = 20
  - ○ Optimal (Opt) = 23
  - ○ Maximum (Max) = 26
- Primer annealing temp (T_m)
  - ○ Minimum (Min) = 58
  - ○ Optimal (Opt) = 60
  - ○ Maximum (Max) = 62
- Primer GC percentage (GC%)
  - ○ Minimum (Min) = 20
  - ○ Maximum (Max) = 50
- Maximum (Max) Self Complementarity = 8
- Maximum (Max) 3′ Self Complementarity = 3

2.1.3 Procedures for verification of polymorphisms

In dbSNP, many polymorphisms have been identified by bioinformatics tools and thus may not be real polymorphisms. In addition, some of the polymorphisms are low-frequency changes that may be unique to the individual in which the SNP was identified. Therefore, to avoid unnecessary genotyping costs and DNA wastage, it is necessary to verify SNPs as polymorphic in a population relevant to your main study, ahead of full-scale genotyping. We routinely use 48 subjects selected from the human variation panel (http://ccr.coriell.org/nigms/comm/order/catprice.html) established for the HapMap project. Even for confirmed polymorphisms, it is often necessary to test them with a selected panel when either the PCR design or genotyping technique being used is new. The purpose of this testing is to verify whether the PCR is unique and whether genotypes can be scored clearly. If unusual results are seen in the testing, such as more than four genotype groups (two homozygote groups, one heterozygote group, and one failed group; see the example in *Fig. 3*, also available in the color section), or there is no clear grouping of the genotype clusters, it is reasonable to suspect that there may be problems in the PCR performance (uniqueness, even amplification, or robustness) and/or the allele-specific reactions. It is important to point out that a normal outcome from the testing procedures does not necessarily guarantee that the PCR is unique and that both alleles are being amplified evenly because:

- Genotyping software does not discriminate hemizygotes from homozygotes.
- There may be some unknown polymorphisms in the region that can influence the performance of the PCR and the allele-specific reaction.

Using the PCR design guidelines outlined above, we have achieved close to 100% PCR success rate for the thousands of markers we have tested over the last several years. Of these markers, 10–15% show zero or low heterozygosity in our testing panel.

2.2 SNP genotyping methods

2.2.1 FP-TDI genotyping protocol

The FP-TDI genotyping method (see *Protocol 2*) is based on single-base extension and fluorescence polarization detection (9). This protocol contains three sequential reactions:

1. PCR.
2. Enzymatic clean-up.
3. Single-base extension.

The genotypes are assigned based on the change of fluorescence polarization of the dyes used to tag ddNTPs.

In this protocol, it is important to consider each step, as all are critical for success.

- PCR. We use amplicons in the range of 100–350 bp, as smaller amplicons tend to be amplified more robustly and minimize the amount of primers and dNTPs in the PCR. It is important to optimize the amount of both primers and dNTPs used in the PCR (so the alkaline phosphatase and exonuclease I are not overwhelmed by unincorporated dNTPs and primers, respectively) as they will interfere with the incorporation of ddNTPs used for the allele-specific reactions.
- PCR clean-up to remove left-over primers and dNTPs. Alkaline phosphatase is added to the PCR product for complete digestion of the dNTPs left over from the PCR. In addition, exonuclease I is also added to remove remaining single-stranded PCR primers as well as single-stranded extensions of the amplicon. The enzymes used are heat sensitive and the protocol takes advantage of this feature to inactivate the enzymes following digestion. This feature also requires that the enzymes be handled with care before being added to the reaction mixture to avoid inactivation or partial inactivation of the enzyme.
- Dye set used for the single-base extension. The fluorescence dyes used to label the ddNTPs are another key for successful genotyping. Although the ddNTPs included in commercially available FP-TDI kits are successful in genotyping a majority of SNPs, there is a fraction of SNPs that do not perform well when using them. This is because the nature and the rules of the interaction between

fluorescence dyes and ssDNA are not fully understood. For this reason, if a marker does not work well with the commercial kits, a practical approach is to use a different set of ddNTPs to carry out the single-base extension reaction (10).

Fluorescence polarization measures both the quantity and the ratio of the allele-specific products of the two alleles. In order to reach the minimal threshold of the fluorescence polarization measurements, PCRs need to be robust to produce a sufficient amount of product. In our experience, PCR products that can be visualized on an agarose gel are sufficient to produce reasonable genotyping results.

2.2.2 TaqMan genotyping protocol

The TaqMan assay (see *Protocol 3*), also called the nuclease cleavage assay, is a single-step method that couples the PCR amplification and allele-specific reaction together (11). This is probably the simplest protocol for SNP typing, and it can be very cost-effective if the sample size is reasonable (800–1000 subjects). This method relies on the 5′ nuclease activity of DNA polymerase to cleave the allele-specific probes, and uses fluorescence resonance energy transfer (FRET) to identify alleles. When a doubly labeled allele-specific probe is matched with the target polymorphism, the 5′ end of the probe will be cleaved during PCR, separating the 5′-end fluorescent group (reporter dye) from the 3′-end fluorescent group (quencher dye). This separation of the two fluorescence dyes leads to the uncoupling of FRET and makes the reporting dye detectable upon excitation at its specific wavelength. Applied Biosystems provide two product lines that use this technique for genotyping, validated assays, and custom-designed assays. For both validated and custom-designed assays, VIC and FAM are used as reporter dyes for alleles 1 and 2. The company designs and tests PCR primers and allele-specific probes using its own proprietary algorithms. To ensure that the assays work robustly, the company provides a ready-to-go mixture containing PCR primers, allele-specific probes, and dNTPs. For this reason, the protocol is quite robust and quality control is relatively easy.

In our experience, two issues can influence the performance of the assay significantly:

1. To achieve the desired scoring rate, it is necessary to quantify the genomic DNAs to be genotyped and to make the concentration equal for all individuals included in a sample panel. This is especially important when the assays are scored by the end-point approach. If the concentration of genomic DNA varies too much, the PCR efficiency will be very different among individuals, which can cause difficulties in scoring of the genotypes.

2. To ensure genotype quality, internal negative controls should be included in sample panels. To score a genotype reliably, a reference base line for a negative-control reaction is required, against which a positive reaction can then be scored with confidence. Without these negative controls, even if some samples can still be scored, the confidence in these scored genotypes will be compromised.

2.3 General considerations

To guarantee consistent and reproducible genotyping, quality control is an important issue. The quality control should cover all materials and protocols, including genomic DNA samples, supplies, and reagents. We recommend:

1. Assessing the quality and quantity of genomic DNA samples. In order to get good genotyping results, it is essential to use DNA samples of similar concentration and quality. Normally, this is done at the beginning of a research project when the sample panel is assembled.
2. Using the same vendors for supplies. Many reagents used for genotyping protocols can be purchased from different vendors. It is not advisable to change vendors frequently, as this makes it difficult to control and compare the quality of these reagents.
3. Testing each new batch of supplies and other reagents.
4. Aliquotting supplies and reagents into smaller volumes. For example, we know that repeated freeze/thawing of primers and dNTPs can be detrimental to PCR efficiency. As there is no easy way of assessing their quality after a number of freeze/thaw cycles, we routinely aliquot the amount of reagent required for 1 week's worth of work and limit their thawing to three times, after which the aliquot will be discarded.
5. Logging supplies and daily activities.

These measurements may seem trivial, but they can have a significant impact on the consistency and reproducibility of high-throughput operations.

2.4 Recommended protocols

As illustrated in *Fig. 1*, most genotyping protocols require a clean-up of the PCR products prior to allele-specific reactions and allele-specific product detection. Because this chapter is focused on PCR, descriptions of other procedures will be brief. However, it is important to point out that PCR is only one step of most genotyping protocols. Successful PCR does not necessarily guarantee the success of a genotyping protocol. In the examples below, I will discuss the steps that are likely to cause problems and describe measures to prevent these problems.

Protocol 1

Standard PCR protocol[a]

Equipment and Reagents

- Genomic DNA (5 ng/μl)
- HotMaster *Taq* DNA polymerase (5 units/μl) and accompanying reaction buffer (Eppendorf)
- PCR primers (stock concentration 200 μM)[b]
- dNTPs (stock concentration 100 mM; Invitrogen)
- Sterile water
- Black 384-well microplates (MJ Research)
- Soft-sealing pad (MJ Research)
- Tetrad PCR machine (MJ Research)
- 1% Agarose gel containing 10 ng/ml ethidium bromide
- 6× Orange loading dye solution (Fermentas)
- Equipment and reagents for agarose gel electrophoresis including 1× TBE agarose gel running buffer (10.8 g/l Tris base; 5.5 g/l boric acid; 4 ml/l 0.5 M EDTA, pH 8.0; diluted from a 10× stock; Sigma)
- DNA size marker (100 bp ladder, Invitrogen)
- UV light source

Method

1. Prepare PCRs in 384-well microplates[c] in a 5 μl volume containing:
   - 5 ng of genomic DNA
   - 100 nM each PCR primer
   - 250 μM dNTPs
   - 0.5 μl of 10× reaction buffer
   - 2.5 mM $MgCl_2$[d]
   - 0.5 units of HotMaster *Taq* DNA polymerase
   - Sterile water up to a final volume of 5 μl

2. Seal the microplates with the soft-sealing pad.

3. Use the following thermal cycling conditions for PCR:
   - Initial denaturation at 95°C for 3 min
   - 30 cycles of 95°C for 30 s, primer-dependent T_m for 30 s, and 65°C for 45 s
   - Final extension at 72°C for 5 min

4. Run the PCR products alongside a DNA size marker on a 1% agarose to visualize the products.

Notes

[a]This method can be used for testing primers.

[b]Primer design for several sequences can performed as a batch process. For this, we use the open-sourced PRIMER 3 program. We have established a protocol to obtain sequences for SNPs from the dbSNP site (http://www.ncbi.nlm.nih.gov/SNP/index.html). The sequences are then masked by the REPEATMASKER program (7) and fed to the PRIMER 3 program (8) using a PERL script. The parameters used are as follows:

- PRIMER_OPT_SIZE = 23
- PRIMER_MAX_SIZE = 26
- PRIMER_MIN_SIZE = 20
- PRIMER_OPT_TM = 60
- PRIMER_MAX_TM = 62

- PRIMER_MIN_TM = 58
- PRIMER_PRODUCT_SIZE_RANGE = 100–350
- PRIMER_MIN_GC = 20
- PRIMER_MAX_GC = 50
- PRIMER_SALT_CONC = 50
- PRIMER_SELF_ANY = 8
- PRIMER_SELF_END = 3
- PRIMER_DNA_CONC = 40
- PRIMER_GC_CLAMP = 0
- PRIMER_MAX_END_STABILITY = 8

Most of the parameters are self-explanatory. For those who do not have bioinformatics support to use automatic procedures, the PRIMER 3 program has a website for manual design (http://frodo.wi.mit.edu/cgi-bin/primer3/primer3_www.cgi) where users can find the definitions of these parameters and change them according to their needs.

^{c}We use Tetrad PCR machines (MJ Research) and therefore use microplates from the same company to ensure robust PCR performance. Other microplates could work well, but testing should be conducted to ensure that their performance is adequate.

^{d}Altering the $MgCl_2$ concentration range between 1.5 and 5 mM can improve PCR efficiency.

Protocol 2

FP-TDI genotyping protocol

Equipment and Reagents

- HotMaster *Taq* DNA polymerase (5 units/μl) and accompanying reaction buffer (Eppendorf)
- PCR primers and one extension primer (stock concentration 200 μM)
- dNTPs (stock concentration 100 mM; Invitrogen)
- Black 384-well microplates (MJ Research)[a]
- Soft-sealing pad (MJ Research)
- Tetrad PCR machine (MJ Research)
- FP-TDI kit (Perkin Elmer) containing:
  - ☐ Fluorescent-labeled ddNTPs
  - ☐ Shrimp alkaline phosphatase
  - ☐ *Escherichia coli* exonuclease I
- Microplate reader (Analyst; Molecular Dynamics Corporation)

Method

1. Prepare PCRs in 384-well microplates[a] in a 5 μl volume containing:
   - 5.0 ng of genomic DNA
   - 100 nM each PCR primer
   - 250 μM dNTPs
   - 0.5 μl of 10× reaction buffer
   - 2.5 mM $MgCl_2$
   - 0.5 units of HotMaster *Taq* DNA polymerase
   - Sterile water up to a final volume of 5 μl

2. Seal the microplates with the soft-sealing pad.

3. Use the following thermal cycling conditions for PCR:
   - Initial denaturation at 95°C for 3 min
   - 10 cycles of 95°C for 30 s, 75°C for 5 s, ramping to 65°C at −0.1°C/s, and 65°C for 45 s
   - 30 cycles of 95°C for 30 s, 60°C for 45 s. Final extension at 72°C for 5 min

4. Add 1 unit of shrimp alkaline phosphatase and 0.1 unit of *E. coli* exonuclease I to each PCR and incubate at 37°C for 1 h. Inactivate the enzymes by incubating at 95°C for 15 min.

5. Perform the single-base extension reactions with the FP-TDI kit[b] according to manufacturer's instructions. The cycling conditions for the reaction are 30 cycles of 95°C for 30 s and 60°C for 30 s.

6. Read the fluorescence with a microplate reader[c].

7. Score genotypes based on fluorescent polarization values[d].

Notes

[a]We use Tetrad PCR machines (MJ Research) and therefore use microplates from the same company to ensure robust PCR performance. Other microplates could work well, but testing should be conducted to ensure that their performance is adequate.

[b]FP-TDI kits are available for all six polymorphisms (A/G, C/T, A/C, G/T, A/T, and C/G).

[c]The filters used for R110 are: excitation 490 ± 10 nm and emission 510 ± 10 nm. The filters used for TAMRA are: excitation 550 ± 10 nm and emission 580 ± 10 nm.

[d]Genotypes are scored using a Microsoft EXCEL template developed in our laboratory (12).

Protocol 3

TaqMan genotyping protocol

Equipment and Reagents
- Black 384-well microplates (MJ Research)
- TaqMan SNP Genotyping kit[a,b] (Applied Biosystems)
- AmpliTaq DNA polymerase (5 units/µl; Applied Biosystems)
- Soft-sealing pad (MJ Research)
- Tetrad PCR machine (MJ Research)
- Microplate reader (Analyst; Molecular Dynamics Corporation)

Method

1. Prepare PCRs in 384-well microplates in a 5 µl volume containing:
   - 2 ng of DNA
   - 0.5× TaqMan assay mix (from the genotyping kit)
   - 0.5 units of AmpliTaq DNA polymerase
   - Sterile water up to a final volume of 5 µl

2. Seal the microplate with the soft-sealing pad.

3. Use the following thermal cycling conditions for PCR:
   - Initial denaturation at 95°C for 5 min
   - 40 cycles of 95°C for 30 s and 60°C for 45 s

4. Following PCR, read the fluorescence intensities for both the VIC and FAM dyes with a microplate reader[c].

5. Score genotypes by plotting the intensity of VIC against that of FAM.

Notes

[a]For a specific region of interest, users can search the Applied Biosystems database using its SNP browser software. Once the SNPs of interest are identified, SNP assays can be ordered from the Applied Biosystems website (https://products.appliedbiosystems.com/ab/en/US/adirect/ab?cmd=catNavigate2&catID=600760). User registration is required. The assays are mixtures containing PCR primers, allele-specific probes, and dNTPs. DNA polymerase can be ordered separately.

[b]There are two different protocols that can be used for these TaqMan assays: real-time PCR and an end-point assay. The real-time PCR protocol requires a real-time PCR machine capable of measuring VIC and FAM fluorescence. The end-point assay protocol does not require a real-time PCR machine, but requires a fluorescence plate reader. In my laboratory, we have carried out comparative studies and found that the results from the two protocols are comparable. As the end-point assay protocol is more flexible and allows a higher throughput, we tend to use this protocol for our research.

[c]The filters used for reading VIC are 520 ± 10 nm for excitation and 550 ± 10 nm for emission. The filters for reading FAM are 490 ± 10 nm for excitation and 520 ± 10 nm for emission.

2.5 Typical results

2.5.1 FP-TDI genotyping

Fig. 2 (also available in the color section) presents some examples of typical results produced using this protocol. In the figure, data from two markers are presented. The top two panels are results from the marker rs31251. We designed PCR primers (primer 1: 5′-AAAAATTAATGAACCTGAACCATCA-3′, and primer 2: 5′-ATACCAGAGCTTTATTTCAGTTTTTG-3′) and an extension primer (5′-AGCATGCTAAAAGTGGAC-3′) for this marker. Genotypes of the subjects can be scored by either simply plotting the fluorescence polarization values of the two alleles or by cluster analysis (see *Fig. 2*). These two panels show similar cluster patterns (left panel: family samples; right panel, case–control samples), despite being obtained from two different sources with varying DNA concentrations and quality.

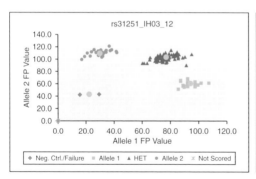
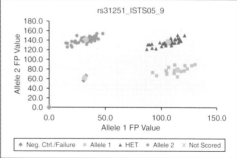
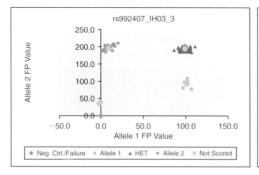
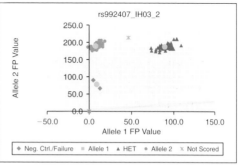

Figure 2. Some examples of SNPs typed by the FP-TDI protocol (see page xxi for color version).
Each panel contains 94 subjects and two negative controls. The top two panels show the results for the SNP rs31251 typed for two different samples. The results from a family sample are shown on the left and those from case–control samples on the right. Both demonstrate very similar results. The bottom two panels show the results for the SNP rs992407, typed for case–control samples. The panel on the right misses the minor allele homozygote group. Comparison with the left panel allows determination of the genotypes of the right panel.

The bottom two panels are results obtained from the marker rs992407. We designed PCR primers (primer 1: 5'-CAAAGGAGCTCAGGATTAGAACA-3', primer 2: 5'-TTGAGTGAAATTGAGGTGTTTCA-3') and an extension primer (5'-GTATATTATTTACAGACAGACCTCAA-3') for this marker. Two different sample plates were typed for this marker: the left panel has three groups of genotypes (homozygotes 1/1, heterozygotes and homozygotes 2/2). However, the right panel has only two genotype groups: homozygotes 1/1 and the heterozygote group. The 2/2 homozygote group is missing. Comparison with the left panel allows determination of the genotypes for the right panel.

It should be noted that the distances between the three genotype groups are different for the two markers. Typically, this pattern is characteristic of a marker and remains unchanged across different samples. These characteristics are very useful in determining which group comprises heterozygotes when the homozygotes for the minor allele are not observed.

2.5.2 TaqMan genotyping

The TaqMan assay is a robust technique and has a low failure rate in our hands. Like the FP-TDI genotyping protocol, there are different ways of scoring genotypes. In *Fig. 3* (also available in the color section), the top-left panel is a simple scatter plot of the intensities of FAM and VIC dyes. This approach is normally sufficient when the gaps between the groups are clear, as is the case here. When the data quality is suboptimal, data transformation can improve the scoring substantially. The top-right panel shows the results obtained using tangent transformed values (a ratio of intensities between FAM and VIC) of the data from the top-left panel.

The bottom two panels of *Fig. 3* show some unusual genotyping results for the SNP rs244738, which indicates a problem with either the PCR or the allele-specific reaction. In this example, the cause is a second polymorphism, rs244737, located 8 bp upstream of the target SNP, rs244738. This second polymorphism creates a differential annealing affinity of the two allele-specific TaqMan probes among individuals, resulting in some subtle but consistent changes in cleavage efficiencies of the allele-specific probes. Without this knowledge, it is difficult to score the genotypes for this marker.

2.6 Summary

Robust and unique PCR is a prerequisite for many SNP typing protocols. For those protocols where the user designs and controls all aspects of the procedure, such as the FP-TDI protocol, a systematic approach in design, operation, and quality control is essential for consistent and reliable genotyping. For those protocols where the company provides the 'ready-to-go' reagents, such as the TaqMan assays, users need to understand the company's guidelines. Sometimes the scientific judgement and rationale underlying these guidelines may not be apparent. Only when users truly understand the principles can they decide when and where to alter the protocol to suit their situation and accomplish efficient and reliable genotyping.

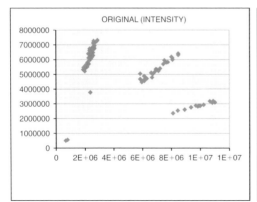
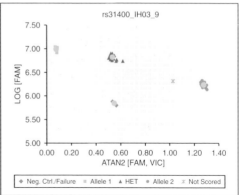

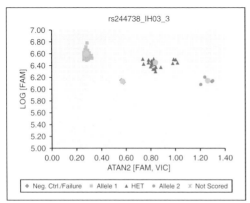
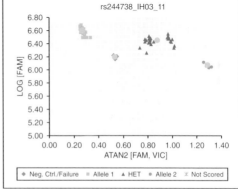

Figure 3. Examples of SNPs typed by the TaqMan protocol (see page xxii for color version).
The top two panels are results for rs31400 for the same sample plate scored by different algorithms. On the left is a scatter plot of fluorescence intensities of the two report dyes (FAM on the x-axis and VIC on the y-axis) as recorded by the fluorescence plate reader. On the right is a plot of the tangent-transformed value (the ratio of FAM/VIC) versus the intensity of FAM. Whilst the two methods produce identical results where the data quality is good, a tangent transformation can make the scoring more objective when the data are suboptimal. The bottom two panels show an unusual clustering pattern, where the heterozygotes are divided into two distinct subgroups, which is consistent across different sample plates. The results are a consequence of a second polymorphism, rs244737, located 8 bp upstream of the target SNP, rs244738. This second polymorphism creates a differential annealing affinity of the two allele-specific TaqMan probes between individuals, resulting in a consistent change in cleavage efficiencies of the allele-specific probes.

3. TROUBLESHOOTING

- Genotyping failures

   There are many factors that can contribute to the failure of both the FP-TDI and the TaqMan genotyping methods. These failures can be divided into two groups.
   - ○ Complete failure, i.e. affecting a whole plate, where none of the subjects can be assigned a genotype. This is likely to be caused by operational errors

and/or reagents and supplies. If this happens in the testing procedure for new SNPs, errors in design of primers and probes should also be considered. To find out where the errors occur, the best approach is to repeat the protocol with a smaller number of subjects and verify the results of each step.

○ Partial failure, where some of the subjects typed in an experiment can be scored, whilst others can not. Partial failures are more common and the causes are more complicated. In essence, this kind failure is caused by factors varying among individual samples. These include uneven sample quantity and quality, partially degraded reagents (dNTPs, primers, or enzymes), physical location in a 96- or 384-well plate, pipetting errors, and thermal cycling conditions. For these reasons, it is more practical to take preventive measurements than to figure out the exact causes for different samples. This is one of the reasons why we recommend the use of consistent parameters and strict quality control measurements (outlined above). For different projects, the acceptable success rate (i.e. the percentage of subjects genotyped) may vary. In general, we expect 95% or better.

- **Complete genotyping failure for the FP-TDI protocol**
  The following approaches can be used to verify where problems may be occurring:

  ○ The aim of the PCR step (step 1, section 2.2.1) is to amplify the target DNA sequence, and this can be verified by agarose gel electrophoresis. In our experience, if a PCR product can be visualized on an agarose gel, then there is sufficient to produce genotyping results.

  ○ The aim of the clean-up step (step 2, section 2.2.1) is to remove excess dNTPs and PCR primers left over from the PCR. We can find out whether the alkaline phosphatase and exonuclease treatments have worked efficiently by running the product from step 3 (single-base extension) on a polyacrylamide gel capable of differentiating primers of different length. If multiple bands are seen on the gel, this is an indication of failed phosphatase treatment. A failed nuclease treatment will produce product bands one base longer than the PCR primers. Failure of either the phosphatase or the nuclease would lead to non-specific primer extension. I should emphasize that it is difficult to verify the failure of the phosphatase and nuclease without performing the single-base extension step. This is because it is not easy to detect small amount of dNTPs and unlabeled PCR primers. After the third step, these unintended products, along with the intended products, will be labeled with fluorescent dyes. This makes detection much easier. Once the problematic step is found, focus on this step and test the reagents used for it. If primer design is suspected to be causing the problem, design new primers and optimize them for use.

- **Complete genotyping failure for the TaqMan method**
  Results can be viewed by running the amplification products on an agarose or polyacrylamide gel to check for the presence of PCR products. If there are no

PCR products, test the genomic DNA, DNA polymerase, assay kit, and cycling conditions (the annealing temperature) to eliminate these factors as the cause of reaction failure. If there are PCR products (a single band of the expected size), it is likely that probe cleavage has not occurred, which may be a consequence of using the wrong type of DNA polymerase. The TaqMan assay depends on the 5′→3′ nuclease activity of DNA polymerase to cleave probes. Some commercially available DNA polymerases do not have this property.

- **Difficulty in scoring Taqman genotyping results**
  This can be a consequence of varying DNA concentration of the samples under investigation, which can make it difficult to distinguish a negative reaction from a weakly positive one (due to suboptimal genomic DNA concentration). In this situation, continuing thermal cycling for another 10–15 cycles (50–55 cycles rather than 40 cycles) may improve the results significantly, as this gives those samples with lower DNA concentration an opportunity to produce more PCR products, narrowing the gap with those samples with higher DNA concentration.

- **Unexpected genotyping results**
  It is often possible to generate unexpected genotyping results, i.e. Hardy–Weinberg equilibrium failures. Whilst it is not possible to give a comprehensive list of possible causes, some key factors to consider are as follows:
  - Ensure that there are no additional polymorphisms in the primer and probe binding sites.
  - Ensure that there has been no mistake in the orientation of the plate, i.e. the samples are correctly assigned to each well and the plate has not been transposed during the genotyping process. This can be achieved by including spatially unique negative control samples on each plate (i.e. the pattern of the negative controls will clearly identify any gross errors in plate formatting/handling).
  - We recommend including positive controls of known genotype (i.e. include samples from the SNP verification step, see section 2.1.3), which can be cross-checked to ensure that the correct assay has been used and/or that the assay is performing correctly.

  When performing case–control analysis, ensure that you have a mixture of cases and controls on each plate so that failure/error of one plate does not lead to bias in the overall result.

4. REFERENCES

1. Weber JL (1990) *Curr. Opin. Biotechnol.* **1**, 166–171.
2. Weber JL & Broman KW (2001) *Adv. Genet.* **42**, 77–96.
★★ 3. Chen X & Sullivan PF (2003) *Pharmacogenomics J.* **3**, 77–96. – *A description of the biochemistry and protocols of methods commonly used for SNP typing.*
4. Kwok PY (2001) Annu. Rev. *Genomics Hum. Genet.* **2**, 235–258.
5. Syvanen AC (2001) *Nat. Rev. Genet.* 2, 930–942.
★★ 6. Pompanon F, Bonin A, Bellemain E & Taberlet P (2005) *Nat. Rev. Genet.* **6**, 847–859. – *A description of the common errors in genotyping and methods used to detect them.*

7. Bedell JA, Korf I & Gish W (2000) *Bioinformatics*, **16**, 1040–1041.
8. Rozen S & Skaletsky H (2000) *Methods Mol. Biol.* **132**, 365–386.
9. Chen X, Levine L & Kwok PY (1999) *Genome Res.* **9**, 492–498.
10. Chen X (2003) *Comb. Chem. High Throughput Screen.* **6**, 213–223.
11. Livak KJ (1999) *Genet. Anal.* **14**, 143–149.
12. van den Oord EJ, Jiang Y, Riley BP, Kendler KS & Chen X (2003) *Biotechniques*, **34**, 610–20, 622.

CHAPTER 6

Using PCR and linkage mapping to identify single genes and quantitative trait loci for livestock traits

Jillian F. Maddox, Imke Tammen, and Sonja Dominik

1. INTRODUCTION

PCR and linkage mapping are tools that have been used frequently in conjunction to map simple and complex inherited traits using pedigrees from a range of livestock and other species (1–4). The general procedure for trait mapping entails: phenotyping animals for the trait(s) of interest; collecting DNA from the phenotyped animals; genotyping all, or a subset, of the phenotyped animals, with markers covering the whole genome; and linkage analysis to localize the trait to a specific region on a chromosome. At this point, other fine-mapping tools such as linkage disequilibrium analysis, association analysis, comparative map analysis, and positional candidate gene mapping can also be used in association with linkage analysis to refine the mapped position further (see section 2.3). The ultimate aim is to refine the chromosomal location for the trait so that either an indirect test can be developed that enables marker-assisted selection, or the causative mutation is identified, enabling the use of a direct DNA test.

1.1 Genetic traits

Traits of interest can be controlled by a single locus, e.g. most simple Mendelian inherited disorders, or, as is the case with many livestock production or quantitative traits, by several loci, known as quantitative trait loci (QTL). Single-locus traits are much easier to map than QTL and can be mapped with fewer animals. Over 75 loci with simple Mendelian inheritance have had their causative mutations identified in livestock (5). This has allowed the development of direct DNA tests that can be used for the selection of animals with favorable alleles, or the removal of animals with deleterious alleles from livestock breeding populations. QTL mapping studies have been performed for all major livestock

PCR: *Methods Express* (S. Hughes and A. Moody, eds.)
© Scion Publishing Limited, 2007

species, and the results from large numbers of studies have been summarized in species-specific QTL map databases (6–9). Whilst only a small number of causative mutations for livestock QTL have been identified to date (10–13), the frequency of identification of mutations responsible for QTL is increasing.

1.2 Genetic markers

Microsatellites, also known as simple tandem repeats or simple sequence repeats, are repeats of a small motif (usually one to six bases) with the most common mammalian type being a $(CA)_n$. or $(GT)_n$ dimer, often referred to as a CA or GT repeat (14, 15). The likelihood of a microsatellite being polymorphic increases with increasing length of the repeat, and the majority of repeats where n is at least 10 are polymorphic. CA repeats of this length are found every 30–50 kb in mammalian genomes. The small size of microsatellites means that they can readily be amplified by PCR, and their high degree of polymorphism means that they have a high likelihood of being informative (i.e. able to identify which allele came from each parent) for genotyping. These properties have made microsatellites the markers of choice for livestock trait mapping studies for many years.

Recently, as a consequence of whole genome or large-scale expressed sequence tag sequencing, large numbers of single-nucleotide polymorphisms (SNPs) have been identified in several livestock species. Whilst SNP markers are about 100-fold more abundant than microsatellites, they are generally less informative, as SNP markers usually have only two alleles, which means that parents are less likely to be heterozygous than for markers with more alleles. In addition, it makes it harder to determine unambiguously which progeny allele is derived from which parent, particularly if only the sire and progeny are genotyped, as is often the case with livestock populations. Although the recent development of cost-effective commercial high-throughput SNP genotyping platforms (16) has made them the markers of choice for genome-wide association studies in humans (17), the equivalent level of technology is still being developed for livestock. In the future, it is likely that SNP genome mapping will largely replace microsatellite mapping for livestock species where appropriate SNP genome mapping 'chips' are available.

Strategies for identifying mutations in genomic regions of interest include both physical methods, such as library screening, and *in silico* methods. The recent availability of large amounts of sequence information for many livestock species, together with a big reduction in sequencing costs, has altered the strategy for developing new markers for regions of interest. Previously, for microsatellite detection, a common strategy was to use a labeled DNA probe for physical screening of a large insert DNA library, or to use PCR with specific primers for a gene or a product likely to map to a region of interest, if the library was in a format suitable for PCR screening. Clones that were identified as belonging to the region of interest then had their inserts cut up into small fragments by digestion with a restriction enzyme, and the resulting fragments were then subcloned. One or more labeled microsatellite repeats (e.g. [^{32}P]dCTP-tailed $(GT)_{20}$) were then used to screen the fragments, and clones containing inserts were then sequenced with vector-specific primers flanking the insert. Whilst this strategy is still used for species without much sequence information, it has largely been replaced in sequence-rich species by an *in silico* strategy whereby a region of interest can be investigated with a genome map viewing tool that also shows microsatellites and SNPs that have been identified previously. For species without these resources, a common strategy for SNP discovery nowadays is to PCR amplify fragments of 500–1000 bp (the frequency of SNPs in many livestock species is at least one polymorphism per 500 bp of sequence) from a region of interest from several animals using region-specific primers, and then to sequence the products to identify mutations. A range of other cheaper physical methods including single-strand conformational polymorphism, denaturing gradient gel electrophoresis, or enzymes, such as the CEL-1 endonuclease, can also be used as a screening mechanism for physical identification of fragments containing mutations that can then be sequenced.

This chapter focuses on laboratory and computer methods that assist in moderate to high-throughput genotyping of microsatellite markers and their subsequent linkage analysis. This is followed by a brief discussion of some of the more commonly used computer packages for QTL analysis.

2. METHODS AND APPROACHES

2.1 Genotyping

Linkage mapping for a population requires genotyping tests to be performed on the same sets of a large number of individuals for many markers. The work is highly repetitive with error levels of between 0.5 and 1% being common, whilst error levels of more than 15% per locus have been reported (18). Error levels of 1% can result in incorrect map orders, inflated map distances, significant loss of linkage information for a trait, and also a decrease in the power of detecting associations (18, 19). Consequently, it is important to observe procedures undertaken within one's laboratory objectively, in order to recognize where errors are likely to occur, and to develop and implement strategies that minimize errors. The automation derived from the use of robotic systems helps reduce error rates; however, many laboratories lack the resources to purchase these systems. The protocols described below give examples of practical, cost-effective ways in which genotyping can be made more systematic to reduce error levels.

2.1.1 DNA extraction

There are many methods of DNA extraction from livestock. In the past, large amounts of highly purified DNA have been required for whole genome scans, and the preferred option has been to extract DNA from blood. FTA (Fast Technology for Analysis of nucleic acids; Whatman) paper/card is a convenient vehicle for collecting, transporting, and long-term storage of blood (or other DNA-containing) samples. On average, a single 2.5 cm blood spot on a standard FTA card can provide sufficient DNA (~4–8 µg) for more than 3000 PCR tests, but currently provides insufficient amounts of DNA for high-density SNP-based genome scans. Similarly, milk, frozen semen, paraffin-embedded tissue, tail hair, and feces have also been used as sources of livestock DNA, but, with the exception of semen, often provide only limited amounts of DNA. Whole genome amplification (20) may be used to overcome the problem of only limited amounts of DNA being available for livestock studies (21) (also see Chapter 18).

2.1.2 Genotyping methods

Protocols 1–6 represent streamlined ways of scaling up PCR assays for laboratories without robotic equipment by using 96- or 384-well plates in a semi-automated fashion.

- *Protocol 1* is a modification of the standard FTA DNA extraction method for a single PCR. It represents a quick and easy method for processing the type of blood samples that are commonly collected from livestock by farmers, and yields sufficient DNA for 60 or more PCR tests. The protocol can easily be scaled up to obtain larger yields of DNA.
- *Protocol 2* describes the creation of DNA master plates, which are used for the transfer of DNA samples to DNA test plates using some form of replicating device, such as a 96-well transfer device or a multipipette, and is a simple way to reduce variation among test plates. If an error is made in setting up the master plate, then it occurs on multiple test plates and is consequently easy to detect.
- *Protocol 3* describes primer design.
- *Protocol 4* represents an example of radioactively labeled PCR and manual gel electrophoresis (see *Figs 1* and *2*) and tailors the PCR cocktail and cycling conditions to fit specific primer pairs. This approach provides the option of multi-loading, which entails staggered loading of samples with electrophoresis between each loading. Silver staining after gel electrophoresis (22) can be used as an alternative to radioactive methods to visualize the PCR products for genotyping. However, silver staining is much less sensitive than radiolabeled or fluorescent methods (see *Protocol 5*), and increased amounts of PCR products need to be produced for this approach to succeed.
- *Protocol 5* demonstrates an indirect method that uses a M13 tailed microsatellite primer in combination with a fluorescently labeled M13 primer (23) and a standard touchdown PCR protocol to reduce costs and simplify the PCR procedures being performed within a laboratory (see *Fig. 3*). This protocol allows the use of a single fluorescently labeled primer for all primer pairs. Synthesis and purification of fluorescently labeled primers is costly, especially as some primers fail to amplify the desired PCR products and others may only be used infrequently for specific mapping projects. There have been several instances where primer pairs that fail to amplify products using *Protocol 5* work well using *Protocol 4*.

- *Protocol 6* describes PCR multiplexing, which can decrease time and increase cost-efficiency, but requires special attention to the primer design (for guidelines on multiplex primer design, see 24–27). As with multi-loading approaches, multiplexing requires prior knowledge of either the expected allele size ranges or the migration positions relative to a reference point such as a tracking dye in order to avoid different fragments overlapping.

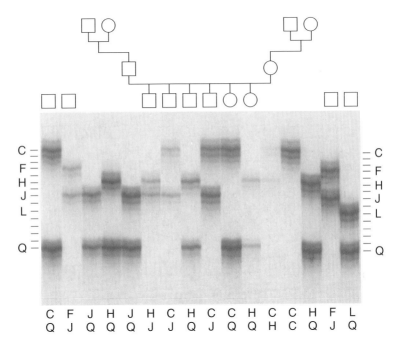

Figure 1. Genotyping the ovine CD5 microsatellite.
The microsatellite was amplified and visualized using the method described in *Protocol 4*. Assignment of alleles and genotyping was performed as described in *Protocol 7*. The allele codes are indicated to the left and right of the autoradiograph (labeled C to Q from largest to smallest), and the genotypes for each individual are indicated below the autoradiograph. The autoradiograph shows the microsatellite alleles for 16 individuals consisting of a family of 12 individuals (top row, four grandparents; middle row, two parents; and bottom row, six offspring) and four unrelated animals (far left and far right). Males are indicated by squares, females by circles, and lines are used to show the relationships between family members. Both parents are heterozygous, with the genotypes for the sire being JQ and for the dam being CH. Three of the grandparents are heterozygous and the maternal grandsire is homozygous (CC). The microsatellite shows a relatively common 'stutter' pattern with the major band (to which the allele code is assigned) being associated with a ladder of smaller less–intense products whose size differs by increments of 2 bp and a faint larger product. Stutter bands are an artifact of PCR that are thought to be caused by slipped-strand mispairing during PCR.

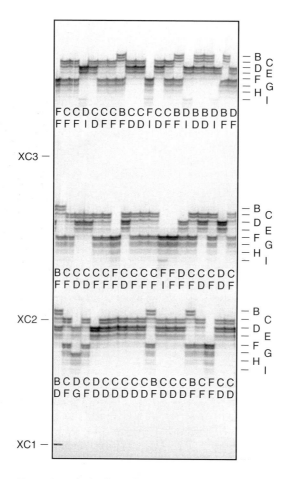

Figure 2. Multi-loading of an ovine microsatellite.
The first set of samples was loaded onto the gel and then electrophoresed until the xylene cyanol band had migrated 5 cm into the gel. The second set of samples was loaded and electrophoresis was performed until the second xylene cyanol band had migrated 5 cm into the gel. The third set of samples was loaded and the gel was run until the first xylene cyanol band (XC1) was 0.5 cm from the bottom of the gel (the positions of the second xylene cyanol band (XC2) and the third xylene cyanol band (XC3) are also indicated on the autoradiograph). The XC bands migrate at the same position as a 110 base fragment in a denaturing 6% acrylamide gel. The allele codes are indicated to the right of the figure (labeled B to I from largest to smallest) and the genotypes for each individual are indicated below each lane.

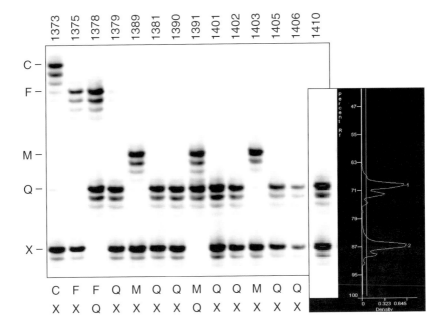

Figure 3. Fluorescent M13 genotyping of the ovine ILSTS28 microsatellite.
The microsatellite was amplified using the method described in *Protocol 5*. Semi-automated gel electrophoresis was performed on a LI-COR machine. The gel image was analyzed using RFLP-SCAN (Scanalytics), which allows visualization of individual genotypes as a chromatogram (as shown in the insert for individual 1410 which has genotype QX, where Q = band 1 and X = band 2 in the track profile shown to the right of the gel picture) to assist with allele calling. The allele codes are indicated to the left of the image (labeled C, F, M, Q, and X from largest to smallest) and the genotypes for each individual are indicated below the image.

Protocol 1

DNA extraction from blood on FTA paper (3.1 mm paper punch)

Equipment and Reagents

- Blood sample on FTA paper (Whatman[a])
- 3.1 mm Paper punch (Fiskars[b])
- 0.2 ml Thin-walled PCR tubes or strips of 0.2 ml PCR tubes and caps (Art Robbins Instruments)
- FTA Purification Reagent (Whatman)[c]
- Aspirator system[d]
- Autoclaved Tris/EDTA (10 mM Tris/HCl (pH 8.0), 0.1 mM EDTA)
- Sterile (autoclaved) Milli-Q water or equivalent
- Thermal cycler (preferably with heated lid)
- Microfuge or plate centrifuge
- 50–200 µl Micropipette or multipipette or repeat dispenser[e]
- −20°C (or −70°C) freezer (manual defrost)[f]

Method

1. Position a clean 3.1 mm paper punch above a 0.2 mm thin-walled PCR tube and punch a 3.1 mm sample from the middle of the blood sample on the FTA paper so that it drops into the tube.

2. Clean the punch by making a punch of clean FTA paper (or blotting paper) and/or blowing compressed air over the tip of the punch (with the punch in ejection mode).

3. Make punches as described in steps 1 and 2 for all samples. Make a punch of clean FTA paper into a tube to use as a negative control. It is also a good idea to do a punch of a sample that has previously extracted well as a positive control.

4. Add 150 µl of FTA Purification Reagent to each tube[e]. Allow the tube to sit for 5 min at room temperature. Use an aspirator system to remove and discard as much as possible of the purification reagent from the tube, being careful not to suck up the paper pellet.

5. Repeat step 4 a further three times (a total of four additions of FTA Purification Reagent).

6. Add 150 µl of Tris/EDTA to each sample tube. Allow the tube to sit for 5 min at room temperature. Use the aspirator to remove as much as possible of the liquid from the tube, being careful not to suck up the paper pellet.

7. Repeat step 6.

8. Add 120 µl of sterile Milli-Q water to each sample tube. Place the lids on the sample tubes and heat the tubes at 100°C for 15 min on the PCR block (with the heated lid option enabled if the thermal cycler has that option).

9. Centrifuge the tubes briefly (in a microfuge at 10 000 g for 2 s or a plate centrifuge for 500 g for 1 min) to remove the liquid from the lids (this step may not always be needed if the thermal cycler has a heated lid). Pipette the liquid up and down twice to ensure the DNA is evenly distributed in the liquid before transferring the DNA in H_2O to a fresh tube leaving the paper punch behind.

10. Aliquots (2–4 µl) of the DNA samples can now be dried onto plates for use as templates for PCR (Protocol 2), or the samples can be frozen and kept for subsequent use[g].

Notes

[a]There is a range of options in terms of FTA paper formats that can be used including 2.5 cm spot (MiniCard, Classic Card), 96-well (CloneSaver) and 384-well (EasyClone) formats (all from Whatman). It is generally cheaper to purchase larger formats (e.g. four-spot Classic Card) and cut the FTA paper to a suitable size, and then staple a piece of FTA paper between a suitably sized piece of folded paper that can be used to record the animal's identity and act as a shield. FTA paper is also suitable for use with unclotted blood samples that have been frozen and thawed.

[b]Whilst Whatman supplies Harris punches for FTA work, a cheaper alternative is to use a hand paper punch such as a Fiskars (www.fiskars.com) hand punch (note that you will need to remove the paper catcher before use). Fiskars (and other hand) punches have a range of punch formats (e.g. 3.1 mm circle, 1.6 mm circle, etc.).

[c]Zhou and colleagues (28) recently proposed replacing the FTA Purification Reagent steps with a single 20 mM NaOH step for 1.2 mm pellets (200 μl, 30–50 min). The authors have successfully tested this approach with 3.1 mm pellets replacing the four FTA Purification Reagent steps (4 and 5) with two 15 min 200 μl 20 mM NaOH steps (or a single 30 minute 20 mM NaCH step).

[d]It is better to use an aspirator than a pipette as it removes more of the liquid and is quicker. If no pump-based aspirator system is available, a water-tap-based system is also suitable. If the aspirator sucks too strongly, then the paper pellet tends to attach to the end of the needle. If the pellet escapes from the tube whilst there are still FTA Purification Reagent addition steps or sticks to the needle, then put the pellet back into the tube. To clean the aspirator system prior to use, wipe the needle with 70% ethanol and suck a small amount of 70% ethanol through it.

[e]If rows of tubes are to be done, then it is easiest to use a multipipette and an appropriate reagent reservoir to dispense reagents. The reagent reservoir needs to be sterile for the Tris/EDTA and H_2O steps, and tips need to be changed between wells for these steps.

[f]Automatic defrost freezers use heating cycles for defrosting. This reduces the longevity of many biological reagents (including DNA, oligonucleotides, and restriction enzymes). Hence, it is recommended that either a manual defrost freezer is used, or that samples are placed within a container that protects against temperature fluctuations (such as a Biocooler; RPI Corporation) in an automatic defrost freezer.

[g]It is not possible to quantitate accurately the amount of DNA at this stage as the yield is too low. However, using this approach, we routinely obtain sufficient DNA for PCR.

Protocol 2

Preparation of DNA master and replicate plates

Equipment and Reagents

- Sterile (autoclaved) Milli-Q water or equivalent
- 96-Well thermal cycler plates or thin-walled PCR tubes (Greiner)
- 96-Well dispensing device (Intelligent Bio-Instruments) or multipipette (Eppendorf)
- Thermal cycler
- Laminar flow hood or other clean area for plate preparation
- –20°C (or –70°C) freezer (manual defrost)

Method

1. Design DNA plate template layouts making sure that they include suitable positive[a] and negative controls in appropriate positions on the plate.

2. Set up a DNA master plate with DNA diluted to 2–8 ng/µl or DNA from *Protocol 1*.

3. Remove the thermal cycler plates from their plastic shells/sheaths (retain the shell).

4. Use a 96-well dispensing device or a multipipette to dispense 2 µl[b] of DNA into the bottom of each well on the replicate plate.

5. Dry the DNA onto the plate[c].

6. Return the dried DNA plate to the plastic shell and seal the shell with a bag sealer[d,e].

7. Keep the DNA master plate with the remaining DNA at –20°C.

Notes

[a]It is a good idea to have at least one standard reference control DNA sample on all plates so that loadings from different plates can easily be aligned in terms of relative allele sizes for a marker. You should also include at least one negative control per plate. The negative controls test for PCR contamination and help with sample alignment when labeling autoradiographs. It is best to position negative controls in the middle of the plate surrounded by samples rather than at the edges of the plate.

[b]If the product yield after PCR is too low from FTA-extracted DNA, then dispense 4 µl per well.

[c]Suitable drying regimes include heating on a thermal cycler (lid up) (or on a plate heat block) at 65°C for 15 min, or placing in a laminar flow hood at room temperature for several hours to overnight.

[d]Alternatively use a self-adhesive plate sealer to cover the plate.

[e]The plates can be kept at room temperature in the dark for up to 3 months. The plate life depends on both the quality and the amount of DNA dried onto the plate. Plates prepared from FTA-extracted DNA, or those with less than 16 ng of DNA or with DNA of poor quality (some degradation or contaminants), are likely to have a shorter shelf-life. Keeping the plates in a freezer does not extend the plate life.

Protocol 3

Microsatellite primer design, synthesis, and storage

Equipment and Reagents
- PRIMER3 program[a]
- Microsatellite sequence for species of interest or for closely related species

Method

1. Paste the source sequence into the sequence box at http://frodo.wi.mit.edu/cgi-bin/primer3/primer3_www.cgi

2. Adjust the relevant PRIMER3 criteria to those suitable for your experiment[b] to design primers either side of the microsatellite sequence.

3. Resuspend the synthesized primers[c] to 200 µM in sterile Milli-Q water (stock tube). Store the primer stocks in liquid form or after freeze drying at –20°C (or colder) in a manual defrost freezer (do not use an automatic defrost freezer)[d]. Create working concentration primers from the stocks by combining the forward and reverse primers with sterile Milli-Q water so that the final concentration of each primer in the mix is 5–10 µM[e].

4. PCR conditions will need to be optimized for each primer pair. Please refer to Chapter 1 for suggested steps.

5. Store aliquots of working concentration primers at –20°C or colder in a manual defrost freezer.

Notes

[a]For a single sequence use, PRIMER3 is available via the Internet at http://frodo.wi.mit.edu/primer3/. The software can be downloaded and installed locally for use in batch mode. Consult the online help or the README file for details on the various user-definable parameters.

[b]We have found that the following guidelines work well for microsatellite primer design:
- T_m of 66°C for both primers (do not worry if this means that the primers are different lengths).
- Avoid long runs of Gs or Cs where possible.
- Where possible, avoid repetitive elements or, at most, position one primer within the repeat.
- Avoid including the microsatellite itself as part of either primer. However, the PCR will often work if one primer is adjacent to the microsatellite and includes a couple of bases from the microsatellite repeat.
- Primers based on sequences from closely related species where both primers have a T at the 3' end produce good results.

Sometimes no 'ideal' primers can be developed. Primers which do not meet the recommended design criteria may still work and are worth testing if no optional primers are available.

[c]When designing primers for fluorescent detection, unlabeled primers need only to be of sequence quality, whereas fluorescently labeled primers should be HPLC purified.

[d]Stock and working concentration unlabeled primers can be kept at –20°C for more than 10 years provided the tubes do not become contaminated with DNases. Fluorescently labeled primers are light sensitive and hence should be stored in a light-proof container. Whilst fluorescently labeled primers are reported to be sensitive to freeze/thaw cycles, the authors have successfully used the same batch of fluorescently labeled M13 primer for 12–18 months until laboratory stocks ran out. It is good practice to apportion primers into appropriate-sized aliquots before being first frozen so that individual tubes are not handled more than five times. When the same fluorescent primers are to be used repeatedly within a short period of time (e.g. 1 week), they can be kept in a refrigerator at 4°C or on wet ice rather than being repeatedly frozen and thawed.

[e]A range of concentrations may need to be tested for primers that are to be used for fluorescent work or for markers that are to be multiplexed in a single PCR tube (see *Protocol 6*).

Protocol 4

Radioactive PCR, denaturing polyacrylamide gel electrophoresis, and autoradiography

Equipment and Reagents

- Microplate or tubes with dried DNA (see *Protocol 2*)
- $[\alpha\text{-}^{33}P]$dATP[a] (Perkin Elmer)
- 25 mM $MgCl_2$[f]
- 2 mM dNTPs[b] (0.2 mM dATP, 2 mM dGTP, 2 mM dCTP, and 2 mM dTTP; Roche)[f]
- *Taq* DNA polymerase (New England Biolabs)[c] and accompanying 10× PCR buffer
- TaqStart antibody (CloneTech)
- Primers (see *Protocol 3*)
- Multipipette plus adjustable repeat pipetter plus Combitips (Eppendorf)
- Sterile paraffin oil (Sigma)
- Self-adhesive microplate sealer or tube lids
- Thermal cycler (MJ Research)
- 1× TBE gel running buffer (0.9 M Tris, 0.9 M boric acid, 0.02 M EDTA)
- Gel (7 M urea, 6% acrylamide[a] (acrylamide:bis-acrylamide 19:1; BioRad), 1× TBE running buffer)
- Tetramethylethylenediamine (TEMED), Sigmacote, ammonium persulfate, (Sigma)
- Sharktooth combs: microtiter 3× format, 0.4 mm thickness, made from Teflon (Owl Separation Systems S2S-MT3 – cut in two to help insertion) and 0.4 mm spacers
- Other polyacrylamide gel apparatus for casting and running gels
- Denaturing loading dye (98% formamide[a], 10 mM EDTA, 0.025% xylene cyanol, 0.025% bromophenol blue – filter through 0.2 μm filter and store frozen or at 4°C)
- Vacuum grease (e.g. glisseal®; Borer Chemie)
- Eight-channel multi-loading gel device (Hamilton)
- Temperature-controlled power pack with temperature probe (BioRad)
- Gel drier (BioRad or Labconco)
- Vacuum pump (Eyela A-3S aspirator[d]) and vacuum gauge
- Dark room
- BioMax MR film (Kodak)
- Film cassette (Sigma)
- Developer and fixer (Kodak or Agfa)

Method

1. Calculate the amounts of each reagent needed for the PCR cocktail[e].

PCR cocktail	Final concentration	Volume per reaction (μl)
Taq polymerase (5 U/μl)	0.1 U/4 μl	0.02
1.1 μg/μl TaqStart antibody	0.022 μg/4 μl	0.02
2 mM dNTPs[b,f]	100 μM	0.20
25 mM $MgCl_2$[g]	1.5–4.5 mM	0.24–0.72
7 μM Primers (forward and reverse)	0.7 μM each	0.40
10× PCR buffer[e]	1×	0.40
Sterile Milli-Q water	–	Make volume up to 4 μl
10 μCi/μl $[\alpha\text{-}^{33}P]$dATP	0.1–0.4 μCi/4 μl	0.01–0.04 (amount depends on age of the $[\alpha\text{-}^{33}P]$dATP)

2. Combine the *Taq* polymerase and the TaqStart antibody in an appropriately sized tube and let them sit at room temperature for 5 min. Add the rest of the reagents[f] in the order above, mix the cocktail by vortexing, and centrifuge the tube to consolidate the sample and reduce bubbles.

3. Dispense 4 μl of PCR cocktail onto the side of each well of a plate containing dried DNA with a repeat dispenser (or single pipette tip)[h].

4. Add 1 drop of sterile paraffin oil to each well[i] to prevent sample evaporation during PCR.

5. Amplify the DNA using the following PCR profile:
   - 1 cycle of 95°C for 2 min 55 s (initial denaturation)
   - 30 cycles of denaturation at 95°C for 5 s, annealing at primer-specific temperature[g] for 30 s, and extension at 72°C for 30 s.
   - 1 cycle of 72°C for 2 min 30 s (final elongation)
   Following PCR, the plate should be stored at –20°C until required.

6. Add 2 vols (8 μl) of denaturing loading dye to each PCR well with a multipipette and gently pipette to mix. Use a fresh set of tips for each set of wells. The samples may be stored in a –20°C freezer before or after this step.

7. For polyacrylamide gel electrophoresis, siliconize one glass plate with Sigmacote before assembling the gel apparatus according to the manufacturer's instructions. Combine:
   - 60 ml of gel mix (sufficient for a gel 40 cm wide × 27 cm long, 0.4 mm thick)
   - 575 μl of 10% ammonium persulfate
   - 28 μl of TEMED

8. Gently swirl to mix, draw the solution up into a 50 ml syringe, and gently expel the gel liquid continuously between the glass plates moving the syringe from side to side. Leave the gel for 1–2 h at room temperature to set[j].

9. Prepare the gel apparatus for running according to the manufacturer's instruction. Turn on the power (~130 W for a gel containing ~45 ml of gel solution) and run it to pre-warm the gel and running buffer[k] to 50°C[k]. Use a syringe and appropriate gauge needle to wash the top surface of the gel with running buffer (from gel buffer chamber). Lightly smear vacuum grease onto the combs and insert them into the gel, until the shark teeth are just below (approximately 1 mm) the surface of the gel. Re-apply the power.

10. Thaw the samples and then denature the samples on a heat block at 95°C for at least 5 min.

11. Turn off the power to the gel (the gel should be at 50°C[m]), rinse the wells thoroughly with gel running buffer using a syringe and needle to remove any bubbles and leached urea, and load 1 μl of the samples directly from the heat block to the gel with the multi-loading device as rapidly as possible. Re-apply the power to the gel. Replace the cover on the sample plate and return it to the freezer.

12. Run the samples the requisite distance into the gel (use the position of the xylene cyanol or bromophenol blue dye bands as an indicator of size: xylene cyanol runs at 110 bp on a 6% denaturing gel) to avoid fragment overlaps and load the next batch of samples if another batch of samples is to be loaded. Note that steps 10 and 11 need to be performed for each set of samples loaded. Step 10 should be performed such that the end of the 5 minute sample denaturation corresponds to when the xylene cyanol or bromophenol blue band has migrated to the position at which the next batch of samples are to be loaded.

13. Repeat step 12 as many times as necessary for multiloading.

14. Run the bottom set of samples the requisite distance (usually 1 or 2 cm from the bottom of the gel) using the loading dyes as markers.

15. Lay down the glass plates containing the gel flat and remove the top plate. Apply blotting paper (cut to slightly larger than the gel) to the gel surface and use the blotting paper to gently peel the gel from the bottom plate. Apply cling film to the surface of the gel and transfer the gel with blotting paper and cling film (cling film up, gel and blotting paper towards porous surface) to a gel drier connected to a vacuum pump. Place an extra piece of blotting paper beneath the gel and blotting paper. Heat at 80°C, with a vacuum level of greater than 27 inches of mercury for sufficient time to dry the gel (this will depend on the vacuum pump; usually between 20 and 80 minutes).

16. Remove the cling film from the surface of the gel and in a dark room place the dried gel in a cassette with a sheet of film[n,o].

17. Expose the autoradiograph overnight (or longer if the bands are likely to be faint) and develop the film. Results for single and multiple loadings can be seen in *Figs 1* and *2*, respectively.

Notes

[a]Appropriate safety precautions need to be taken for handling [α-^{33}P]dATP, formamide, and acrylamide.

[b]Purchase a set of individual dNTP stocks and make up a mix where the dNTP concentration that corresponds to the radioactive dNTP is 0.2 mM and the others are 2 mM.

[c]Any good-quality commercial *Taq* polymerase and 10× buffer can be used. Some 10× buffers already contain $MgCl_2$. Different buffers may perform slightly differently in PCR.

[d]Faster pumps are considerably more expensive and often require expensive maintenance. The pump should pull a vacuum of at least 27 inches of mercury.

[e]The multiplier for the cocktail will depend on the number of samples and an allowance for pipette discrepancies and wastage (e.g. set up 105 reactions worth of reagents for 96 samples).

[f]The dNTPs, 25 mM $MgCl_2$, and 10× PCR buffer should be aliquotted and stored at –20°C. It is a good idea to vortex these after thawing (especially the 10× PCR buffer).

[g]Start by PCR testing the primers with 1.5 mM $MgCl_2$ in the PCR cocktail at an annealing temperature (T_{ann}) at 5–8°C below the T_m (e.g. for a T_m of 66°C, try a T_{ann} of 58°C). If this doesn't work, try a range of $MgCl_2$ concentrations between 0.5 and 4.5 mM, and a range of annealing temperatures between 50 and 65°C. Other factors that may improve the PCR are to alter the PCR cocktail so that it includes up to 10% dimethyl sulfoxide or 10% glycerol or to omit the TaqStart antibody (if there are likely to be mismatches between primer and template sequence).

[h]Ensure that you do not touch the DNA area of the well as this avoids having to change the pipette tip between wells.

[i]Oil is necessary for PCR volumes of less than 20 μl, even if the thermal cycler has a heated lid. It is quicker to add one drop of oil per well with a 1 ml micropipette than to try and add a specified amount to each well. One milliliter of oil will dispense about 32 drops. There is no need to centrifuge the plate/tube after adding the PCR cocktail and oil. The plastic on the plates is sufficiently slippery that the 4 μl of PCR cocktail slides rapidly to the bottom of the well. The oil slides down on top of the cocktail and acts as a barrier. We do not seal the plate during PCR – we tried this but found it too difficult to remove the seal. After the PCR has finished, put on an acetate plate sealer before placing the sealed plate in the freezer.

[j]If the gel is to be used on the following day, then soak some paper towels in 1× TBE and apply them to the open gel surfaces, and wrap the gel and paper towels in cling film. The wrapped gel should be stored at room temperature.

[k]To reduce the time needed to warm the gel, the gel running buffer can be preheated to ~45°C in the microwave. Do not heat the buffer more than this, as it may cause the glass plate in contact with the buffer to crack.

[l]It is important that the room in which the gel is being run is kept at a temperature of about 21°C. If the room temperature is warmer than this, then the gel will run more slowly and the band resolution will be less sharp. If the gel is run at temperatures above 50°C, the glass plates are more likely to crack. Toughened glass is not suitable for use with running gels as the surface is not flat (the plates should be made from float glass).

[m]The temperature of the gel can be read from the display on the temperature-controlled power pack. An alternative to a temperature-controlled power pack is to use an adhesive temperature indicator (e.g. Owl Separation Systems or Thermographic Measurements Ltd) and attach it to the front gel plate.

[n]BioMax MR film has emulsion on only one surface (the non-shiny side) and this surface needs to be in contact with the dried gel (shiny side away from the dried gel).

[o]This protocol can be modified to be used for single-strand conformational polymorphisms or PCR-restriction fragment length polymorphism analysis, or for running samples on agarose gels using ethidium bromide or silver to detect the DNA.

Protocol 5

M13 tailed microsatellite primer, fluorescent M13 primer, and touchdown PCR for semi-automated gel electrophoresis

Equipment and Reagents

■ Microplate or tubes
■ DNA (50 ng per reaction, plated and dried in 96- or 384-well plates)
■ 25 mM MgCl$_2$
■ dNTPs (Roche)
■ *Taq* DNA polymerase and accompanying 10× PCR buffer[a]
■ Microsatellite-specific forward and reverse primers[b] (20 μM) with one primer having a 5′-CACGACGTTGTAAAACGAC tail
■ Fluorescently labeled M13 primer[c] (5′-fluorescent label-CACGACGTTGTAAAACGAC-3′)
■ Multipipette plus adjustable repeat pipetter with combitips (Eppendorf)
■ Sterile paraffin oil (Sigma)
■ Thermal cycler (MJ Research)
■ Semi-automated gel electrophoresis equipment (e.g. LI-COR or ABI)

Method

1. Calculate the amounts of each reagent needed for the PCR cocktail.

PCR cocktail	Final concentration	Volume per reaction (μl)
Taq polymerase (10 U/μl)[a]	0.5 U/10 μl)	0.05
5 mM dNTPs	0.2 mM	0.4
25 mM MgCl$_2$[d]	2.5 mM	1
20 μM microsatellite primers	0.08 μM each	0.04 each
1 μM M13 primer	0.02 μM	0.2
10× PCR buffer	1×	1
Sterile Milli-Q water	–	Make volume up to 10 μl

2. Thaw all ingredients; keep them on ice and mix well before use.

3. Prepare the PCR cocktail, mix thoroughly by vortexing, and centrifuge briefly.

4. Dispense 10 μl of PCR cocktail onto the side of each well of a plate containing 50 ng of dried DNA with a repeat dispenser (or single pipette tip)[e].

5. Add 1 drop of sterile paraffin oil to each well[f].

6. Amplify the DNA using the following PCR program:
    ■ 1 cycle of 95°C for 5 min (initial denaturation)
    ■ 5 cycles of denaturation at 95°C for 45 s, annealing starting at 68°C for 90 s (with the annealing temperature being reduced by 2°C per cycle), and extension at 72°C for 60 s
    ■ 4 cycles of denaturation at 95°C for 45 s, annealing starting at 58°C for 60 s (with the annealing temperature being reduced by 2°C per cycle), and extension at 72°C for 60 s
    ■ 25 cycles of denaturation at 95°C for 45 s, annealing at 50°C for 60 s, and extension at 72°C for 60 s
    ■ 1 cycle of 72°C for 5 min (final elongation)

7. Add denaturing loading dye to each PCR product and either store upright in a dark box in a –20°C freezer until use or run on a semi-automated polyacrylamide gel apparatus[g] (typical results can be seen in *Fig. 3*)

Notes

[a]Any commercial *Taq* polymerase/10× buffer can be used. If the *Taq* polymerase or buffer do not include a hot-start component, then it is advisable to prepare the PCR cocktail on ice to reduce nonspecific amplification and primer-dimer formation.

[b]One of the microsatellite-specific primers (e.g. the forward primer) needs to be synthesized with the additional nucleotides CACGACGTTGTAAAACGAC at the 5′ end. This incorporates the binding site for the fluorescently labeled M13 primer into the PCR product. The stepped program is a variant of a touchdown protocol that reduces nonspecific bands by getting the PCR started at a high annealing temperature that is reduced in steps and thus is likely to suit a range of different primer combinations without further optimization. The final annealing temperature of 50°C is suitable for use with the fluorescently labeled M13 primer.

[c]The type of fluorescent label depends on the equipment used for semi-automated electrophoresis. The fluorescently labeled M13 primer can be resuspended and diluted in either Milli-Q water or 10 mM Tris/1 mM EDTA (pH 7.5) and needs to be stored in a light-proof container. As this is the only fluorescently labeled primer required, the primer can be ordered in bulk to reduce costs. The fluorescent M13 primer needs to be aliquotted appropriately before use. The authors resuspend the dried oligonucleotide to make a 'concentrated' stock solution of 200 μM (200 pmol/μl), make a 20 μM 'diluted' stock solution from this, and finally prepare a 1 ml 1 μM working solution. All solutions are stored frozen and an aliquot of concentrated stock is only thawed to make more diluted stock when previous aliquots of diluted stocks have been used up. Likewise, an aliquot of diluted stock is only thawed to make a working solution when previous aliquots of working solution have been used up.

[d]The $MgCl_2$ concentration and the number of cycles in the PCR program can be varied to optimize the PCR.

[e]Providing that the DNA in the well is not touched and that a negative control is in the final well, there is no need to change the pipette tip between wells.

[f]Dispense the oil with a 1 ml micropipette but avoid contact between the pipette tip and the PCR cocktail.

[g]Gel electrophoresis and image analysis differ among different apparatuses and the manufacturer's instructions should be followed.

Protocol 6

Multiplex PCR

Equipment and Reagents
- Primer pairs for the appropriate microsatellites[a]
- Other equipment and reagents as for *Protocol 4*

Method

1. Determine the relative sizes of the markers that will be used in the multiplex PCR. Gaps of 15–20 bp should be left between adjacent marker size ranges in case there are extra alleles[b].

2. Double the normal amount of *Taq* polymerase and TaqStart antibody for combinations of two to five sets of primers. Try the primers in various combinations and determine what microsatellite combinations will co-amplify without problems.

3. Calculate the amounts of each reagent needed for the PCR cocktail:

PCR cocktail	Final concentration	Volume per reaction (μl)
Taq polymerase (5 U/μl)	0.2 U/4 μl	0.08
1.1 μg/μl TaqStart antibody	0.044 μg/4 μl	0.04
2 mM dNTPs	100 μM	0.20
25 mM MgCl$_2$	1.5–4.5 mM	0.24–0.72
7 μM Primers (forward and reverse)	0.35 μM each	0.20
10× PCR buffer[e]	1×	0.40
Sterile Milli-Q water	–	Make volume up to 4 μl
10 μCi/μl [α-^{33}P]dATP	0.1–0.4 μCi/4 μl	0.01–0.04 (depending on age of [^{33}P]dATP)

4. Follow from step 2 onwards in *Protocol 4*, omitting steps 12 and 13 (typical results can be see in *Fig. 4*).

Notes

[a]The PCR must amplify under the same conditions with known size ranges. In addition, they should have tight allele ranges and not have big gaps between adjacent alleles (between 75 and 450 bp, depending on the equipment). This size range should allow four to six microsatellite markers.

[b]If the gaps are smaller than this, then either redesign the primers for the microsatellite with the problem size range so that the primers amplify a product with an appropriate size range or add a stuffer sequence to one or both primers for the problem microsatellite to adjust the size range. The stuffer sequence could be the M13 tail (see *Protocol 5*) or a 5′ G-type tail. Suitable 5′ G tails include GTTTCTT, GTGTCT, GTGTCTT, and GTTTCT (29).

[c]Primer concentrations may need to be further adjusted, especially for fluorescent genotyping. Generally one needs to use more primer for larger product sizes to compensate for the decreased efficiency of amplification.

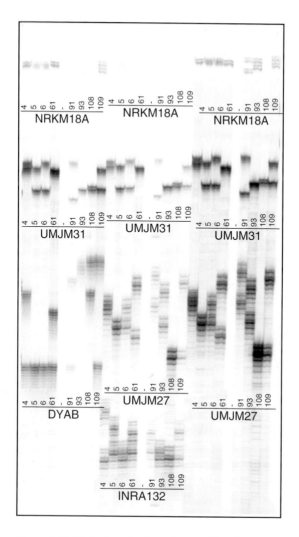

Figure 4. Multiplexing of ovine microsatellites (see *Protocol 7*).
Different combinations of primers were tested to determine their multiplexing ability
(annealing temperature of 59°C and 1.5 mM MgCl$_2$) and their relative size ranges. The
primers tested were: INRA132 (smallest, 152–178 bp), UMJM27 (not previously sized,
GenBank allele: 193 bp), DYAB (181–227 bp), UMJM31 (not sized, GenBank allele: 267
bp), and NRKM18A (largest, not sized, GenBank allele: 369 bp) microsatellites. The
multiplex testing showed that the UMJM27 and DYAB microsatellites have overlapping
size ranges; hence, these microsatellites are not suitable for multiplexing together. The
gap between the largest INRA132 allele and the smallest UMJM27 or DYAB allele is less
than 20 bp; hence, INRA132 should not be multiplexed with either UMJM27 or DYAB
without the size of the smaller or larger microsatellite being adjusted. There is room for an
additional microsatellite that is smaller than INRA132 in the multiplex as there are no
fragments in the 75–132 bp range, and also room for one that is between the sizes of
UMJM31 and NRKM18A.

Protocol 7

Manual genotyping

Equipment and Reagents
■ Autoradiograph or dried silver-stained gel
■ Database system for storing genotypes[a]

Method
1. Label the autoradiograph or dried gel with the date, gel identifier, marker, and sample numbers.
2. Enter the gel layout details (gel identifier, gel date, marker, sample numbers) into the database[b].
3. Determine an allele scoring system for the marker[c] (see *Fig. 1* for a scoring example) and enter the genotypes into the database.
4. Get a second person to score and enter the genotypes independently into the database.
5. Compare the two sets of scoring and print out any discrepancies.
6. Get both scorers to check the discrepancies and correct any mistakes in their scoring.

Notes
[a]There is a range of free (e.g. MYSQL or POSTGRESQL) or commercial database tools (e.g. Microsoft ACCESS) that could be used that are preferable to storing the data in Microsoft EXCEL spreadsheets. One simple table structure could have fields for gel identifier, gel scorer code, animal identifier, marker name, first allele, and second allele. It is important to use appropriate table indexes and database integrity mechanisms to assist with database performance and data integrity issues.
[b]For plate templates that are to be used multiple times, it is easiest to create a template to eliminate the need to type in sample numbers again.
[c]Look at all autoradiographs or dried gels for a marker. Determine which allele has the largest number of base pairs (top allele) and which allele is the smallest (bottom allele). Count the number of different alleles. It is easier to use a single character for each allele and assign alleles in alphabetic order from top to bottom (or vice versa) than to record actual sizes. Allele codes can readily be converted to allele sizes via an allele size-conversion table for a marker. If there are fewer than 26 alleles and the overall size distance between the top allele and the bottom allele is less than 25 times the microsatellite repeat motif size, then assign a letter for each potential position of an allele starting from the top allele. If the size range is greater than this but there are fewer than 26 alleles, then do not assign alleles for some of the positions where no allele is seen. When there are fewer than 26 alleles, it is often better to position the alleles in the middle of the A–Z range in case larger or smaller alleles are subsequently revealed.

2.1.3 Data management

The next step after size separation of alleles for a marker is to enter the genotypes into a database system. This can be done either manually or automatically (30), depending on the type of size-separation equipment used and the appearance of the microsatellite. *Protocol 4* is suited to manual genotyping, as outlined in *Protocol 7*. Two independent scorers should do the genotype scoring, and computer software should be used to identify any scoring discrepancies and to test the data for Mendelian inheritance. Where equipment with the capability for automatic calling is used, the microsatellite appearance needs to be appropriate for this type of capture method, and the allele calling and quality scores should be checked by a person, as an experienced human genotyper is often better at resolving difficult microsatellite stutter patterns than the algorithms commonly used by genotyping software.

2.2 Data analysis: linkage analysis and trait mapping

Two software tools, one for linkage analysis (CRI-MAP), and the other for QTL data analysis (QTL EXPRESS), are described in *Table 1*.

Linkage analysis can be used for both:

- Construction of whole genome or chromosome linkage maps prior to single gene or QTL mapping.
- Mapping of single-gene traits following appropriate transformation of the phenotype to a genotype (e.g. for a recessive trait, carrier 1 2, affected 2 2, normal 1 1).

CRI-MAP (the basic steps are outlined in *Table 1*) is suited for both linkage analysis and single-gene mapping of common livestock pedigrees. These and many other software tools can be obtained free from the Internet (http://www.nslij-genetics.org/soft/), e.g. LINKAGE (35) and MERLIN (36). However, most do not allow for the complex pedigree structures that are commonly used in livestock mapping experiments.

For QTL mapping (the basic steps for QTL EXPRESS are outlined in *Table 1*), phenotypes for traits of interest are measured and recorded in a database, along with other relevant information (e.g. day of measurement, experimental group, sex, birth date, birth type, age at weaning, etc.). Many QTL software packages also incorporate the ability to carry out statistical adjustments, and these packages would be more suited to novice users than packages that require the statistical adjustments to be performed in order to create suitable input data prior to use of the QTL package.

Procedures need to be developed within a laboratory whereby individuals and their samples are consistently identified, in order to allow correct correlation of genotype and phenotype data for all mapping work. Electronic identifiers, bar coding, and computer-based recording and database systems are all useful technologies that help to reduce error rates. Information on mapping algorithms and strategies for livestock mapping can be found elsewhere (37–43).

Although we recommend the use of CRI-MAP and QTL EXPRESS software for novice users, there are several other software packages that have been found to be useful by the authors:

- MULTIMAP (44) is a program used to automate multilocus linkage map construction. It uses a modified version of CRI-MAP. The software and manual can be obtained from http://compgen.rutgers.edu/multimap/ and additional instructions for using MULTIMAP with livestock pedigrees can be found at http://www1.angis.org.au/jmaddox/pages/multimap.htm.

Table 1. Basic steps for the use of CRI-MAP and QTL EXPRESS software packages

Construction of multilocus linkage maps using CRI-MAP	QTL analysis using QTL EXPRESS
1. Prepare a gen file according to the instructions in the manual and save as a text file. • The CRI-MAP software and manual (31) can obtained from the CRI-MAP home page (http://compgen.rutgers.edu/multimap/crimap/). • Gen files can be merged together using the merge option of CRI-MAP. A suite of extra programs that complement CRI-MAP has been developed (32) and is available from (http://www.genlink.wustl.edu/software/). 2. Run the *prepare* option of CRI-MAP (*crimap n prepare*), where *n* is the chromosome number of the gen file, to generate the other input files (e.g. par, dat, etc.) that CRI-MAP will need for subsequent operations. • Different actions can be undertaken from the command line (*crimap # action*). In general, it takes several runs with repeats of actions and comparison of the results before the final map is established. • The par file can be edited with a text editor. 3. Run the *twopoint* option of CRI-MAP (*crimap n twopoint*) to generate information (two-point LOD scores) that will enable grouping of loci into chromosomal groups. This step may be unnecessary if loci are already in chromosomal groups so that each gen file corresponds to a single chromosome. 4. Run the *build* option of CRI-MAP (*crimap n build*) to create a chromosomal map. Initially build a framework map (LOD 3). 5. Use the *all* option of CRI-MAP (*crimap n all*) to determine the best positions for the remaining loci (do the loci one at a time). 6. Inspect the final map with the *chrompic* option of CRI-MAP (*crimap n chrompic > chrn.chrompic*; where the > redirects the output to a file for subsequent browsing) to determine whether any genotyping looks suspicious. Suspicious genotypes are those where either (i) there is a single entry for a particular parental source flanked by the opposite parental source within a short (less than 10 or 20 cM) distance (e.g. 101 or oio), or (ii) there is a parental phase change for both copies of a chromosome for the end informative locus (e.g. 0001) where the distance between it and the adjacent informative locus is short.	1. Set up a phenotype, genotype, and map file as described in the online documentation. • QTL EXPRESS (33) is used online on http://qtl.cap.ed.ac.uk/. Data is submitted over the Internet and results generated instantly. Online introduction and documentation is available on the website. • The general format is similar for all analysis options: *Genotype file* – includes marker names, information on the data coding, the animal identifiers and pedigree, sex, generation, and the marker genotypes. *Map file* – information on the interval for genotype probability calculation, marker names, and mapping distances in cM. Map information on multiple chromosomes can be contained in a single file. *Phenotype file* - information on fixed effects and traits, animal ID, and records on fixed effects and production traits. 2. Choose the appropriate analysis option for your experimental design and follow the interactive input screen. • Five different analysis approaches can be chosen. Data from separate experiments can be combined in a single analysis. QTL EXPRESS uses Haley–Knott regression (34) for QTL mapping. • The user enters the paths to all of the input files on their computer or clicks on the QTL EXPRESS browse button to locate them. The data is then uploaded to the QTL EXPRESS web site and subjected to automatic data-checking procedures, which generate error messages for genotype or other data input errors. The web site also provides the options of performing (a) permutation tests to provide significance levels at 1 and 5%, and (b) bootstrapping to generate confidence intervals. The information on the significance of the results is provided as an F-statistic and as a likelihood ratio. 3. View the complete results or a summary online. • The output can be copied and pasted into a file on the user's computer.

- QTL CARTOGRAPHER (45) (http://statgen.ncsu.edu/qtlcart/index.php) and R/QTL (46) (http://www.rqtl.org), although not as suitable for novice users, allows for a greater range of analysis approaches and more user options.

2.3 Definition of terms

- **Association analysis**: Association analysis is a method of genetic analysis that compares the frequency of alleles among individuals with traits that can be categorized into classes. A given allele is considered to be associated with a trait if that allele occurs at a significantly higher frequency among individuals expressing the trait.
- **Covariate:** An independent continuous variable that affects response in the dependent variable and is not a class effect.
- **Linkage or genetic linkage:** Genetic linkage occurs when alleles for loci that are sufficiently close on the same chromosome are inherited together more often than would be expected if the loci were assorting independently. The further two loci are apart from each other on a chromosome, the greater the chance that crossover (or recombination) occurs between the two loci during meiosis (gamete formation).
- **Linkage mapping or linkage analysis:** Linkage mapping is the process by which linkage information can be used to construct a linkage map from a set of polymorphic genetic (DNA, protein, or trait) markers.
- **Linkage disequilibrium and linkage equilibrium:** Loci are said to be in linkage disequilibrium when specific haplotypes for the loci occur at a higher frequency than that predicted from the product of their respective allele frequencies. In contrast, loci are said to be in linkage equilibrium when haplotypes for the loci occur at the frequency predicted from the product of their respective allele frequencies. For linkage disequilibrium mapping purposes, loci need to be on the same chromosome. However, it is not necessary for loci to be on the same chromosome to be in linkage disequilibrium, as favorable allele loci on different chromosomes that are selected in concert will appear to be in linkage disequilibrium.
- **Marker–assisted selection**: This involves the use of DNA markers that are closely linked to a single locus or a QTL in a breeding programme. The intention of such an approach is to increase the accuracy of predicting the breeding value of animals when a direct DNA test is not available.
- **Positional candidate mapping**: An experimental approach in which knowledge of the location of a candidate gene is used as a starting point for linkage mapping.
- **Spurious logarithm of odds (LOD) peaks:** Standard approaches to QTL mapping can produce false significant results (visible as large peaks in the LOD profile) in regions of low information content or if the phenotype data is not normally distributed.

Additional terms can be found in on-line glossaries such as http://ghr.nlm.nih.gov/ghr/glossary/ and http://www.fao.org/biotech/index_glossary.asp.

3. TROUBLESHOOTING

- **Variable results are seen for different individuals performing the PCRs**

  It is well known that PCR works better for some individuals than others. It is important that each user within a laboratory has their own complete set of reagents. Allocate the task of making up aliquots of the reagents only to those individuals whose PCRs always work well. It is important that laboratory practices ensure that people are not interrupted or distracted when performing PCR-related tasks. Check that all pipettes have recently had the calibration tested and verified.

- **The PCR works poorly, fails, or shows contamination**
  - If both the primers and the DNA have previously been shown to work well in PCR, then the most likely cause of faint or no bands is degradation of one of the reagents, with $MgCl_2$ being the most likely. The 25 mM $MgCl_2$ stock solution should be aliquotted prior to the first use into amounts that are likely to be suitable for a maximum of five freeze/thaw cycles. It has been reported that low (25 mM) concentrations of $MgCl_2$ do not freeze/thaw well, in contrast to higher concentrations (250 mM or 1 M) and the authors have found that if the PCR product yield deteriorates, then using a fresh $MgCl_2$ stock often fixes the problem. In the authors' experience, the second most likely cause of PCRs failing is if the TaqStart antibody has degraded. Unopened tubes of TaqStart antibody will deteriorate if kept at room temperature, suggesting that at least some batches of the commercial TaqStart antibody are contaminated with proteases or bacteria.
  - Another reason for faint bands could be a low starting amount of DNA (with consequent insufficient levels of product amplification for detection), or DNA that is insufficiently pure (where contaminants inhibit the PCR), poor-quality DNA (where the DNA has been partially or totally degraded by DNases resulting in too low a starting amount of DNA of the correct fragment length), or too much DNA (usually has too many PCR inhibitors). If one is using FTA paper, ensure that an appropriate amount of blood has been placed on the paper according to the manufacturer's instructions. The blood can be fresh (straight from the animal) or from a collection tube containing an anticoagulant. It is important not to use heparin as an anticoagulant, unless the batch of heparin has been tested previously and shown not to cause problems. Some forms of heparin bind to DNA and need to be removed by digestion with a heparinase prior to PCR. Primer synthesis problems can also cause PCR problems such as low yields, background bands and an increased number of stutter bands. Note that primers are synthesized from the 3′ end.
  - The authors have found that the most frequent cause of sample contamination comes from sample loading problems, where either the multi-loader is not rinsed appropriately between samples (it is recommended to carry out three 10 µl rinses with H_2O), or there is sample leakage between adjacent wells.

○ Whilst aerosols have been widely reported as a cause of PCR contamination, the authors have found them not to cause much of a problem with the amount of template DNA and the number of amplification cycles (30–35) used in the above protocols. Instead, a more common source of PCR contamination results from the use of contaminated pipette tips where a user fails to eject a tip and cross-contaminates samples or buffers. Other common sources of PCR failure include omitting to add an ingredient, adding an ingredient twice, or using a faulty batch of an ingredient. A simple way to control these types of error is to prepare a check list for each PCR and to record batch numbers (or batch preparation date) of all reagents used together with the assay date and who performs the assay. Reagents should be ticked on the sheet immediately after they have been added.

- **The PCR gives lots of background bands**
  If the PCR has previously worked well for the primers under the conditions being used, then the most likely cause, if using a TaqStart antibody, is degraded or insufficient TaqStart antibody or an inadequate hot-start procedure. The TaqStart antibody is supplied in glycerol and needs to be mixed thoroughly before being aliquotted.

  If this is the first time a set of primers have been tested, then try increasing the annealing temperature, adjusting the primer concentration, or decreasing the $MgCl_2$ concentration (for some markers, the concentration of $MgCl_2$ needs only to be 0.5 mM). (See also earlier comments about primer synthesis problems.)

- **The marker maps to the wrong position**
  If a marker maps to the wrong position, then it is possible that there has been a primer synthesis error or that the marker is in a repeat region (this may be in the current or in the original map) and the wrong repeat has been genotyped. This can be resolved by sequencing the amplified product and comparing it with the target sequence.

- **Genotyping results are not as expected**
  Failure to find a significant LOD score probably means that the study has too few animals to detect this sort of trait, or not all of the genome has been scanned, or the level of genotype or phenotype errors is too high.

  Stringent protocols and double-checking procedures of individual identities need to be followed in building the data file to reduce the occurrence of errors.

4. REFERENCES

★★ 1. Andersson L & Georges M (2004) *Nat. Rev. Genet.* **5**, 202–212. – *A review providing an overview of the genetics of complex traits in domestic animals.*

2. Schwerin M (2001) *Reprod. Domest. Anim.* **36**, 133–138.

3. Flint J, Valdar W, Shifman S & Mott R (2005) *Nat. Rev. Genet.* **6**, 271–286.

4. Ott J (1999) *Analysis of Human Genetic Linkage*, 3rd edn. The Johns Hopkins University Press, Baltimore.

★ 5. OMIA – Online Mendelian Inheritance in Animals. Available at: http://omia.angis.org.au/ – *online database of genes, inherited disorders and traits with links to NCBI, PubMed and Entrez Gene.*

6. Wang J, He X, Ruan J, *et al.* (2005) *Nucleic Acids Res.* **33**, D438–D441.

★★ 7. Khatkar MS, Thomson PC, Tammen I & Raadsma HW (2004) *Genet. Sel. Evol.* **36**, 163–190. – *Review and meta-analysis of QTL mapping in cattle.*

8. Hu ZL, Dracheva S, Jang W, *et al.* (2005) *Mamm. Genome,* **16**, 792–800.

9. Bovine QTL Viewer. Available at: http://bovineqtl.tamu.edu/

★ 10. Grobet L, Martin LJ, Poncelet D, *et al.* (1997) *Nat. Genet.* **17**, 71–74. – *Positional candidate gene mapping of causative mutation.*

11. Freking BA, Murphy SK, Wylie AA, *et al.* (2002) *Genome Res.* **12**, 1496–1506.

12. Barendse W, Bunch R, Thomas M, Armitage S, Baud S & Donaldson N (2004) *Aust. J. Exp. Agric.* **44**, 669–674.

13. Grisart B, Coppieters W, Farnir F, *et al.* (2002) *Genome Res.* **12**, 222–231.

14. Weber JL & May PE (1989) *Am. J. Hum. Genet.* **44**, 388–396.

15. Buchanan FC, Littlejohn RP, Galloway SM & Crawford AM (1993) *Mamm. Genome,* **4**, 258–264.

16. Syvanen AC (2005) *Nat. Genet.* **37**, S5–S10.

17. Hirschhorn JN & Daly MJ (2005) *Nat. Rev. Genet.* **6**, 95–108.

18. Pompanon F, Bonin A, Bellemain E & Taberlet P (2005) *Nat. Rev. Genet.* **6**, 847–859.

19. Buetow KH (1991) *Am. J. Hum. Genet.* **49**, 985–994.

20. Dean FB, Hosono S, Fang L, *et al.* (2002) *Proc. Natl. Acad. Sci. U. S. A.* **99**, 5261–5266.

21. Hawken RJ, Cavanagh JA, Meadows JR, *et al.* (2006) *J. Dairy Sci.* **89**, 2217–2221.

22. Bassam BJ, Caetano-Anolles G & Gresshoff PM (1991) *Anal. Biochem.* **196**, 80–83.

23. Oetting WS, Lee HK, Flanders DJ, Wiesner GL, Sellers TA & King RA (1995) *Genomics,* **30**, 450–458.

24. Schoske R, Vallone PM, Ruitberg CM & Butler JM (2003) *Anal. Bioanal. Chem.* **375**, 333–343.

★★★ 25. Markoulatos P, Siafakas N & Moncany M (2002) *J. Clin. Lab. Anal.* **16**, 47–51. – *Description of the benefits of multiplex PCR.*

★★★ 26. Butler JM (2005) *Methods Mol. Biol.* **297**, 53–66. – *Description of the approaches required to design a multiplex PCR assay.*

27. Butler JM, Buel E, Crivellente F & McCord BR (2004) *Electrophoresis,* **25**, 1397–1412.

28. Zhou H, Hickford JG & Fang Q (2006) *Anal. Biochem.* **354**, 159–161.

29. Brownstein MJ, Carpten JD & Smith JR (1996) *Biotechniques,* **20**, 1004–1006, 1008–1010.

30. Idury RM & Cardon LR (1997) *Genome Res.* **7**, 1104–1109.

31. Lander ES & Green P (1987) *Proc. Natl. Acad. Sci. U. S. A.* **84**, 2363–2367.

32. Weaver R, Helms C, Mishra SK & Donis-Keller H (1992) *Am. J. Hum. Genet.* **50**, 1267–1274.

33. Seaton G, Haley CS, Knott SA, Kearsey M & Visscher PM (2002) *Bioinformatics,* **18**, 339–340.

34. Haley CS & Knott SA (1992) *Heredity,* **69**, 315–324.

35. Lathrop GM, Lalouel JM, Julier C & Ott J (1984) *Proc. Natl. Acad. Sci. U. S. A.* **81**, 3443–3446.

36. Abecasis GR, Cherny SS, Cookson WO & Cardon LR (2002) *Nat. Genet.* **30**, 97–101.

37. Broman KW, Wu H, Sen S & Churchill GA (2003) *Bioinformatics,* **19**, 889–890.

38. Baret PV, Knott SA & Visscher PM (1998) *Genet. Res.* **72**, 149–158.

39. Kerr RJ, McLachlan GM & Henshall JM (2005) *Genet. Sel. Evol.* **37**, 83–103.

40. Meuwissen TH & Goddard ME (2004) *Genet. Sel. Evol.* **36**, 261–279.

41. Lathrop M, Cartwright P, Wright S, Nakamura Y & Georges M (1991) In *Gene-mapping Techniques and Applications,* pp. 177–198. Edited by LB Schook, HA Lewin & DG McLaren. Marcel Dekker, New York.

42. Soller M (1991) In *Gene-mapping Techniques and Applications,* pp. 21–50. Edited by LB Schook, HA Lewin & DG McLaren. Marcel Dekker, New York.

43. Lynch M & Walsh B (1998) *Genetics and Analysis of Quantitative Traits.* Sinauer Associates, Sunderland, MA.

44. Matise TC, Perlin M & Chakravarti A (1994) *Nat. Genet.* **6**, 384–390.

45. **Basten CJ, Weir BS & Zeng Z-B** (2002) *QTL Cartographer: a Reference Manual and Tutorial for QTL Mapping*. Department of Statistics, North Carolina State University, Raleigh, NC.
46. **Broman KW, Boyartchuk VL & Dietrich WF** (2000) *Technical Report MS00-04*. Department of Biostatistics, Johns Hopkins University, Baltimore, MD.

CHAPTER 7

PCR restriction fragment length polymorphism analysis for genotyping of single-nucleotide polymorphisms

Simon Hughes

1. INTRODUCTION

The focus of this chapter is PCR restriction fragment length polymorphism (RFLP) analysis, an amalgam of PCR, which permits amplification of a specific region of DNA defined by two primers, and RFLP, a technique by which individuals can be differentiated by analysis of DNA fragment patterns generated by cleavage of their DNA by restriction enzymes. PCR-RFLP is a quick, easy, and cheap methodology for genotyping known single-nucleotide polymorphisms. The methodology does not require expensive fluorescently labeled oligonucleotides and is ideal for laboratories that do not routinely perform genotyping, as it does not require specialist equipment.

1.1 Restriction enzymes

Restriction enzymes are enzymes that cut dsDNA. Since the identification of the first type II restriction enzyme, *Hind*II (isolated from *Haemophilus influenzae*), by H.O. Smith, K.W. Wilcox, and T.J. Kelley in 1968, restriction enzymes have become commonplace in the laboratory due to their use in a number of applications, such as physical mapping of DNA, cloning of DNA, Southern blotting, and RFLP.

More than 900 restriction enzymes, isolated from over 230 different bacterial strains, are currently known and can be divided in to four types, I to IV. Types I, III, and IV comprised a single large enzyme complex that has both restriction and methylase activities. All three identify specific recognition sequences, but cut at sites that can be hundreds of bases up or downstream of this site. For type II

PCR: *Methods Express* (S. Hughes and A. Moody, eds.)
© Scion Publishing Limited, 2007

enzymes, the restriction enzyme is independent of the methylase and cuts at specific sites within or close to the recognition sequence. This chapter will focus on the use of type II restriction enzymes.

Type II restriction enzymes do not cleave DNA randomly, unlike DNAse I, but scan the length of the dsDNA for a particular 4–12 nt recognition sequence. Once identified, the enzyme cuts both strands of the DNA, generating either blunt ends or staggered 'sticky' ends. Recognition sites vary from enzyme to enzyme; however, many are palindromic (the DNA sequence reads the same in both directions on the sense and antisense strands). Details of restriction enzyme sequence motifs can be found at http://www.neb.com/nebecomm/ EnzymeFinderSearchByName.asp.

1.2 RFLP

Traditional RFLP analyses use one or more restriction enzymes to cut genomic DNA at specific recognition sites (1). Once cut, the DNA fragments are separated by agarose gel electrophoresis according to their molecular size, transferred to a nylon membrane, and hybridized with a labeled DNA probe that will only bind to alleles for a specific locus. When comparing DNA from different individuals, size differences result from base additions, deletions, substitutions, or sequence rearrangements that change the enzyme recognition sites. As a consequence, the size and position of DNA fragments will differ among individuals. This approach has several applications and has been used for forensic applications (2–4) as well as genetic linkage studies (5, 6). Despite the advantage of being highly discriminatory, RFLP analysis has a number of disadvantages:

- It requires several milligrams (5–10 mg) of high-quality/high-molecular-weight DNA.
- It may require the use of radioactive materials for visualizing results.
- Analysis is slow and generally cannot be automated.

These problems obviously limit the use of RFLP, and the emergence of alternative genetic markers, i.e. microsatellites and single-nucleotide polymorphisms (SNPs), has led to this technique been largely superseded in humans.

1.3 PCR-RFLP

In contrast to standard RFLP, PCR-RFLP has many advantages, which include:

- Analysis of restriction sites on a site-by-site, locus-by-locus, or gene-by-gene basis.
- Requirement for nanogram quantities of DNA.
- Works well on degraded samples.
- No requirement for specialist equipment as results can be visualized using a standard agarose gel.
- Analysis is quick and can easily be automated.

This chapter describes the use of PCR-RFLP, where the presence of a SNP either generates or eliminates the recognition sequence for a particular restriction enzyme. If a restriction enzyme can be identified that only cuts the DNA when one allele of the SNP is present, this enzyme can be used for SNP identification. Screening for such alterations is made possible by designing primers that:

1. Flank a restriction site where a SNP, within the site, either generates or eliminates the recognition site (standard PCR-RFLP).
2. In conjunction with a SNP, generate or eliminate a restriction site (engineered PCR-RFLP or ePCR-RFLP).

2. METHODS AND APPROACHES

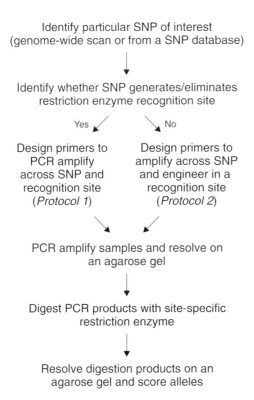

Figure 1. Flow diagram depicting the steps involved in PCR-RFLP.
Primers are designed that will generate an amplicon of <200 bp that flank either the restriction site containing the SNP or the engineered restriction site and the SNP. The PCR product is generated and visualized on a 1% agarose gel to determine that the PCR product is the expected size. The PCR product is digested using the enzyme specific for the recognition site. The digest products are visualized on a 4% agarose gel and the alleles scored.

2.1 Restriction enzyme identification

The selection of restriction enzymes is restricted by the sequence context of the SNP you want to interrogate. *Fig. 1* displays a flow diagram outlining the experimental design for any given SNP of interest. In some cases, the SNP will disrupt a naturally occurring restriction enzyme site, and *Protocol 1* describes a method for identifying whether the SNP of interest alters a restriction site (see *Fig. 2*). However, in many cases, the SNP of interest will not naturally disrupt a restriction enzyme site and *Protocol 2* describes the method for generating engineered restriction sites using the SNP and modified primers (see *Figs 3* and *4*).

Protocol 1

Restriction enzyme site identification for standard PCR-RFLP

Equipment
- DNA sequence containing SNP (5–8 nt)
- Computer with internet access

Methods
1. Open the New England NEBcutter v2.0 website (http://tools.neb.com/NEBcutter2/index.php).

2. Paste in the sequence containing the SNP with allele 1 (using the sequence from *Fig. 2*, this would be with the A allele). Click submit.

3. If this identifies a restriction site or sites, record which enzymes cut within the sequence.

4. Paste in the sequence containing the SNP with allele 2 (using the sequence from *Fig. 2*, this would be with the G allele). Click submit.

5. If this identifies a restriction site or sites, record which enzymes cut within the sequence.

6. Identify which enzymes differentially digest and only cleave once within the PCR product[a].

7. Design primers either side of the recognition sequence containing the SNP[b,c,d].

Notes

[a]Although this approach will work for several SNPs, it will not work for all, as a SNP will not always generate or eliminate a restriction enzyme recognition site. In this instance, it is possible to engineer in a site; this is described in *Protocol 2*.

[b]Ensure that when the PCR products are digested, the two alleles differ in size by at least 10 bases to enable ease of scoring of the polymorphism.

[c]Example primers are shown below in *Protocol 3*.

[d]Desktop DNA analysis programs such as VECTOR NTI or MAPDRAW also have restriction enzyme sequence identification functionality and could also be used.

(a)

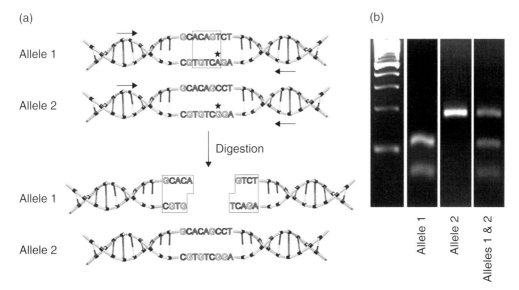

(b)

Figure 2. Standard PCR-RFLP.
(*a*) The location of the SNP is indicated by a star. In this case, the SNP leads to the substitution of a single nucleotide, which can either generate or eliminate the recognition site for a particular restriction endonuclease. In the example provided, the SNP can be either A or G. The substitution of G for A eliminates the enzyme recognition site. If a PCR product is generated (primer indicated by arrows) that contains the SNP and recognition site, and the product is then digested and resolved on an agarose gel, the different alleles can be detected. (*b*) Homozygote A individuals will give two bands (allele 1), as the restriction site has been retained and can thus be cut. Homozygote G individuals will give a single band (allele 2), as the restriction site has been eliminated and cannot be cut. Heterozygote A/G individuals (allele 1 and 2) will have all three bands, as one restriction site has been retained and one has been eliminated.

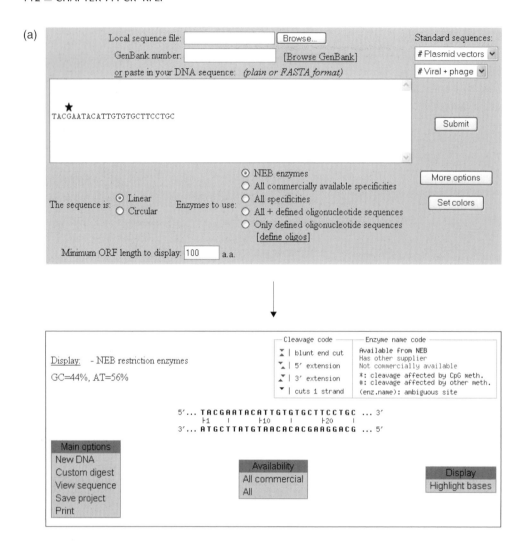

Figure 3. Engineering a restriction enzyme site for use in ePCR-RFLP.
This example uses the New England Biolabs NEBcutter v2.0 website
(http://tools.neb.com/NEBcutter2/index.php). (*a*) The sequence containing the SNP (G, indicated by a star) is pasted into the text box, as indicated in the top panel. The output (lower panel) will show which restriction enzyme recognition sites are present. In this example, there are no restriction sites identified.

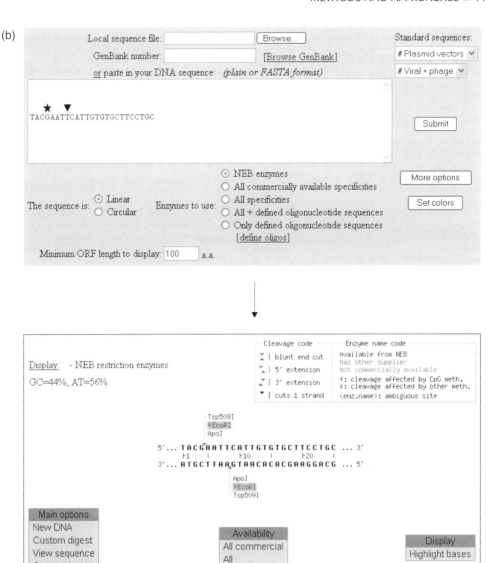

Figure 3 (*cont'd*). Engineering a restriction enzyme site for use in ePCR-RFLP.
(*b*) In the top panel, if the A 4 bases downstream of the SNP is substituted by a T (indicated by an arrowhead), this generates recognition sites for a series of restriction enzymes: *Apo*I (5′-RAATTY-3′ where R = A or G), *Eco*RI (5′-GAATTC-3′) and *Tsp*509I (5′-AATT-3′), shown in the lower panel.

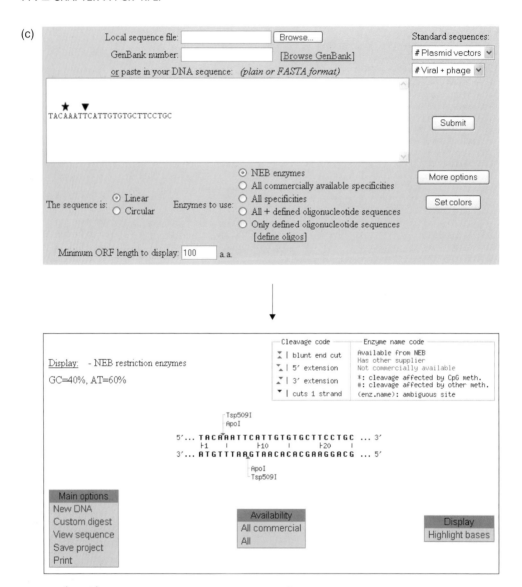

Figure 3 (*cont'd*). Engineering a restriction enzyme site for use in ePCR-RFLP.
(*c*) In the top panel, if the G of the SNP is substituted for the A (indicated by a star), whilst leaving the substituted T in place (indicated by an arrowhead), the *Eco*RI site is eliminated (bottom panel). Therefore, *Eco*RI can be used to discriminate between the two SNP alleles (G or A) due to the generation or elimination of the restriction site.

Once a sequence surrounding the SNP is determined that can help identify allele 1 or 2, this sequence can then be incorporated into either the forward or reverse primer. However, remember to not include the base that corresponds to the SNP in the primer (see *Fig. 4*).

Protocol 2

Engineering a restriction enzyme site for use in ePCR-RFLP

Equipment
- DNA sequence containing SNP (5–8 nt)
- Computer with internet access

Methods

1. Open the New England NEBcutter v2.0 website (http://tools.neb.com/NEBcutter2/index.php).

2. Paste in the sequence containing the SNP with allele 1 (using the sequence in *Fig. 3a* this would be with the G allele). Click submit.

3. If no restriction sites are observed, alter a single base adjacent to the SNP by replacing it with either A, G, C, or T (a substitution of an A to T is shown in *Fig. 3b*)[a]. Click submit.

4. If this generates a restriction site or sites, record which enzymes cut within the sequence.

5. Paste in the sequence containing the SNP with allele 2 (using the sequence from *Fig. 3c* this would be with the A allele). Alter the base adjacent to the SNP to the same nucleotide as in step 3[a]. Click submit.

6. If this generates a restriction site or sites, record which enzymes cut within the sequence.

7. Identify which enzymes differentially digest.

8. If no differentially digesting enzymes are identified, repeat steps 2–6[b]. It may be necessary to alter more than one base to generate a restriction site[c].

9. Once the sequence surrounding the SNP that generates a restriction site for identification of either allele 1 or 2 is determined, include this sequence into either the forward or reverse primer but *do not* include the base that corresponds to the SNP[c,d,e,f,g] (see *Fig. 4*).

Notes

[a]Although this step may generate restriction sites, it is possible that when comparing this sequence with that for the alternate allele/base for the SNP (as in step 5), no restriction enzyme will be identified that can differentiate between alleles. Alteration of adjacent bases may help to solve this.

[b]If the altered base was first changed to an A, try altering this to a G or C to see if this generates a restriction site.

[c]We would not recommend altering more than three bases, as this may adversely affect the PCR. We do not recommend altering the base at the extreme 3′ end of the primer, as this will also adversely affect the PCR.

[d]It is important not to include the nucleotide corresponding to the SNP in the primer, as this will prevent allele discrimination.

[e]For the primer containing the engineered restriction site, ensure that the primer is 20–30 bases in length and that the restriction site is at the 3′ end of the primer. These considerations are important in order to guarantee that the size difference between the two alleles is >10 bases, which will enable ease of scoring of the polymorphism.

[f]If the altered base is to the 5′ side of the SNP, engineer the change into the forward primer; however, if the altered base is on the 3′ side of the SNP, engineer the change into the reverse primer.

[g]Example primers are shown below in *Protocol 3*.

(a)

(b)

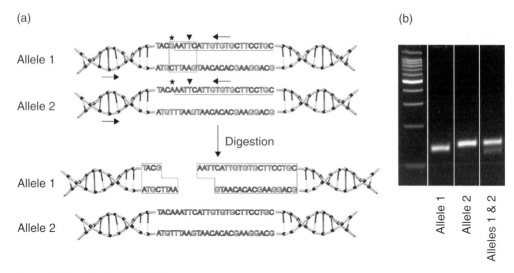

Figure 4. Example of ePCR-RFLP.
The location of the SNP is indicated by a star and the location of the altered nucleotide by an arrowhead. The nucleotide has been altered from A to T. In this example, the reverse primer (underlined) has been designed to lie directly adjacent to the SNP. The altered nucleotide in combination with the SNP leads to either the generation or elimination of the recognition site for a particular restriction endonuclease. In this example, the SNP can be either A or G. The substitution of A for G eliminates the enzyme recognition site. If a PCR product is generated (primer indicated by arrows) that contains the SNP and recognition site, and the product is then digested and resolved on an agarose gel, the different alleles can be detected. (*b*) Homozygote G individuals will give a single smaller band (allele 1), as the restriction site has been retained and can thus be cut. Homozygote A individuals will give a single larger band (allele 2), as the restriction site has been eliminated and cannot be cut. Heterozygote A/G individuals (alleles 1 and 2) will have all three bands, as one restriction site has been retained and one has been eliminated. (The small band is difficult to visualize on an agarose gel and generally will not be seen.)

2.2 DNA extraction and PCR amplification

The DNA that can be analyzed using PCR–RFLP (see *Protocols 3* and *4*) can come from virtually any source, such as fresh tissue, cultured cells, fixed tissue, DNA obtained from macro- or microdissected tissue sections, etc. However, it is important to ensure that the DNA is free of any contaminants that might adversely affect the PCR. Therefore, it is recommended that the DNA is isolated using a kit or protocol that will generate DNA of high purity. The source of DNA polymerase used for PCR is up to the user; however, we would recommend a polymerase that has proofreading capacity to ensure correct incorporation of bases.

Protocol 3

PCR

Equipment and reagents
- Genomic DNA (25 ng/μl)
- Platinum *Taq* DNA polymerase (5 units/μl), plus accompanying 10× PCR reaction buffer and 50 mM MgCl$_2$ (Invitrogen)
- 10 mM dNTPs (Invitrogen)
- Nuclease-free water (Promega)
- Oligonucleotides
  - □ Test primer 1a: 5′-TGCTTGAGTGATGGACTAGAT-3′ (2 μM)
  - □ Test primer 1b: 5′-TTCTAGCCTAAGTTCCTGAACTGT-3′ (2 μM)
  - □ Test primer 2a: 5′-CTGAATGATACCTATGAGAGCAGT-3′ (2 μM)
  - □ Test primer 2b: 5′-GCAGGAAGCACACAATGAATT-3′ (2 μM)[a]
- Thermal cycler
- 1% Agarose gel containing 10 ng/ml ethidium bromide
- 6× Orange loading dye solution (Fermentas)
- Equipment and reagents for agarose gel electrophoresis including 1× TBE agarose gel running buffer (10.8 g/l Tris base; 5.5 g/l boric acid; 4 ml/l 0.5 M EDTA, pH 8.0; diluted from a 10× stock; Sigma)
- DNA size marker (100 bp ladder; Invitrogen)
- UV light source

Method
1. Combine per reaction:
   - 2 μl of 10× PCR reaction buffer
   - 1.5 μl of dNTPs
   - 1 μl of MgCl$_2$
   - 1 μl of test primer 1a and 1 μl of test primer 1b, or 1 μl of test primer 2a and 1 μl of test primer 2b
   - 0.1 μl of Platinum *Taq* DNA polymerase
   - 1 μl of genomic DNA
   - Nuclease-free water up to a final volume of 20 μl

2. Mix briefly by vortexing or pipetting. Centrifuge at 12 000 ***g*** for 5–10 s to consolidate the sample.

3. Amplify the DNA using the following PCR profile:
   - Initial denaturation at 95°C for 2 min
   - 35 cycles of denaturation at 95°C for 30 s, annealing using a 60–55°C 'touchdown PCR[b]' approach for 30 s, and elongation at 72°C for 1 min
   - Final elongation: 72°C for 10 min

4. Analyze the PCR products by mixing 5 μl of the reaction mix with 1 μl of 6× orange loading dye solution and resolving the sample by agarose gel electrophoresis alongside a DNA size marker[c].

Notes

[a]The underlined base in test primer 2b refers to an engineered base change in the primer, which substitutes a T for an A, thus generating an *Eco*RI restriction site.

[b]Touchdown PCR uses a cycling program where the annealing temperature decreases by 0.2–0.5°C

every cycle. The first cycle annealing temperature should be approximately 2–5°C above the maximum melting temperature (T_m) of the primers, whilst the final cycle annealing temperature should be 2–5°C below the T_m of the primers. The primers will anneal at the highest temperature that is permitted by the primer sequences, and the product produced will most likely be the sequence of interest. Touchdown PCR improves the specificity of primer–template formation and with it the specificity of the final PCR product.

[c]For some restriction enzymes, it may be necessary to clean up the PCR products using a PCR clean-up kit such as those available from Qiagen due to PCR and restriction digest buffer incompatibility.

Protocol 4

Restriction digestion

Equipment and Reagents

- Sequence-specific restriction enzyme and accompanying 10× digestion buffer (New England Biolabs)
  - ☐ Test primer set 1 – *Hpy*CH4III (10 000 units/ml)
  - ☐ Test primer set 2 – *Eco*RI (20 000 units/ml)
- Nuclease-free water (Promega)
- Thermal cycler
- 4% Agarose gel containing 10 ng/ml ethidium bromide
- 6× Orange loading dye solution (Fermentas)
- Equipment and reagents for agarose gel electrophoresis including 1× TBE agarose gel running buffer (10.8 g/l Tris base; 5.5 g/l boric acid; 4 ml/l 0.5 M EDTA, pH 8.0; diluted from a 10× stock) (Sigma)
- DNA size marker (100 bp ladder, Invitrogen)
- UV light source

Method

1. Combine per reaction:
   - 15 µl of PCR product from *Protocol 3*
   - 2 µl of 10× digestion buffer[a]
   - 0.1 µl of restriction enzyme
   - Nuclease-free water up to 20 µl

2. Mix briefly by vortexing or pipetting. Centrifuge at 12 000 ***g*** for 5–10 s to consolidate the sample.

3. Incubate the sample at 37°C for 3 h, then deactivate the restriction enzyme by incubating at 65°C for 15 min.

4. Analyze the digestion products by mixing the 20 µl reaction mix with 3.5 µl of 6× orange loading dye solution and resolving the sample by agarose gel electrophoresis alongside a DNA size marker.

Notes

[a]The addition of bovine serum albumin may be necessary for the activity of some restriction enzymes.

2.3 Results and data interpretation

If you are planning to use DNA of lower quality, i.e. DNA extracted from formalin-fixed paraffin tissue, it is advisable to optimize the protocols above using DNA extracted from fresh tissue or cells, where both alleles of the SNP of interest are known to be present. One possible source is DNA from cell lines. This is advisable to make sure that the assays work and provide scorable results.

The typical results obtained for the two test primer sets can be seen in *Figs 2* and *4*.

- **Test primer set 1** (see *Fig. 2b*): homozygotes for allele 1 generate two bands of approximately 160 and 90 bp (due to retention of restriction sites and complete digestion), whilst homozygotes for allele 2 retain the original PCR product size of approximately 250 bp (due to loss of restriction sites and no digestion). Heterozygotes, with one copy of allele 1 and one copy of allele 2, generate three bands of 250, 160, and 90 bp (due to retention of one restriction site and loss of the other).
- **Test primer set 2** (see *Fig. 4b*): homozygotes for allele 1 generate two bands of approximately 102 and 32 bp (due to retention of restriction sites and complete digestion), whilst homozygotes for allele 2 retain the original PCR product size of approximately 134 bp (due to loss of restriction sites and no digestion). Heterozygotes, with one copy of allele 1 and one copy of allele 2, generate three bands of 134, 102, and 32 bp (the small band is difficult to visualize on an agarose gel and generally will not be seen) due to retention of one restriction site and loss of the other.

The two primer pairs interrogate their associated SNPs using slightly different approaches.

- Test primer set 1: the SNP generates or eliminates a restriction site (see *Fig. 2*).
- Test primer set 2: the SNP in combination with an alteration in the primer sequence generates or eliminates a restriction site (see *Fig. 4*).

Both approaches allow the fast and simple screening of large patient cohorts. Due to the nature of the experimental work involved – preparation of the PCR and restriction digests, which involve multiple pipetting steps – the whole screening approach lends itself to automation.

3. TROUBLESHOOTING

- **Unable to identify restriction sites using the method described in either *Protocol 1* or *Protocol 2***
  Unfortunately, not all SNPs can be scored using the methods described here. When this is the case, alternative approaches will need to be used, i.e. primer extension, allele-specific hybridization, or sequencing.
- **There is no amplification in the PCR**
  Have the primers being used been optimized? Have the primers being used

been used successfully before? If the answer to either of these question is no, then it is important to optimize the primers, and methods for this can found in Chapter 1. If the answer to the questions above is yes, then consider the following. It is always important to include a positive-control sample that will always amplify by PCR. If the positive control has worked and the test samples have not, then it suggests a problem with the test DNA samples. However, if the positive control has not amplified, it points to a problem with the reactions. If so, it is important to consider the following:

○ Have all reaction constituents been added? (Ticking off each constituent from a list once it has been added to the reaction will ensure this.)

○ Have the correct concentrations of primers and/or $MgCl_2$ have been added?

○ Has a sufficient concentration of DNA been added to the reaction? The minimum advisable concentration is approximately 10 ng/μl.

○ dNTPs and oligonucleotides should be stored in small aliquots, as repeated freeze/thaw cycles of a single stock can affect the integrity of these reagents and thereby affect the efficiency of PCR.

○ Are the thermal cycling conditions correct for the primer pair of interest? If these suggestions do not work, then it is advisable to change reaction reagents and try again.

- **Amplification of the negative control**
  Due to the sensitivity of PCR, the reactions can easily be contaminated. If negative controls produce an amplification product, a number of steps can be tried to eliminate this:

  ○ Repeat the PCR using fresh reagents.

  ○ Use filtered tips to avoid the introduction of contaminants via aerosol from the pipette.

  ○ Physically separate the areas in the laboratory where reactions are set up. Prepare and pipette the PCR mixture at one bench and then add the DNA to the reaction in a different location in the laboratory. Also use different pipettes for reaction preparation and pipetting of DNA.

  ○ It is strongly advisable to aliquot all reaction constituents; if an aliquot becomes contaminated, only that aliquot will be lost and not the entire stock.

- **Inconsistent amplification**
  The most likely cause of inconsistent amplification is degraded DNA, for example when working with DNA from fixed tissue. As a consequence of degradation, the average size of the DNA will be decreased, which will affect the PCR. When working with low-quality DNA, the use of small amplicons (<200 bp) is recommended, as these generally work better. This problem can also be overcome by moving the location of one or more of the primers.

- **It is not possible to discriminate between alleles**
  There are a number of possible solutions:

  ○ Increase the concentration of the gel or alternatively use a mixture (50:50) of standard agarose and low-melting-temperature agarose (as this can provide improved resolution of closely sized bands).

- Use a high-resolution agarose such as Metaphor (Cambrex Bioscience), which can resolve 2 bp differences.
- If using the detection method outlined in *Protocol 1*, increase the size of the PCR product by increasing the distance between forward and reverse primers, whilst keeping the restriction site and SNP equidistant for the two primers. This will make a bigger size difference between the two alleles once they have been digested.
- If using the detection method outlined in *Protocol 2*, redesign the primer that lies adjacent to the SNP by making it larger. This will make a bigger size difference between the two alleles once they have been digested.
- Add a nonspecific sequence tag to increase the size of the primer.

- **The distribution of genotypes is not as expected**
  There a few factors to consider:
  - First check whether the genotype distribution is in Hardy–Weinberg equilibrium (whereby genotype frequencies and gene frequencies of a large, randomly mating population do not vary, provided immigration, mutation, and selection do not take place) for the population you are studying. If the genotype data is not in Hardy–Weinberg equilibrium, a possible explanation is the presence of a SNP in one of the primer-binding sites, which can lead to preferential amplification of one allele over the other. In this case, check public databases to ensure that the primers are not affected by known polymorphisms.
  - For human studies in particular, it is important to consider the ethnic make-up of the patient cohort, as this can affect results.
  - If neither (1) nor (2) is a confounding factor, then it is possible that the assay is not working optimally. This was observed for my work when studying two genes that required a primer design that followed *Protocol 2*. These primer pairs were being used to study fixed samples. Thus, it is possible that the combination of these two factors affected the results. In this instance, it is advisable to rescreen a subset of your samples using one or more alternative screening methods, e.g. primer extension or sequencing. If these alternative methods give identical results, then it is safe to assume that the original approach has worked. If not, then a different approach needs to be adopted.

4. REFERENCES

1. Sambrook J & Russell D (2001) *Molecular Cloning: a Laboratory Manual*, 3rd edn. Cold Spring Harbor Laboratory Press, New York.
2. Rankin DR, Narveson SD, Birkby WH & Lai J (1996) *J. Forensic Sci.* **41**, 40–46.
3. Johnson ED & Kotowski TM (1996) *J. Forensic Sci.* **41**, 569–578.
4. Cali F, Le Roux MG, D'Anna R, *et al.* (2001) *Int. J. Legal Med.* **114**, 229–231.
5. Spike CA, Bumstead N, Crittenden LB & Lamont SJ (1996) *J. Hered.* **87**, 6–9.
6. Emi M, Takahashi E, Koyama K, Okui K, Oshimura M & Nakamura Y (1992) *Genomics*, **13**, 1261–1266.

CHAPTER 8

Forensic genetic DNA typing with PCR-based methods

Claus Børsting, Juan J. Sanchez, and Niels Morling

1. INTRODUCTION

Invention of PCR (1, 2) and DNA fingerprinting (3, 4) revolutionized forensic investigations around 1985 and stimulated an entirely new field of research known as forensic genetics. Twenty years later, PCR is essential for all forensic genetic investigations, and the term 'DNA fingerprint' is still widely used to describe the unique DNA profile of an individual, even though the technique has been replaced by more sensitive and efficient methods.

1.1 PCR and forensic genetics

PCR is a highly efficient method for amplification of trace material, and the introduction of PCR made it possible to identify a person from just a few picograms of DNA. However, PCR was not put to immediate use in forensic laboratories, because the extreme sensitivity of PCR also made the technique susceptible to contamination problems (5). A typical forensic sample is collected from dead bodies or surfaces where the DNA has been exposed to various environmental or chemical conditions that may degrade or destroy the DNA. Thus, forensic DNA samples are often scarce and a poor target for PCR. Addition of just a few intact DNA molecules from another source may be amplified more efficiently and cause false results. PCR products are especially efficient targets in PCR and there are numerous examples in the literature where PCR products have contaminated reagents, laboratory equipment, surfaces, ventilation shafts, and laboratory workers (5). Early on, it was clear that laboratories performing PCR routinely needed new laboratory protocols to avoid and check for cross-contamination (6, 7), and today these protocols are included in the accredited laboratory standards for testing laboratories (ISO 17025) recommended by the International Society of Forensic Genetics (ISFG) (8).

In highly degraded trace materials, the majority of DNA fragments are 200 base

PCR: *Methods Express* (S. Hughes and A. Moody, eds.)
© Scion Publishing Limited, 2007

pairs (bp) or less, and consequently only very short segments (<200 bp) of DNA can be amplified efficiently. This has forced the forensic genetics community to search for relevant DNA loci with a size of 150 bp or less. Fortunately, PCR amplification of short segments of DNA is more efficient than amplification of long segments, and thus maximal sensitivity of the genetic analyses can be achieved.

A locus is only relevant for forensic genetic testing if there is more than one known variant (allele) in the human population. Short tandem repeats (STRs) are short regions of DNA where a sequence of 2–7 bp is repeated several times. STRs do not have any known genetic function and it is not known how they were introduced into the genome, but they have high mutation rates (approximately 1 mutation per 300 generations) and, consequently, the number of alleles in any given population is high. STRs with 4 bp repeat units are currently the preferred loci for forensic genetic testing. The typical number of repeats in a STR locus varies from 5 to 30 (9) and thus most STR alleles are shorter than 150 bp. In the USA, the FBI selected 13 STRs located on different chromosomes for the Combined DNA Indexing System (CODIS) database (10); seven of these STRs are also recommended by Interpol and are used by all European countries (11). The international forensic community has worked closely with companies in the development of multiplex PCR kits, and today there are several commercial products available that allow amplification of all CODIS loci in a single multiplex PCR. These kits are validated and used by forensic laboratories all over the world (12, 13) and have facilitated the development of standardized databases, which have proved to be a highly valuable tool for national and international law enforcement.

1.2 Mitochondrial DNA (mtDNA) analysis

The high copy number of mtDNA in human cells makes mtDNA an obvious target for forensic genetic investigations, especially when the amount of nuclear DNA is limited or the DNA is highly degraded. Mitochondria are inherited maternally, and there is no recombination between mitochondrial genomes because there is only one chromosome per mitochondrion. Therefore, the variation of mtDNA genomes among humans is relatively small and the power of discrimination is limited compared with STR analyses. In 1999, the European DNA Profiling (EDNAP) Group's mitochondrial DNA population database project (EMPOP; www.empop.org) was initiated with the purpose of creating a common forensic standard for mtDNA sequencing and an on-line mtDNA database with high-quality mtDNA sequences from different human populations. Only forensic laboratories qualified by successful participation in EMPOP collaborative exercises have permission to import mtDNA sequences (14). The initiative is a reaction to the significant number of errors in public mtDNA databases (15). Most of these errors are clerical errors caused by manual recording of the results at some point in the data handling (14) and, consequently, EMPOP have issued a number of recommendations for mtDNA sequencing and implemented additional control mechanisms for import of mtDNA sequences into the EMPOP database (14, 16).

1.3 Single nucleotide polymorphisms (SNPs)

The search for informative loci is still ongoing, and during the last few years, SNPs have attracted the attention of the forensic community (17). SNPs have a number of advantages for forensic genetics. First, SNP analysis can be performed on PCR products with a size of 50 bp or less (18), and can be performed on highly degraded samples where STR analysis may fail (19). Secondly, the mutation rate of SNPs is very low, which is an important advantage in situations where immediate relatives are investigated (e.g. case work involving missing persons, mass disasters, paternity, and immigration issues). It is not likely that SNPs will replace STRs in routine forensic case work (17) but, just like mtDNA sequencing has proved its use in special cases, SNPs may become a valuable addition to the forensic arsenal of loci in the future.

2. METHODS AND APPROACHES

2.1 PCR fragment analysis

PCR begins with an initial denaturation of the target DNA and the PCR primers, followed by three steps, each of which is defined by a specific temperature (see *Fig. 1*):

1. Denaturation at 94–96°C.
2. Annealing of sequence-specific PCR primers to the DNA target at a temperature close to the melting temperature of the PCR primers (usually around 60°C).
3. Elongation of the PCR primers at 65–74°C.

The three thermal steps are repeated a number of times (usually 25–40) and under optimal conditions the number of target DNA molecules is doubled with each repeated cycle. The PCR products themselves are targets in subsequent PCR cycles and consequently the target sequence is amplified exponentially. In theory, ten copies of a DNA target are turned into $10 \times 2^{25} = 335\,544\,320$ copies after 25 PCR cycles. In reality, the efficiency of the reaction is decreased as the substrates (nucleotides and PCR primers) are used up and the DNA polymerases are degraded by the high temperature. In a well-optimized PCR, an average of 90% of the DNA targets are copied in each cycle, and thus, ten copies are turned into $10 \times 1.9^{25} = 93\,076\,490$ copies after 25 cycles.

2.2 PCR optimization

Optimization of a PCR begins in front of the computer with selection of the DNA target and careful design of the PCR primers, and it ends in the laboratory by adjusting the annealing and elongation temperatures, the concentrations of PCR primers, dNTPs, and Mg^{2+}, and possibly by selecting the most appropriate DNA

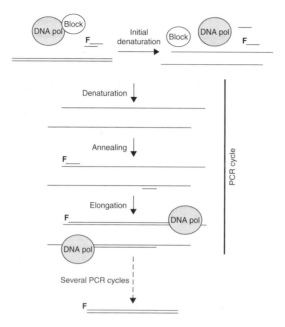

Figure 1. PCR.
PCR involves an initial denaturation step followed by a number of PCR cycles (denaturation, annealing, and elongation). During the initial denaturation step, dsDNA is separated into single DNA strands and the DNA polymerase becomes active as the chemical moiety blocking the enzyme activity is removed. Each PCR cycle begins with denaturation of the double-stranded PCR products into single DNA strands. In the annealing phase, the PCR primers bind to their targets and the PCR primers are subsequently extended by the DNA polymerase during the elongation phase. F, fluorescent label.

polymerase (20). It is possible to amplify several different DNA targets in the same PCR tube (known as multiplex PCR). This puts more constraints on the selection of DNA targets and the design of PCR primers (see Chapter 9 for more details), because the average amplification efficiency of each fragment needs to be similar. If some of the fragments are amplified more efficiently than other fragments, the difference in number of PCR products after completion of the PCR will be so large that it complicates downstream applications (a difference of 15% in average amplification efficiency gives rice to a ten-fold difference in the number of PCR products after 25 cycles).

2.3 PCR multiplexing

Multiplexing is essential for forensic genetic investigations because there is often very little DNA and it is not possible to obtain more of the sample later on. Typically, a stain (e.g. blood) collected from a crime scene is divided into two and DNA is purified from each half. The DNA from both preparations is quantified (see *Protocol 3*) and analyzed at least twice by PCR (see *Protocols 4* and *5*) before the results are compared. Therefore, it is pivotal that each analysis step is as sensitive and informative as possible.

With the development of PCR-based quantification methods, minimal amounts of DNA (<100 pg) are used to determine the concentration, and with the highly sensitive PCR multiplexes developed for forensic investigations, 200 pg or less is usually sufficient to obtain a complete STR profile (see *Protocol 4*). Thus, a stain collected from a crime scene need only contain 100 human cells (equivalent to 700 pg of genomic DNA) to allow successful identification of a person.

2.4 Recommended protocols

All of the PCR protocols described in this chapter employ hot-start polymerases, which are chemically modified enzymes that are inactive until the chemical moiety attached to the enzyme is removed by heating. During PCR set-up and the first ramp of thermal cycling, the annealing temperature is not optimal and the PCR primers may hybridize to each other (known as primer dimers) or to non-specific positions in the target DNA. During this period, the hot-start polymerases are inactive and therefore misprimed PCR primers are not extended, which improves the sensitivity of the PCR significantly and reduces the formation of non-specific PCR products.

All successful PCRs, and especially multiplex PCRs, depend on a careful balance between the concentration of DNA targets and PCR primers. Usually, the amount of DNA in a well-optimized PCR may vary 10–100-fold without affecting amplification efficiency significantly, and in standard situations it is not necessary to quantify the DNA sample prior to setting up the PCR. However, forensic samples often contain DNA from nonhuman sources and this DNA will also influence the PCR. Furthermore, forensic samples are often collected from unusual sources that may extricate unexpected inhibitors of the PCR. Therefore, there are three relevant factors for all typical forensic samples:

1. The total concentration of DNA.
2. The concentration of human DNA that may be amplified in the PCR.
3. The level of inhibition.

There are several standard methods for determination of the total concentration of DNA in a sample (21). *Protocol 3* is a PCR-based method for simultaneous determination of the level of inhibition and the concentration of human DNA that may be amplified by PCR (22). In addition, there are commercial kits (Applied Biosystems, Promega, Invitrogen) and locally developed kits (23, 24) available for determination of the concentration of male DNA or the concentration of mtDNA, which may be important for the investigation of certain samples.

2.4.1 DNA preparation

Quality and purity of the prepared DNA is important for all PCR-based applications. *Protocol 1* is a crude but fast and efficient way of extracting DNA from blood, semen, hair roots, and epithelial cells. Phenol/chloroform extraction (21) is recommended for more difficult samples, such as hair shafts or trace samples collected from materials that are known to release PCR inhibitors.

The number of reference samples collected from persons appearing in paternity or immigration cases, or as victims or suspects in crime cases, has increased dramatically in the last few years. These samples are fresh and they rarely cause problems. However, the sheer number of samples is so high that efficient protocols for sampling, storage, and extraction is pivotal for many forensic laboratories. FTA cards (Whatman) are an excellent medium for collection, shipping, and storage of DNA samples. Cells lyse when they come into contact with the coating on the FTA cards and any biohazards are eliminated. The DNA is released from the cells and binds irreversibly to the filter matrix. After washing of the FTA card, PCR can be performed directly on the card. *Protocol 2* is a manual protocol for washing of FTA cards, but the washing protocol and set-up of PCR are easily automated on liquid-handling robots (25).

Protocol 1

Chelex extraction of genomic DNA

Equipment and Reagents
- Chelex 100 resin (Bio-Rad)
- Double-distilled water (ddH$_2$O)
- Microcentrifuge
- Block heater
- Vortex mixer

Method

1. Transfer 3 µl of blood into 1 ml of ddH$_2$O in a 1.5 ml tube and mix for 15 s on a vortex mixer.

2. Incubate the mixture at room temperature for 30 min.

3. Mix for 15 s on a vortex mixer.

4. Spin the mixture in a microcentrifuge for 5 min at 10 000 r.p.m. and remove the supernatant.

5. Add 200 µl of 5% Chelex 100 resin and mix for 5 s on a vortex mixer.

6. Incubate the mixture at 56°C for 1 h.

7. Mix for 15 s on a vortex mixer.

8. Incubate the mixture at 100°C for 8 min.

9. Mix for 15 s on a vortex mixer and spin the mixture in a microcentrifuge at 10 000 r.p.m. for 5 min.

10. Transfer the supernatant into a clean 0.5 ml tube[a].

Notes

[a]The concentration of genomic DNA in the supernatant is approximately 1 ng/µl.

Protocol 2

Washing and preparation of FTA cards for PCR

Equipment and Reagents
■ BSD PowerPunch (BSD Robotics)
■ Double-distilled water (ddH$_2$O)
■ Platform shaker
■ Block heater

Method

1. Cut out a 1.2 mm disk from the sample FTA card using the BSD PowerPunch[a] and transfer the disk to a PCR tube.

2. Add 200 µl of ddH$_2$O and turn the tube upside down a few times.

3. Place the tube on a platform shaker and incubate at room temperature for 15 min at 200 r.p.m.[b]

4. Remove the supernatant and incubate the tube for 30 min at 50°C with the lid open to dry the FTA disk[c].

Notes

[a]Semiautomatic solutions for punching are available from BSD Robotics.
[b]An additional washing step (steps 2 and 3) may be necessary if the sample contains a lot of PCR inhibitors.
[c]PCR can be performed directly on the dried disk by adding PCR primers, PCR buffer, and *Taq* DNA polymerase to the tube.

2.4.2 Real-time PCR

In *Protocol 3*, the PCR products are detected by TaqMan probes as the PCR progresses (known as real-time or quantitative PCR; see Chapter 4 for more details). TaqMan probes have a fluorescent label at the 5′ end and a nonfluorescent quencher at the 3′ end, and they hybridize to the PCR products during the annealing phase (see *Fig. 2*). As the PCR primers are extended during the elongation phase, the TaqMan probes hybridized to the PCR products are degraded by the 5′→3′ exonuclease activity of the *Taq* DNA polymerase and the fluorophore is separated from the quencher. During the PCR, a tungsten halogen lamp directs light to the microtiter plate and the fluorophores released from the degraded TaqMan probes will emit fluorescence, which can be detected using a charge-coupled device (CCD) camera on the instrument. The number of released fluorophores is proportional to the number of PCR products and thus the amount of light emitted is a measure of the number of PCR products in the reaction mix.

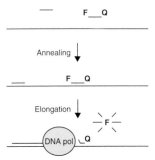

Figure 2. TaqMan probe technology.
The TaqMan probe is a short oligonucleotide with a fluorophore (F) at the 5′ end and a quencher (Q) at the 3′ end. The quencher absorbs emission from the fluorophore as long as the probe is intact. The 5′→3′ exonuclease activity of the DNA polymerase cleaves the TaqMan probe during the elongation phase of the PCR and emitted fluorescence is detected.

Protocol 3

Quantification of human DNA

Equipment and Reagents
- Quantifiler Human DNA Quantification kit (Applied Biosystems)
- Human genomic DNA (Promega)[a]
- Tris/EDTA/glycogen (TEG) buffer (10 mM Tris/HCl (pH 8.0), 0.1 mM EDTA (pH 8.0), 2 µg/ml glycogen)
- Double-distilled water (ddH$_2$O)
- Spectrometer
- Microtiter plate centrifuge
- ABI PRISM optical adhesive covers (Applied Biosystems)
- ABI PRISM 7000 Sequence Detection System (Applied Biosystems)[b]

Method[c]
1. Confirm the concentration of the human genomic DNA standard by spectrophotometric analysis (A_{260} = 1 for 50 µg/ml).

2. Make 80 µl aliquots of the human genomic DNA at a concentration of 50 ng/µl in TEG buffer[d].

3. Make the following serial dilutions in TEG buffer from a 50 ng/µl aliquot of the human genomic DNA standard: 16.70 ng/µl, 5.56 ng/µl, 1.85 ng/µl, 0.62 ng/µl, 0.21 ng/µl, 0.07 ng/µl, and 0.02 ng/µl.

4. Combine the following reagents (per sample) and mix well by pipetting up and down in each well of a microtiter plate (final volume 25 µl):
   - 10.5 µl of Quantifiler Human PCR Primer Mix[e] (Applied Biosystems)
   - 12.5 µl of Quantifiler PCR Reaction Mix[f] (Applied Biosystems)
   - 2 µl of sample or 2 µl of diluted human genomic DNA standard[g] or 2 µl of a negative control (ddH$_2$O)

5. Seal the microtiter plate with an ABI PRISM optical adhesive cover and spin the microtiter plate in a plate centrifuge for 30 s at 1000 r.p.m. to consolidate the sample.

6. Place the microtiter plate in an ABI PRISM 7000 Sequence Detection System and import or type in the sample list.

7. Run the following PCR program:
   ■ 95°C for 10 min
   ■ 40 cycles of 95°C for 15 s and 60°C for 1 min
   ■ Hold at room temperature

8. Analyze the results using ABI PRISM 7000 sds software (Applied Biosystems)[h].

Notes

[a]If the human DNA standard in the Quantifiler kit is used for creation of the standard curve, the concentration of the samples will be overestimated by a factor of two (26).

[b]The CCD camera on the ABI PRISM 7000 Sequence Detection System can detect four different dyes simultaneously.

[c]Steps 1–5 of this method should be performed in an area free of PCR products, preferably in a laminar air flow hood.

[d]The aliquots may be stored at –20°C for 6 months.

[e]The primer mix contains PCR primers to amplify the human DNA target in the sample, TaqMan probes (labeled with FAM dye) to detect the human PCR products, an internal PCR control system consisting of a synthetic DNA template, PCR primers to amplify the synthetic DNA template, and TaqMan probes (labeled with VIC dye) to detect the synthetic DNA template. If the sample contains PCR inhibitors, amplification of the internal PCR control will be reduced or completely absent.

[f]The reaction mix contains AmpliTaq Gold DNA polymerase, dNTPs, a passive reference (ROX dye) used for normalization of the fluorescent signal, and optimized buffer components.

[g]Each dilution of the human genomic DNA standard must be analyzed twice.

[h]The software generates a standard curve based on the dilutions of the human genomic DNA standard and calculates the fit (R^2 value) between the regression line and the individual data point of the each dilution. If $R^2 > 0.98$, the standard curve is acceptable. The concentration of human DNA in each sample is calculated based on the standard curve. The concentration of human DNA in the negative control must be zero.

2.4.3 Fluorescent PCR

In *Protocol 4*, the PCR products are labeled via a fluorophore attached to the 5′ end of one of the PCR primers (see *Fig. 1*). During electrophoresis, the fluorophores are excited by a laser light and detected by a CCD camera on the electrophoresis instrument. The sizes of the PCR products are determined by the use of an internal size marker with a different fluorescent label.

Protocol 4

STR typing of human DNA

Equipment and Reagents
- AmpF/STR Identifiler PCR Amplification kit (Applied Biosystems)[a]
- AmpliTaq Gold DNA Polymerase (5 units/µl; Applied Biosystems)
- Double-distilled water (ddH$_2$O)
- Microcentrifuge
- Thermal cycler

Method

1. Combine the following reagents in a 0.2 ml PCR tube (final volume 10 µl) and mix by pipetting up and down[b]:
   - 4 µl of AmpF/STR PCR reaction mix (Applied Biosystems)[c]
   - 2 µl of AmpF/STR Identifiler primer set (Applied Biosystems)[d]
   - 0.2 µl of AmpliTaq Gold DNA Polymerase (Applied Biosystems)
   - 2.8 µl of ddH$_2$O[e]
   - 1 µl of sample (e.g. from *Protocol 1*) or 1 µl of negative control (ddH$_2$O) or 1 µl of positive control[f]

2. Spin the tubes briefly in a microcentrifuge to consolidate the samples and place them in the thermal cycler.

3. Run the following PCR program:
   - 95°C for 11 min
   - 27 cycles of 94°C for 1 min, 59°C for 1 min, and 72°C for 1 min
   - 60°C for 1 h[g]
   - Hold at 4°C

Notes

[a]The Identifiler kit amplifies all CODIS loci (CSF1PO, D3S1358, D5S818, D7S820, D8S1179, D13S317, D16S539, D18S51, D21S11, vWA, FGA, TH01, and TPOX), two additional STR loci (D2S1338, D19S433), and the amelogenin sex marker, which differentiates between a male and female sample.
[b]This step should be performed in an area free of PCR products, preferably in a laminar air flow hood.
[c]The reaction mix contains dNTPs and optimized buffer components.
[d]The primer mix contains PCR primers to co-amplify 15 STRs and amelogenin.
[e]When the target DNA is placed on FTA cards (see *Protocol 2*), use 3.8 µl of ddH$_2$O to make the PCR mix and add 10 µl of the mix to each FTA disk.
[f]A positive control should always be included to confirm that the PCR mix contains all of the necessary components and that the PCR cycle program has completed successfully.
[g]*Taq* DNA polymerases often, but not always, synthesize dsDNA with 1 nt overhangs (A/T overhangs), resulting in ssDNA products of two different sizes, both of which will be detected as individual fragments by electrophoresis (see *Protocol 6*). The incubation at 60°C for 1 h is included to ensure that all PCR products have 1 nt overhangs and thus are the same size.

METHODS AND APPROACHES ■ 133

2.4.4 mtDNA sequencing

DNA sequencing is a DNA synthesis reaction performed with deoxy- (dNTP) and dideoxynucleotides (ddNTPs) (27, 28). The sequencing primers are extended with dNTPs, whereas incorporation of a ddNTP prevents further extension of the new DNA strand and the DNA synthesis reaction is terminated. The concentrations of dNTPs and ddNTPs are balanced carefully to ensure that DNA strands of different lengths are synthesized. In *Protocol 5*, the four ddNTPs are labeled with different fluorophores (FAM, JOE, TAMRA, and ROX dyes), revealing the identity of the last base on each synthesized DNA strand. The different-sized sequencing products are separated by capillary electrophoresis and the emitted fluorescence creates a sequence of color signals representing the complementary sequence of bases of the original DNA strand. The sequencing reaction described in *Protocol 5* is performed as a cycle sequencing reaction with 25 consecutive cycles of denaturation, annealing, and DNA synthesis to increase the sensitivity of the assay. However, as the sequencing products are not substrates for subsequent sequencing reactions, the amplification is linear as opposed to the exponential amplification of PCR.

In the mtDNA control region, there are at least three hypervariable (HV) regions (15), and typical targets for forensic and anthropological investigations are HV1 (nt 16024–164365 and HV2 (nt 73–340) (14). In *Protocol 5*, the HV1 and HV2 sequences are analyzed by making four sequence reactions (a forward and reverse sequence reaction for each region) using the PCR primers as sequencing primers (see section 4).

Protocol 5

DNA sequencing of mtDNA

Equipment and Reagents
- 10 mM dNTP mix (Invitrogen)
- 25 mM $MgCl_2$ (Applied Biosystems)
- 5 units/µl AmpliTaq Gold DNA polymerase and accompanying AmpliTaq Gold reaction buffer (Applied Biosystems)
- 100 µM HV1 forward primer (L15997: 5'-CACCATTAGCACCCAAAGCT-3')
- 100 µM HV1 reverse primer (H16401: 5'-TGATTTCACGGAGGATGGTG-3')
- 100 µM HV2 forward primer (L48: 5'-CTCACGGGAGCTCTCCATGC-3')
- 100 µM HV2 reverse primer (H408: 5'-CTGTTAAAAGTGCATACCGCCA-3')
- MicroSpin S300 HR spin column (Amersham Pharmacia Biotech)
- CentriSep column (Applied Biosystems)
- BigDye Terminator Cycle Sequencing kit (Applied Biosystems)
- Double-distilled water (ddH$_2$O)
- Vortex mixer
- Microcentrifuge
- Vacuum centrifuge
- Thermal cycler

Method[a]
1. Combine the following reagents (final volume 25 µl) in a 0.2 ml PCR tube and mix briefly on a vortex mixer:
   - 2.5 µl of 1 µM forward primer[b]
   - 2.5 µl of 1 µM reverse primer[b]
   - 2.5 µl of 10× AmpliTaq Gold reaction buffer
   - 2.5 µl of 25 mM $MgCl_2$
   - 0.5 µl of 10 mM dNTP
   - 0.2 µl of 5 U/µl AmpliTaq Gold DNA polymerase
   - 12.3 µl of ddH$_2$O
   - 2 µl of sample or 2 µl of negative control (ddH$_2$O) or 2 µl of positive control

2. Spin the tubes briefly in a microcentrifuge to consolidate the samples and place in the thermal cycler.

3. Run the following PCR program:
   - 95°C for 3 min
   - 40 cycles of 95°C for 20 s, 60°C for 10 s, and 72°C for 15 s
   - 72°C for 7 min
   - Hold at 4°C

4. Transfer the reaction to a MicroSpin S300 HR spin column and spin for 3 min at 3000 r.p.m. in a microcentrifuge[c] and collect the flow-through in a 1.5 ml tube.

5. Discard the column.

6. Combine the following reagents in a 0.2 ml PCR tube (final volume 20 μl) and mix briefly on a vortex mixer:
   - 4 μl of BigDye Terminator v1.1 cycle sequencing ready reaction mix
   - 2 μl of BigDye buffer
   - 3.2 μl of 3.2 μM primer[b]
   - 6.8 μl of ddH$_2$O
   - 4 μl of PCR product (from step 4)

7. Spin the tubes briefly in a microcentrifuge to consolidate the sample and place in the thermal cycler.

8. Run the following PCR program:
   - 96°C for 2 min
   - 25 cycles of 96°C for 15 s, 50°C for 5 s, and 60°C for 4 min
   - Hold at 4°C

9. Transfer the reaction to a CentriSep column and spin for 3 min at 3000 r.p.m. in a microcentrifuge[c] and collect the flow-through in a 1.5 ml tube.

10. Discard the column.

11. Dry the reaction in a vacuum centrifuge for 35 min at 60°C.

Notes

[a]Step 1 should be performed in an area free of PCR products, preferably on a laminar air flow bench. The remaining steps should be performed in post-PCR work areas.

[b]The PCR primers may be synthesized with a universal tag (i.e. M13F and M13R; see Chapter 1). Sequences close to the sequencing primer are often difficult to analyze, and by using a sequencing primer that is complementary to the tag, the first 20 bases of the sequencing reaction will originate from the PCR primer and not from the sample. In this way, the entire amplified region may be sequenced. However, tagged primers are often less efficient in the PCR and, consequently, it is not recommended to use tagged primers on difficult samples.

[c]Primers and nucleotides are retained in the column.

Protocol 6

Capillary electrophoresis for fragment analysis

Equipment and Reagents
- Hi-Di formamide (Applied Biosystems)
- GeneScan 500 LIZ size standard (Applied Biosystems)
- AmpF/STR Identifiler Allelic Ladder (Applied Biosystems)
- ABI 3130xl DNA Genetic Analyzer (Applied Biosystems)
  - ☐ Performance Optimized Polymer 4 (Applied Biosystems)
  - ☐ 10× Genetic Analyzer Buffer (Applied Biosystems)
- Vortex mixer
- Microtiter plate centrifuge
- Fume hood

Method
1. Mix 1.5 ml of Hi-Di formamide and 25 µl of GeneScan 500 LIZ size standard in a fume hood.
2. Transfer 15 µl to each well of a 96-well microtiter plate.
3. Add 1.5 µl of PCR product (from *Protocol 4*) or 1 µl of AmpF/STR Identifiler Allelic Ladder to each well.
4. Add the lid and spin the plate briefly in a plate centrifuge to consolidate the sample.
5. Incubate the samples at 95°C for 2 min.
6. Place the plate into the ABI 3130xl DNA Genetic Analyzer and import or type in the sample list.
7. Run the electrophoresis with the following settings:
   - Oven temperature: 60°C
   - Pre-run voltage: 15 kV
   - Pre-run time: 180 s
   - Injection voltage: 3 kV
   - Injection time: 10 s
   - Run voltage: 15 kV
   - Run time: 1300 s
8. Analyze the data file from the DATA COLLECTION v3.0 software using GENESCAN v3.7 and GENOTYPER v3.7 software (Applied Biosystems)[a]. See *Fig. 3* (also available in the color section) for sample results.

Notes

[a]The GENESCAN v3.7 software normalizes the fluorescent signals and translates laser scan numbers into nucleotide lengths based on the internal GENESCAN 500 LIZ size standard. The GENOTYPER v3.7 software translates nucleotide lengths into STR alleles based on the AmpF/STR Identifiler Allelic Ladder.

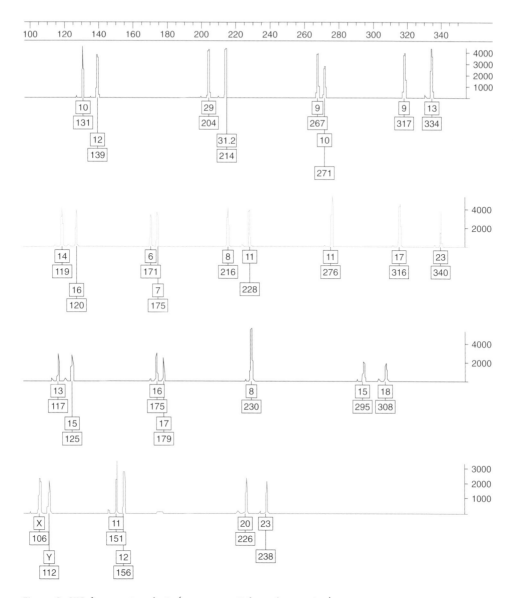

Figure 3. STR fragment analysis (see page xxiii for color version).
A typical electropherogram obtained from one sample amplified with the AmpFℓSTR Identifiler PCR Amplification kit is shown. Each peak represents a STR allele. The number of STRs and the length of the PCR product in nucleotides is shown in boxes below the peak. Four, five, and four marker STR systems are detected in the results shown in blue, green, and black, respectively, and two marker STR systems and the sex marker amelogenin are shown in the results in red. The DNA profile identifies the person as a male with two alleles in 13 STR systems (heterozygous genotype) and one allele in two systems (homozygous genotype). The DNA profile of this person is shown in table format in *Table 1* (Suspect 1).

Table 1. Case analysis

The case scenario described below is fictitious. A woman was found murdered in her apartment and the police collected numerous samples from the crime scene. DNA from these stains and a sample taken from the woman were typed using the AmpFℓSTR Identifiler PCR Amplification kit (see *Protocols 4* and *6*). The DNA purified from one of the stains contained more than two alleles in several STR systems (the DNA profile named Stain) and it was concluded that the stain contained cells from more than one person. There was not found to be more than four alleles in any of the STR systems, indicating that the stain originated from two people, and not from three or more. The woman's DNA profile was also determined (the DNA profile named Victim), and by comparing the DNA profiles of the woman and the stain, it was clear that the woman was one of the contributors to the stain, because all of her STR alleles were found in the DNA profile of the stain. While these samples were analyzed in the forensic laboratory, two suspects were arrested by the police, and samples were taken from the suspects and delivered to the laboratory. The samples from the suspects were also typed using the AmpFℓSTR Identifiler PCR Amplification kit (the DNA profiles named Suspect 1 and Suspect 2). By comparing the DNA profiles from the woman, the stain, and either Suspect 1 or Suspect 2, the following conclusions were made. Suspect 1 was excluded as a contributor in seven STR systems (D21S11, D7S820, CSF1PO, TH01, D2S1338, D19S433, and vWA), because Suspect 1 had an STR allele not found in the stain. Furthermore, the stain contained one allele in seven STR systems (D7S820, TH01, D13S317, D19S433, vWA, TPOX, and D5S818) that was not found in the DNA profiles of either the woman or Suspect 1. In contrast, Suspect 2 was not excluded in any of the STR systems, and the combined STR profile of the woman and Suspect 2 match perfectly with the DNA profile of the stain. Statistical calculation based on these findings were made and reported to the police. Suspect 2 was charged with murder and later convicted.

STR locus	Victim	Stain	Suspect 1	Suspect 2
D8S1179	9, 10	9, 10, 12	10, 12	12
D21S11	29, 30	29, 30	29, 31.2	29, 30
D7S820	8, 9	8, 9, 11	9, 10	8, 11
CSF1PO	9	9	9, 13	9
D3S1358	14, 15	14, 15, 16	14, 16	15, 16
TH01	7	7, 9, 10	6, 7	9, 10
D13S317	8, 10	8, 9, 10, 11	8, 11	9, 11
D16S539	12, 13	11, 12, 13	11	11
D2S1338	21	21, 23	17, 23	21, 23
D19S433	12, 13	12, 13, 14	13, 15	13, 14
vWA	16	16, 18	16, 17	16, 18
TPOX	7, 8	7, 8, 9, 11	8	9, 11
D18S51	15	15, 18	15, 18	18
Amelogenin	X	X, Y	X, Y	X, Y
D5S818	9, 11	9, 11, 12, 13	11, 12	12, 13
FGA	20, 23	20, 23	20, 23	23

Protocol 7

Capillary electrophoresis for DNA sequencing

Equipment and reagents
■ Hi-Di formamide (Applied Biosystems)
■ HPLC-grade water (Applied Biosystems)
■ ABI 3130xl DNA Genetic Analyzer (Applied Biosystems)
    □ Performance Optimized Polymer 4 (Applied Biosystems)
    □ 10× Genetic Analyzer Buffer (Applied Biosystems)
■ Vortex mixer
■ Microtiter plate centrifuge
■ Fume hood

Method

1. Resuspend the sequencing reaction (from *Protocol 5*) in 20 μl of Hi-Di formamide and mix briefly on a vortex mixer.

2. Transfer the reaction to a microtiter plate.

3. Add the lid and spin the plate briefly in a plate centrifuge to consolidate the sample.

4. Place the plate in the ABI 3130xl DNA Genetic Analyzer and import or type in the sample list.

5. Run the electrophoresis with the following settings:
   ■ Oven temperature: 55°C
   ■ Pre-run voltage: 15 kV
   ■ Pre-run time: 180 s
   ■ Injection voltage: 1.2 kV
   ■ Injection time: 6 s
   ■ Run voltage: 15 kV
   ■ Run time: 1700 s

6. Analyze the data file from the DATA COLLECTION v3.0 software using DNA SEQUENCING ANALYSIS v5.2 software (Applied Biosystems)[a].

7. Analyze the quality of the sequences using PHRED (CodonCode)[b] and align the sequences using SEQUENCHER v4.5 (Gene Codes)[c]. See *Fig. 4* (also available in the color section) for sample results.

Notes

[a] The DNA SEQUENCING ANALYSIS v5.2 software normalizes the fluorescent signals and makes base calls.

[b] The PHRED software examines the peaks around each base call to assign a quality score to each base call. Quality scores are linked logarithmically to error probabilities and range from 4 to about 60. A quality score of 10 is equivalent to 90% accuracy of the base call, and a quality score of 20 is equivalent to 99% accuracy of the base call, etc. (30).

[c] The SEQUENCHER v4.5 software compares the imported sequences (forward and reverse orientations) to assemble the best possible contig. If the quality score is 20 or higher for a base call in all imported sequences, the base call is accepted. If the quality score is less than 20 in one or more of the imported sequences, manual inspection of the sequences is recommended.

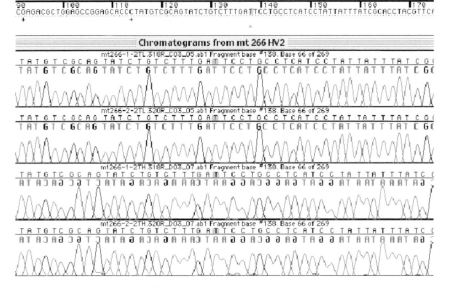

Figure 4. Sequence analysis (see page xxiv for color version).
A typical alignment of two forward and two reverse DNA sequencing reactions is shown. At the top is shown the Cambridge reference sequence (29) and below are the results from the four sequencing reactions. A light blue background indicates that the PHRED quality score was higher than 20. A red letter indicates that manual correction of the sequence was performed. The proposed consensus sequence is shown above the electropherograms.

3. TROUBLESHOOTING

- **PCR negative controls demonstrate contamination**

  It is essential to separate pre-PCR work areas physically from post-PCR work areas and to ensure that all traffic of reagents and, if possible, laboratory workers is unidirectional from the pre-PCR to the post-PCR work area. Each designated work area must be fully equipped with the necessary instruments, laminar air flow benches, laboratory disposables, laboratory coats, etc. Good laboratory behavior must be conducted in all areas; thus, work stations must be cleaned with 10% freshly prepared bleach before and after each laboratory step, reusable trays and racks must be immersed in 10% bleach before they are cleaned and reused, the laboratory workers must wear laboratory coats and gloves at all times, and used laboratory disposables and reagents must be removed from the designated areas in closed containers.

It is recommended that pre-PCR work areas are built with positive air pressure and post-PCR work areas with negative air pressure.

In all of the protocols described in this chapter, dUTP is used in the PCR instead of dTTP. The enzyme uracil glycosylase removes the uracil base from the phosphodiester backbone of uracil-containing DNA, whereas DNA without uracil (i.e. natural thymine-containing DNA) is left unharmed. If contamination of a sample or laboratory is observed or suspected, incubation of PCR mixes with uracil glycosylase for 10 min at 50°C will destroy any contaminating PCR products prior to the PCR. The uracil glycosylase enzymes are subsequently inactivated during the initial denaturation step of the PCR (31).

- Are the test samples contaminated?
  It is recommended that all laboratory workers, pathologists, and police technicians who may come into contact with crime case samples are typed for all of the markers used in the laboratory, and that the DNA profiles are collected in a local database. Every crime case profile should be screened against the database to look for possible contamination.

- Results are not reproducible with DNA of low concentration
  If a sample contains very few copies of genomic DNA, amplification imbalances or allele dropouts are frequently observed and the DNA profiles from two independent experiments may differ. This is usually caused by a stochastic effect, e.g. random selection of chromosomes whilst pipetting from the sample or random variation in primer hybridization during the first few PCR cycles. Therefore, it is recommended that every PCR laboratory decide on a lower threshold for the amount of DNA that can be reproducibly amplified in PCR, and set up rules for analysis of PCRs where the concentration is below the threshold (known as low-copy-number PCR).

- mtDNA sequencing produces unexpected results
  An individual may have two different species of mitochondria, either because two different species were inherited from the mother or because a mutation occurred early in the embryonic stage. If an individual has more than one species of mitochondria, the mtDNA sequences obtained from *Protocol 5* will be difficult to analyze. Usually, the difference is found in the number of C/G nucleotides in the long C/G stretches of HV1 and HV2 (known as heteroplasmy), and the mtDNA sequence of each species can be deduced by performing additional sequencing reactions using nested sequencing primers (32).

4. REFERENCES

★★★ 1. **Saiki RK, Scharf S, Faloona F,** *et al. Science,* 230, 1350–1354. – *The original publication describing PCR.*
★★★ 2. **Saiki RK, Gelfand DH, Stoffel S,** *et al. Science,* 239, 487–491. – *The original publication describing the use of* Taq *DNA polymerase in PCR.*
★★★ 3. **Jeffreys AJ, Wilson V & Thein SL** (1985) *Nature,* 316, 76–79. – *The original publication describing the DNA fingerprinting technique.*

4. Gill P, Jeffreys AJ & Werrett DJ (1985) *Nature,* **318**, 577–579.
5. Borst A, Box ATA & Fluit AC (2004) *Eur. J. Clin. Microbiol. Infect. Dis.* **23**, 289–299.
★ 6. Kwok S & Higuchi R (1989) *Nature,* **339**, 237–238. – *The first guidelines for PCR laboratories.*
7. Victor T, Jordaan A, du Toit R & van Helden PD (1993) *Eur. J. Clin. Chem. Clin. Biochem.* **31**, 531–535.
8. Morling N, Allen RW, Carrecedo A, *et al.* (2002) *Forensic Sci. Int.* **129**, 148–157.
9. Goldstein DB & Schlötterer C (2000) *Microsatellites: Evolution and Applications.* Oxford University Press, Oxford.
★ 10. Budowle B, Moretti TR, Baumstark AL, Defenbaugh DA & Keys KM (1999) *J. Forensic Sci.* **44**, 1277–1286. – *The original publication describing the selection of the 13 CODIS STR loci.*
11. Schneider PM & Martin PD (2001) *Forensic Sci. Int.* **119**, 232–238.
★ 12. Gill P, Sparkes R & Kimpton C (1997) *Forensic Sci. Int.* **89**, 185–197. – *The first validation of STR multiplexes in forensic genetics.*
13. Hallenberg C & Morling N (2001) *Forensic Sci. Int.* **116**, 23–33.
★ 14. Parson W, Brandstätter A, Alonso A, *et al.* (2004) *Forensic Sci. Int.* **139**, 215–226. – *The first report on EMPOP collaborative exercises.*
15. Brandstätter A, Peterson CT, Irwin JA, *et al.* (2004) *Int. J. Legal Med.* **118**, 294–306.
16. Bandelt HJ, Salas A & Lutz-Bonengel S (2004) *Int. J. Legal Med.* **118**, 267–273.
17. Gill P (2001) *Int. J. Legal Med.* **114**, 204–210.
18. Børsting C, Sanchez JJ & Morling N (2004) *Int. J. Legal Med.* **118**, 75–82.
★ 19. Sanchez JJ, Phillips C, Børsting C, *et al.* (2006) *Electrophoresis,* **27**, 1713–1724. – *The original publication describing the development of a 52-SNP multiplex assay designed for forensic applications.*
20. Sanchez JJ, Børsting C & Morling N (2005) *Methods Mol. Biol.* **297**, 209–228.
★★ 21. Ausubel FM, Brent R, Kingston RE, *et al.* (eds) (1995) *Current Protocols in Molecular Biology.* John Wiley & Sons, New York. – *A comprehensive collection of protocols for molecular biology.*
22. Green RL, Roinestad IC, Boland C & Hennessy LK (2005) *J. Forensic Sci.* **50**, 809–825.
23. Andreasson H, Gyllensten M & Allen M (2002) *BioTechniques,* **33**, 402–404, 407–411.
24. Walker JA, Hedges DJ, Perodeau BP, *et al.* (2005) *Anal. Biochem.* **337**, 89–97.
25. Hansen AJ, Simonsen BT, Børsting C, Hallenberg C & Morling N (2006) *Progr. Forensic Genet.* **11**, 663–665.
26. Nielsen K, Mogensen HS, Eriksen B, Hedman J, Parson W & Morling N (2006) *Progr. Forensic Genet.* **11**, 759–761.
27. Sanger F, Nicklen S & Coulson AR (1977) *Proc. Natl. Acad. Sci. U. S. A.* **74**, 5463–5467.
28. Sears LE, Moran LS, Kissinger C, *et al.* (1992) *BioTechniques,* **13**, 626–633.
29. Anderson S, Bankier AT, Barrell BG, *et al.* (1981) *Nature,* **290**, 457–465.
30. Ewing B & Green P (1998) *Genome Res.* **8**, 186–194.
31. Longo MC, Berninger MS & Hartley JL (1990) *Gene,* **93**, 125–128.
32. Rasmussen EM, Sørensen E, Eriksen B, Larsen HJ & Morling N (2002) *Forensic Sci. Int.* **129**, 209–213.

CHAPTER 9

Large PCR multiplexes with special reference to forensic single-nucleotide polymorphism typing

Juan J. Sanchez, Claus Børsting, and Niels Morling

1. INTRODUCTION

The introduction of PCR started a revolution in genetic investigations and although other methods for amplification of DNA do exist, e.g. the oligonucleotide ligation assay (1), PCR is by far the most commonly used method. For forensic genetics, PCR made it possible to introduce investigations of heritable DNA markers of identity, such as short tandem repeats (STRs), into routine crime, paternity, and immigration testing.

In crime-case work, it is important to be able to perform extensive genetic investigations on very small amounts of DNA (<200 pg of DNA). The application of PCR, specifically multiplex PCR, means that a number of DNA fragments can be amplified at the same time in one reaction, which is of the utmost importance in forensic investigations. Currently, a number of STR typing kits capable of analyzing 10–20 STR loci are available commercially. Such kits are very valuable, and the majority of forensic genetic cases can be solved by using these kits. In a number of situations, however, further genetic information or other kinds of genetic investigation is needed. One example is the need for more genetic information with the increasing numbers of individuals in DNA databases and the need for shorter amplified DNA fragments in cases of partly degraded DNA (2). The development of new sets of STR markers and short STRs will also rely upon PCR amplification in the near future.

1.1 Genetic markers

STRs are generally typed by means of electrophoretic methods, and the great majority of forensic genetic laboratories use commercially available multicolor fluorescence kits that allow PCR amplification and subsequent detection with

PCR: *Methods Express* (S. Hughes and A. Moody, eds.)
© Scion Publishing Limited, 2007

capillary electrophoresis (CE) with multicolor fluorescence detection, e.g. Applied Biosystems 3100.

Other genetic markers, particularly sets of single-nucleotide polymorphism (SNP) markers, have been developed for forensic genetic typing (3, 4). The advantage of SNPs is that they can be detected based on amplification of very short amplicons, which is a great advantage in forensic genetic cases where the DNA to be investigated is partly degraded, e.g. in DNA typing of victims in mass disasters. However, each SNP gives less information compared with, for example, STRs, and between 50 and 100 SNPs are needed to give a sufficient amount of genetic information in a crime case, compared with only 10–20 STRs (5).

1.2 SNP typing

SNP typing of the amplified DNA can be performed in a number of ways. It is not necessary to combine amplification and SNP typing in one reaction, but it may be beneficial in relation to cost, speed, etc. In our laboratory, we use the single-base extension (SBE) method as a convenient typing method in which the PCR amplification and the SNP typing occur in separate multiplex reactions (6, 7). This approach involves (see *Fig. 1*):

- Designing PCR amplification primers flanking the SNP(s) of interest (see *Protocol 1*).
- Designing SBE primers with variable-length sequence tags for multiplex CE analysis (see *Protocol 1*).

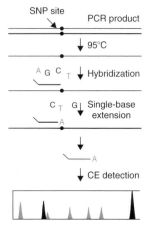

Figure 1. Workflow diagram illustrating the single-base extension (SBE) assay.
SNP flanking primers are designed (see *Protocol 1*) to allow PCR amplification of the sequence surrounding the SNP(s) of interest (see *Protocol 2*). Unincorporated excess primers and dNTPs are removed (see *Protocol 3*) and a secondary SBE PCR is performed using a primer that anneals directly adjacent to the SNP. The addition of a sequence-specific single fluorescently labeled ddNTP (A, G, C, or T) adjacent to each primer allows the SNP to be typed (see *Protocol 3*). Typing is performed by means of CE and subsequent multicolor fluorescence detection (see *Protocol 4*).

- PCR amplification of the sequence surrounding the SNP(s) of interest (see *Protocol 2*).
- Removal of excess primers and dNTPs (see *Protocol 3*).
- Secondary SBE PCR using a primer that anneals directly adjacent to the SNP, whereby the addition of a single fluorescently labeled ddNTP (A, G, C, or T) to each primer allows the SNP to be typed (see *Protocol 3*).
- SNP typing is performed by means of CE and subsequent multicolor fluorescence detection (see *Protocol 4*).

The SBE reaction generally is performed as a multiplex reaction that allows several SNPs to be typed at one time. Other useful methods that combine SBE with mass spectrometry, typically matrix-assisted laser desorption/ionization time-of-flight (MALDI-TOF) mass spectrometry (8, 9), DNA–DNA hybridization between the amplified DNA and DNA probes (10–12) or real-time PCR (13, 14) are informative but are not commonly used in forensic genetic laboratories.

We find the combination of multiplex PCR, SBE, and CE with multicolor detection the most convenient method for a forensic genetic laboratory that wants to set up SNP typing without buying complete commercial kits. A key reason for this choice is because a modern forensic genetic laboratory will already posses the majority of the necessary equipment and expertise needed. Furthermore, the combination of these technologies offers the necessary flexibility concerning design and detection of the PCR multiplexes.

1.3 SNP multiplexing

The advantage of multiplex PCR is obvious in forensic genetics. Multiplex PCR allows investigation of a large number of loci on small amounts of DNA. However, much care and work is needed to establish a sensitive and well-balanced large multiplex PCR. Thus, finding the best strategy for forensic genetic purposes is a trade-off between singleplex and multiplex PCR.

Several formats of multiplex PCR exist and these include:

- The use of universal primer tags on specific primers (3). This method has the advantage that it is a one-step procedure, but the work associated with optimization and the expense of kits should not be underestimated. In addition, the flexibility concerning the design of large multiplexes is limited compared with other multiplex methods.
- Assays based on combinations of the oligonucleotide ligation assay and PCR (e.g. SNP-Plex and Gene-Plex) (1). However, the need for specialized equipment can be limiting, as can the requirement of the SNP-Plex assays for large amounts of DNA.

As an alternative to developing multiplexes to maximize the amount of data generated from limited amounts of DNA, whole genome DNA amplification strategies, e.g. the GenomiPhi kit (15), can be used to generate additional template (see Chapter 18 for further details). However, amplification is not well suited for forensic genetic purposes because of the variability of the amplification efficiency and the difficulties in the detection of genetic markers when small amounts of DNA are available (16, 17).

The focus of this chapter is the development of large PCR multiplexes for forensic purposes, but the approaches described are equally valid for nonforensic applications. The development of small multiplexes with up to 15 amplicons, for example, is usually not a major challenge. However, the development of multiplexes of 30–100 amplicons, which are all amplified to the same degree in a reproducible way, as is needed in forensic genetic case work, is a significant challenge.

Our motivation for developing large PCR multiplexes for SNP typing was primarily the need to study several markers on small amounts of DNA from forensic samples. Multiplex PCR SNP typing also offers advantages by decreasing sample handling and increasing efficiency, speed, and automation. In our SNP work, we have developed multiplex PCR SNP systems for several groups of SNPs, e.g. 35 Y chromosome SNPs (6) and 52 autosomal SNPs (4) for detection with SBE and CE.

In this chapter, we have described the whole series of activities necessary for the establishment of forensic genetic SNP typing for detection of SBE products using CE. In this way, we hope to give the reader an appreciation of the steps required for performing their own SNP typing or other multiplex PCRs.

2. METHODS AND APPROACHES

2.1 Multiplex design

The process of designing a multiplex PCR (see *Fig. 2*) requires careful planning so that the various elements are compatible with each other and so that a reasonable degree of flexibility is obtained in the developmental phase.

The process of designing the multiplex PCR amplification of a number of DNA fragments begins with strategic decisions. It is important to choose a strategy that allows a combination of:

1. Flexibility in choosing primers.
2. Highly specific reactions.
3. Uniform amplification of PCR products.

Once the PCR conditions have been selected and a set of multiplex PCR amplicons have been developed, addition of further amplicons can only be done if the new primers can work under the conditions chosen. If two or more multiplex sets have been developed, interchanging of amplicons between sets can only be performed if the amplicons can be amplified under similar conditions. We use PCR conditions primarily defined by a theoretical primer melting temperature of $60 \pm 5°C$ at a salt concentration (K^+, Na^+, or $Tris^+$) of 180 mM and a purine:pyrimidine ratio close to 1:1, but other conditions may be chosen. It is desirable to be able to detect all of the amplicons in order to analyze the products for homogeneity and estimate the yield of the PCR amplification so that, at the end, a balanced PCR multiplex

Multiplex PCR	Multiplex SBE
1. Compile reference sequences from GenBank. 2. Choose PCR primers: • 35–65% G or C nucleotide • Annealing temperature 60±5°C • 17–26 bp in length 3. Align primer sequences and test for: • Hairpin loops • Primer dimers • Repetitive sequences using BLAST 4. Order HPLC-purified primers. 5. Test primers in singleplex PCR. 6. Combine equimolar amounts of 7–12 primer pairs and perform multiplex PCR. 7. Adjust primer concentrations empirically, based on PCR product yields. 8. Combine the 7–12 multiplex PCR primer sets into one large multiplex PCR.	1. Select SBE primers: • Purine : pyrimidine ratio of 1:1 • Annealing temperature 55±5°C • 17–26 bp in length 2. Align primer sequences and test for: • Hairpin loops • Primer dimers • Repetitive sequences using BLAST • Homology to other multiplex amplicons 3. Order HPLC-purified primers. 4. Test the SBE primers in singleplex PCR. 5. Design SBE primers with 5′ tails. • Two biallelic SNPs can be detected within the same 4 nt window 6. Order HPLC-purified primers with 5′ tails. 7. Test the SBE primer in singleplex PCR. 8. Test the SBE multiplex PCR. 9. Adjust SBE primer concentrations empirically, based on SBE product yields.

Figure 2. Main steps involved in the development of a large multiplex PCR.

consisting of homogeneous amplicons can be obtained. For forensic purposes, the amplicon sizes should be as small as possible in order to be able to analyze partly degraded DNA. However, this makes detection of the individual fragments difficult.

In our experience, the problems of creating a multiplex PCR involve ensuring that:

• The primers do not interact.
• The primers are specific.
• Each primer batch consists of homogeneous molecules.
• The concentrations of the primers are balanced in such a way that the multiplex reaction gives equal amounts of each amplicon.

For optimization of the multiplex PCR, we carry out an analysis involving two multiplex stages, which both need optimization (see *Fig. 2*). The first stage involves the initial amplification PCR where optimization begins with selecting the DNA target, followed by careful design of the amplification PCR primers (see *Protocol 1*), and is continued in the laboratory by adjusting the annealing and elongation temperatures, the concentration of PCR primers, dNTPs, Mg^{2+}, etc. (see *Protocol 2*).

The second multiplex stage is the SBE reaction and it is important to create primers that can be separated by CE in order to obtain clear SNP typing results (~4 nt size difference between products). SBE is performed as a thermal cycling reaction with an annealing step and a denaturation step (see *Protocol 3*). The annealing temperature is carried out at 50–60°C and, consequently, the SBE primers must have a minimal length of 17–18 nt. However, to achieve the required 4 nt size difference between products required for resolution by CE, SBE primers contain a 5′ tail with nucleotides that do not influence the hybridization process (usually from 1 to 40 nt, but can be up to 100 nt). This can be done by adding polymers of A, G, T, or C, or by adding random nucleotide sequences that do not bind to human DNA (6, 18). Thus, the SBE primers will be composed of:

- 17–26 nt complementary to the target sequences
- Nonhuman tails of varying lengths

This strategy can be used to generate large multiplexes. However, the quality of the synthesized oligonucleotides does become limiting and oligonucleotides longer than 80–90 nt are currently costly. In addition, the quality of the primers varies considerably among vendors when the length of the primers is >60–70 nt. Also, although in general it is possible to separate extended SBE primers by CE if the products differ by 4 nt, there may be difficulties in separating some primers that are shorter than 30–35 nt because the electrophoretic mobility of short primers depends strongly on the nucleotide composition.

In our hands, the optimal length of SBE primers is 40–60 nt. If two or more multiplex PCR sets of amplicons have been developed, interchanges of amplicons and SBE primers between the sets can be performed only if both the amplicons and the SBE primers are compatible.

2.2 Recommended protocols

- *Protocol 1* deals with primer design. This is one of the most important elements in multiplex PCR. Much time in the laboratory can be saved by making a careful selection of primers.
- *Protocol 2* describes the multiplex PCR. If one wants to have flexibility to shuffle SNPs from one multiplex set to another, it is helpful if the PCR conditions are kept the same in the various multiplex PCRs.
- *Protocol 3* describes SBE with the SNaPshot kit, but other kits can also be used.
- *Protocol 4* describes CE. We describe an enhancement of the electrophoresis where two sets of samples can be investigated almost simultaneously.

Protocol 1

Primer design

Equipment
- Computer with internet connection

Method

PCR primers

1. Find the sequences surrounding the selected loci in an appropriate database (for human sequences, use, for example, http://www.ncbi.nlm.nih.gov/entrez/query.fcgi?CMD=search&DB=nucleotide).

2. Search the relevant genome database for sequences with high homology to the sequence surrounding the selected loci[a] using the BLAST algorithm (http://www.ncbi.nlm.nih.gov/BLAST/).

3. Select the most suitable primers using the PRIMER 3 software[b] (http://frodo.wi.mit.edu/cgi-bin/primer3/primer3_www.cgi). Apply the following rules for primer design:
   - A theoretical melting temperature of 60 ± 5°C at a salt concentration (K$^+$, Na$^+$, or Tris$^+$) of 180 mM
   - 35–65 % G or C nucleotides in the primer sequence
   - Maximally four G or C nucleotides among the last 7 nt in the 3' end
   - G or C nucleotides at positions 1 and 2 from the 3' end is preferred
   - Keep the amplicon size as small as possible[c]

4. Check the template DNA to which the selected primers will bind for secondary structures using the software available at http://www.idtdna.com/Scitools/Applications/mFold/.

5. Screen the selected primers for primer-dimer and hairpin interactions using, for example, the AUTODIMER program (http://www.cstl.nist.gov/div831/strbase/AutoDimerHomepage/AutoDimerProgramHomepage.htm)[d].

6. Test each primer for homology to other amplicons in the multiplex.

7. Arrange the multiplex assays into sets with compatible primers that lack complementarity to each other, especially at their 3' ends.

SBE primers

1. Design primers that have a length of sequence complementary to the target of 17–26 nt and an annealing temperature of 50–60°C. Tails of varying length can then be included to allow discrimination between different primer sets[e].

Notes

[a]If the locus is located in a sequence with high homology to other sequences in the genome, primer selection can be very difficult and the PCR products may consist of sequences from more than one locus. If this is the case, these primers should be excluded.

[b]Another website is http://www.molbiol.bbsrc.ac.uk/reviews.html.

[c]Short DNA fragments (approximately 100 bp) are amplified more efficiently than long DNA fragments, mainly due to the lower probability of secondary structure interference. DNA in forensic samples may be partly degraded, and short rather than long DNA fragments are more likely to be unbroken and available for amplification (2, 11).

[d]If primer-dimer formation is suspected at the 3' end of primers, the primer-dimer formation must not include more than 4 nt or the free energy (dG) must be above –3.0 kcal/mol. Similarly, hairpin loops where the 3' end is part of the hairpin stem must be avoided. In general, hairpin loops with dG below –5.0 kcal/mol should be avoided.

[e]It is important that the primers can be separated by their lengths under the chosen electrophoretic conditions and, therefore, the lengths of primers must differ by at least 4 nt. Primer design for multiplex SBE-based genotyping can be done using the SBE primer program (http://www.zaik.uni-koeln.de/AFS/Projects/Bioinformatics/sbeprimer.html). In a four-color fluorescence system like the SNaPshot kit, two different biallelic SNPs can be detected within the same 4 nt window. For example, a primer range of 18–100 nt and a 4 nt window can detect approximately 40 SNPs in a single experiment.

Protocol 2

PCR

Equipment and Reagents

- Sterile purified water
- HPLC-purified PCR primers (25 µM per primer pair)
- GeneAmp PCR reagent kit with AmpliTaq Gold DNA polymerase (5 units/µl) containing GeneAmp 10× PCR buffer[a] and 25 mM MgCl$_2$ (Applied Biosystems)
- 10 mM dNTP mix (Amersham Biosciences)
- 25 mM MgCl$_2$
- Nuclease-free water (Promega)
- PCR thermal cycler

Method

1. Extract the genomic DNA or cells spotted on FTA paper (see Chapter 8).

2. Test each primer pair in singleplex PCR using:
   - 1 µl of DNA template (10 ng/µl)
   - 1.25 µl of GeneAmp 10× PCR buffer
   - 0.25 µl of 10 mM dNTP mix
   - 0.5 µl of the primer pair
   - 0.1 µl of AmpliTaq Gold DNA polymerase
   - Nuclease-free water up to a final volume of 12.5 µl

3. Use the following thermal cycling conditions:
   - Denaturation at 94°C for 10 min
   - 30 cycles of 95°C for 30 s, 55°C for 30 s, and 72°C for 30 s
   - 7 min at 72°C

4. Analyze the PCR products by gel electrophoresis in order to check whether a single fragment of the correct size has been amplified.

5. For multiplexing, combine the primers in sets of 7–12 SNPs and perform the PCR using:
   - 1 µl of DNA template (1 ng/µl)
   - 2.5 µl of GeneAmp 10× PCR buffer
   - 6.5 µl of 25 mM MgCl$_2$
   - 1.5 µl of 10 mM dNTP mix
   - Primers to a final concentration of 0.2 µM of each primer pair[b,c]
   - 0.5 µl of AmpliTaq Gold DNA polymerase
   - Nuclease-free water up to a final volume of 25 µl

6. Use the following thermal cycling conditions:
   - Denaturation at 94°C for 5 min
   - 35 cycles of 95°C for 30 s, 60°C for 45 s, and 65°C for 30 s
   - 7 min at 65°C

7. Adjust the primer concentrations in order to get a balanced set of amplicons[c] (concentrations above 0.5 µM may inhibit multiplex PCR, probably by increasing primer–primer annealing).

8. Analyze the PCR products by polyacrylamide gel electrophoresis to check whether a single fragment of the correct size is amplified. It is also possible to use the fluorescent SBE signal detected on an ABI instrument as an indirect estimation of the relative amount of each PCR product that has been amplified (see *Fig. 3a*).

Notes

[a]Consider the use of a PCR multiplex master mix with a balanced combination of NH_4^+ and K^+ (e.g. the Qiagen multiplex PCR kit).

[b]Prepare a PCR primer mix (225 μl total volume) by combining 3.5 μl of each oligonucleotide diluted to 25 μM and then use 13 μl of this stock containing all primers in the PCR. The primer mix must be kept on ice at all times in order to avoid artifacts due to hairpin or primer-dimer formation.

[c]For PCR primer sets, adjust the primer concentrations by adding double the amount of primer pairs amplifying poorly and adding half the amount of primer pairs that amplify most efficiently in order to avoid variation in the total amount of oligonucleotides added to the PCR.

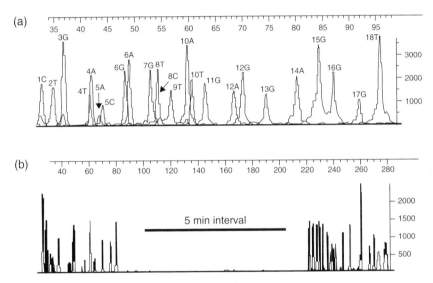

Figure 3. Representative electropherograms of two SNP multiplex assays.
(*a*) Example of an autosomal multiplex extension reaction. (*b*) Results after the injection of two SBE multiplexes immediately after each other in the same ABI 3100 capillary array.

Protocol 3

SBE assay using the SNaPshot kit

Equipment and Reagents

- MinElute PCR purification spin column (Qiagen)
- *Escherichia coli* exonuclease I (Exo I) (10 units/μl; Amersham Pharmacia Biotech)
- Shrimp alkaline phosphatase (SAP) (1 units/μl; Amersham Pharmacia Biotech)
- ABI Prism SNaPshot Multiplex kit (Applied Biosystems)
- 160 mM Ammonium sulfate (Sigma)

Method

1. Remove excess primers and dNTPs by adding the following to 2.5 μl of PCR product:
   - 0.75 μl of SAP[a]
   - 0.023 μl of Exo I[a]

2. Incubate at 37°C for 30 min, followed by 80°C for 15 min.

3. Alternatively, use MinElute PCR purification spin columns and elute the PCR products in 20 μl of Milli-Q water.

4. For the SBE reaction combine:
   - 1 μl of purified PCR product
   - 4 μl of SNaPshot reaction mix
   - 1 μl of SBE primer mix (0.1 μM of each SBE primer diluted in 160 mM ammonium sulfate[b])
   - 2 μl of nuclease-free water

5. Perform the SBE reaction in a thermal cycler with 30 cycles of 96°C for 10 s, 50°C for 5 s, and 60°C for 30 s.

6. Add 1 μl of 1 unit/μl SAP[a] to the SBE mix and incubate at 37°C for 45 min, followed by 75°C for 15 min.

Notes

[a]SAP and Exo I remove unwanted deoxynucleotides and primers to prevent them interfering with downstream applications.

[b]The addition of ammonium sulfate to the SBE primer mix reduces nonspecific peaks, mostly in the green dye electropherograms (4, 19), probably caused by the nontemplate addition of single adenosine molecules to the 3′ end of some of the PCR-amplified fragments (20).

Protocol 4

Standard protocol for CE detection on an ABI 3100 instrument

Equipment and Reagents
- Hi-Di formamide (Applied Biosystems)
- GeneScan 120 LIZ size standard (Applied Biosystems)
- AmpF/STR Identifiler Allelic Ladder (Applied Biosystems)
- Performance Optimized Polymer 4 (Applied Biosystems)
- 10× Genetic Analyzer Buffer (Applied Biosystems)
- HPLC-grade water (Applied Biosystems)
- Sixteen 36 cm × 50 µm ABI 3100 capillary arrays (Applied Biosystems)
- ABI 3100 DNA Genetic Analyzer (Applied Biosystems)
- Vortex mixer
- Microtiter plate centrifuge
- Fume hood

Method

1. Mix 1.9 µl of Hi-Di formamide and 20 µl of GeneScan 120 LIZ size standard.

2. Transfer 19 µl to each well of a 96-well microtiter plate.

3. Add 1 µl of the purified SBE product (from *Protocol 3*) to each well and spin the plate briefly at 900 r.p.m. in a microtiter plate centrifuge.

4. Incubate the samples at 95°C for 2 min.

5. Place the plate in the ABI 3100 DNA Genetic Analyzer and import or type in the sample list.

6. Run the electrophoresis with the following settings (see section 2.3 for multiple-injection conditions):
   - Oven temperature: 60°C
   - Pre-run voltage: 15 kV
   - Pre-run time: 180 s
   - Injection voltage: 3 kV
   - Injection time: 10 s
   - Run voltage: 15 kV
   - Run time: 1000 s

7. Analyze the data file from the DATA COLLECTION v1.1 software using GENESCAN v3.7 and GENOTYPER v3.7 software (Applied Biosystems).

2.3 Multiple-injection protocol

After the development of different multiplex packages, it is useful to have a method of analyzing all of the markers corresponding to the same sample in a single file. This requires multiple subsequent injections on the CE instrument. For this, labeled products from different reactions must be injected one after another in the same capillary without injection of polymer in between and without any carry-over from the previous injection to the next one (a typical result obtained using a 5 min delay between injections is shown in *Fig. 3b*). This approach is not a standard method on the ABI 3100 and requires some modification of the manufacturer's methods (see *Table 1*):

1. Edit an ABI 3100 method. The default methods are located in d:\appliedbio\support files\datacollection support files\method files on the ABI 3100 computer. The files are simple text files that can be modified using a simple text editor such as Notepad. Lines starting with '//' are explanations and the instrument does not execute them. Follow steps 2–9.
2. Copy the method into a text editor, such as Microsoft Notepad, but not Microsoft Word, and rename it (e.g. SNP36_POP4_Multipleinjection.mtd).
3. Modify the method by deleting the parts of the method outlined in step 1 of *Table 1*, which fill the syringe with polymer.
4. Delete the parts of the method outlined in step 2 of *Table 1*, which fill the capillaries with polymer.
5. It is important that the following two lines are not deleted in the edited run module:
   - FLU:BVAL Open
   - SYST:WAIT 3
6. Delete the washing steps after the injection of polymer and the pre-run steps, outlined in step 3 of *Table 1*.
7. Import the new method into the d:\appliedbio\3100\bin directory using MethodImportUtility.bat in the d:\appliedbio\3100\bin directory. Click Browse to find the method you want to import into the database. Once the .mtd file has been imported into the database, you can open the AB 3100 DATA COLLECTION software and the new default run module should then appear in the module editor.
8. Open the new default run module, which contains the edited run parameters that the instrument needs to follow, and make the following changes:
   - For the standard SNP36_POP4_DefaultModule, change the run time to 300 s and save the run module three times as SNP36_POP4_Multi1, SNP36_POP4_Multi3, and SNP36_POP4_Multi5. The run module is saved three times because each injection has to be performed in the right order. Injection 1, 3, or 5 is the first injection for a sample, whereas injection 2, 4, or 6 (see below) is the second injection for the same sample but a different amplification reaction.
   - For the new SNP36_POP4_MultipleinjectionDefaultModule, change:
     - The cap fill volume to 1
     - The pre-run voltage to 0
     - The pre-run time to 1
   - Save the run module three times as SNP36_POP4_Multi2, SNP36_POP4_Multi4, and SNP36_POP4_Multi6. The specific steps using these setting were deleted in the method. Thus, they are not used during the run; this is just information for the user.
9. Different multiplex reactions for the same DNA sample are set up in the plate record in order to ensure that different injections for the same DNA sample are

Table 1. Steps for multi-injection

Step 1	Step 2	Step 3
// Start Syringe Synchronize	//	// Capillaries Washing (water1)
//	SYST:MESS "Filling Array"	SYST:MESS "Washing Capillaries in water"
SYST:MESS "Synchronizing Syringes"	FLU:CAP:FILL $Cap_Fill_Vol	STR:ROB:MOV:SIT Waste
FLU INIT	// 1X fill is 72 step	SYST:WAIT 60
//FLU:POLY:PTHReshold 300 8	FLU:CAP:SYNC	// Capillaries Washing (water1)
FLU:RPOL:PLUN	//	SYST:MESS "Washing Capillaries in water"
FLU:RPOL:SYNC	FLU:POLY:REL Up 100	STR:ROB:MOV:SIT Water2
FLU:POLY:PLUN	FLU:POLY:SYNC	SYST:WAIT 60
FLU:POLY:SYNC		// Capillaries Move to Buffer
//		SYST:MESS "Moving Capillaries to Buffer Site"
SYST:MESS "Filling Polymer Syringe"		STR:ROB:MOV:SIT Buffer
FLU:POLY:FILL 200		SYST:WAIT 5
FLU:POLY:SYNC		// pre-RUN
//		//
FLU:RPOL:INIT		SYST:MESS "Starting Pre-Run"
FLU:RPOL:SYNC		EPS:VOLT:SETT $Pre_Run_Voltage
FLU:RPOL:REL Down 1		EPS On
FLU:RPOL:SYNC		SYST:WAIT $Pre_Run_Time
//		//SYST:MESS "PreRun before reading"
		EPS:VOLT:READ?
		EPS:CURR:READ?
		OVEN:TEMP:READ?
		//
		EPS Off
		EPS:VOLT:SETT 0
		SYST:MESS "End of Pre-Run"
		//

injected in the same capillary (e.g. the well position A1 and A3 in the microtiter plate would allow injection in capillary no. 1). To perform multiple runs, you must assign sequential run module names in the order you wish them to be run. For example:

- From A1 to H2, select the run module called: SNP36_POP4_Multi.1
- From A3 to H4, select the run module called: SNP36_POP4_Multi.2
- From A5 to H6, select the run module called: SNP36_POP4_Multi.3
- From A7 to H8, select the run module called: SNP36_POP4_Multi.4
- From A9 to H10, select the run module called: SNP36_POP4_Multi.5
- From A11 to H12, select the run module called: SNP36_POP4_Multi.6

3. TROUBLESHOOTING

If the multiplex reaction does not work as expected, this can be due to a number of reasons, including:

- *Primer sequences.* If the multiplex PCR does not work or gives weak results, first check whether the targeted locus is located within a DNA sequence that is found in other places in the genome or whether it contains a sequence of low complexity (e.g. proline-rich regions or poly(A) tails). Check whether the primers bind to regions within the template DNA fragment(s) or the other primers.
- *Purity and homogeneity of primers.* The quality of the PCR primers is one of the most import factors for a successful outcome of multiplex PCR. The primers must be purified, e.g. by HPLC purification, and the lengths and composition of the molecules in a primer stock must be uniform. With generally available methods of DNA synthesis, the quality of synthesized oligonucleotides varies considerably when the size of the primers is longer than 60–70 nt, and it is very difficult and expensive to synthesize homogeneous SBE primers longer than 80 nt. Check the quality of the oligonucleotides by mass spectrometry, e.g. MALDI-TOF, or by extending the primers with a fluorescently labeled dNTP and analyzing the extended primer molecules using fluorescent CE (see *Protocol 3*).
- *Balanced signal intensities.* It is important that the signal intensities in the SBE are as uniform as possible. If the signals are out of balance, the concentrations of the primers for amplification of DNA and the SBE reaction must be balanced again.
- *Quality of dNTPs.* dNTPs are sensitive to repeated freezing and thawing and they should be stored in small aliquots at −20 °C, maximally for 8 months. Use fresh dNTPs if the PCR amplification gradually or suddenly fails.
- *Quality and quantity of genomic DNA.* High-purity DNA is needed for multiplex PCR amplification. High-quality DNA should be included in each multiplex run as a positive control. The test DNA should be purified if the results are considerably weaker than those of the positive control. DNA quantification is generally advisable, especially when Chelex or phenol/chloroform DNA extraction methods are used.

- *Nonspecific additional reactions.* If weak, additional nonspecific reactions are repeatedly observed, the touchdown PCR method can be considered. In the first three cycles of the PCR, use a high annealing temperature (e.g. 65 °C for a primer pair with a T_m of 60 °C) and in the remaining cycles, use the T_m as the annealing temperature. The method usually minimizes mispriming and thus increases the proportion of target-specific products. However, redesign of the primers will in most cases be a better solution that will make the multiplex PCR more robust and reliable.
- *Genetic polymorphism in the primer-binding DNA regions.* Polymorphisms in the primer-binding site may cause allelic dropout and other genotyping failures due to inefficient amplification of that specific locus or allele. This leads to lack of results in a SNP system in some individuals. Check whether a single nucleotide mismatch can cause an amplification to fail or reduce the binding efficiency, e.g. by using the BLAST program from the NCBI website. However, a BLAST comparison will only find matches among sequences that have been reported to the database. In addition, check for potential null alleles by comparison of the observed number of homozygotes with the expected number under the assumption of Hardy–Weinberg equilibrium (21).

4. REFERENCES

1. Tobler AR, Short S, Andersen MR, *et al.* (2005) *J. Biomol. Tech.* **16**, 398–406.
2. Gill P, Fereday L, Morling N & Schneider PM (2006) *Forensic Sci. Int.* **156**, 242–244.
3. Dixon LA, Murray CM, Archer EJ, Dobbins AE, Koumi P & Gill P (2005) *Forensic Sci. Int.* **154**, 62–77.
★ 4. Sanchez JJ, Phillips C, Børsting C, *et al.* (2006) *Electrophoresis*, **27**, 1713–1724. – *Original publication describing the development of a 52-SNP multiplex assay designed for forensic applications.*
5. Gill P (2001) *Int. J. Legal Med.* **114**, 204–210.
6. Sanchez JJ, Børsting C, Hallenberg C, Buchard A, Hernandez A & Morling N (2003) *Forensic Sci. Int.* **137**, 74–84.
7. Vallone PM & Butler JM (2004) *J. Forensic Sci.* **49**, 723–732.
8. Mengel-Jørgensen J, Sanchez JJ, Borsting C, Kirpekar F & Morling N (2004) *Anal. Chem.* **76**, 6039–6045.
9. Petkovski E, Keyser-Tracqui C, Hienne R & Ludes B (2005) *J. Forensic Sci.* **50**, 535–541.
10. Heller MJ (2002) *Annu. Rev. Biomed. Eng.* **4**, 129–153.
11. Børsting C, Sanchez JJ & Morling N (2004) *Int. J. Legal Med.* **118**, 75–82.
12. Kemp JT, Davis RW, White RL, Wang SX & Webb CD (2005) *J. Forensic Sci.* **50**, 1109–1113.
13. Ahmadian A, Ehn M & Hober S (2006) *Clin. Chim. Acta.* **363**, 83–94.
★ 14. Whiley DM & Sloots TP (2005) *Pathology*, **37**, 364–370. – *Comparison of SNP detection performance of several multiplex PCR assays.*
15. Dean FB, Nelson JR, Giesler TL & Lasken RS (2001) *Genome Res.* **11**, 1095–1099.
16. Balogh MK, Børsting C, Sánchez-Diz P, *et al.* (2005) *Progr. Forensic Genet.* **11**, 1288.
17. Holbrook JF, Stabley D & Sol-Church K (2005) *J. Biomol. Tech.* **16**, 125–133.
18. Lindblad-Toh K, Winchester E, Daly MJ, *et al.* (2000) *Nat. Genet.* **24**, 381–386.
19. Doi Y, Yamamoto Y, Inagaki S, Shigeta Y, Miyaishi S & Ishizu H (2004) *Leg. Med. (Tokyo)* **6**, 213–223.
20. Brownstein MJ, Carpten JD & Smith JR (1996) *BioTechniques*, **20**, 1004–1010.
21. Chakraborty R, De Andrade M, Daiger SP & Budowle B (1992) *Ann. Hum. Genet.* **56**, 45–57.

CHAPTER 10

Pre-implantation genetic diagnosis of monogenic disease: PCR-based methods for the identification of mutations in single cells

Dagan Wells

1. INTRODUCTION

PCR has the capacity to amplify specific DNA fragments to levels at which they can be detected and analyzed or subjected to chemical/enzymatic manipulations. With careful experimental design and stringent precautions against contamination, it is even possible to amplify a single DNA molecule. This remarkable sensitivity of PCR has permitted the analysis of minute quantities of DNA obtained from various sources. These include DNA retrieved from crime scenes, DNA isolated from archaeological specimens, and DNA derived from extremely small biological specimens in a research and/or clinical context (e.g. microbiopsies, laser-capture microdissected samples, fetal cells isolated from the maternal circulation, etc.).

One of the most challenging applications of PCR amplification is in the context of pre-implantation genetic diagnosis (PGD), a method in which genetic analysis is performed on a single cell. This chapter will discuss appropriate strategies for the analysis of single cells and other minute tissue samples, using PGD methods as an example. New developments in single-cell DNA amplification, such as methods of whole genome amplification, will also be explored.

1.1 Principles of PGD

PGD represents an alternative to pre-natal testing for patients at risk of transmitting an inherited disease to their children (1). The technique involves the generation of multiple embryos using *in vitro* fertilization technology. The

PCR: *Methods Express* (S. Hughes and A. Moody, eds.)
© Scion Publishing Limited, 2007

embryos are then tested in order to determine whether or not specific disease-associated mutations are present. Only those embryos found to be free of causative mutation(s) are transferred to the mother's uterus and consequently any pregnancy that results should be unaffected. Testing is generally carried out 3 days after fertilization of oocytes, at which point the embryos are composed of six to ten cells. In order to ensure that embryo viability is not compromised, a maximum of two cells can be taken for analysis (see *Fig. 1*). As a consequence of the extremely limited amount of genetic material available, the PGD tests employed must display an unusual level of sensitivity and accuracy.

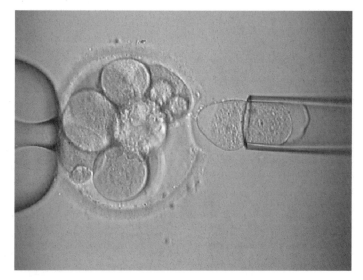

Figure 1. Biopsy of a cleavage-stage embryo for PGD.
The embryo is held in place by gentle suction from the pipette on the left of the image, while a second pipette is used to aspirate a single cell. The cell is removed through a hole in the zona pelucida – the membrane encapsulating the embryo – created using a laser.

A wide variety of PGD protocols for the detection of single-gene disorders have been published (1); however, all remain dependent on the use of PCR. In addition to the problems faced by standard PCR methods, PGD must overcome the greatly increased risk of contamination and phenomena unique to single-cell DNA amplification, such as allele drop-out (the failure of one of the two alleles in a heterozygous cell to amplify).

1.2 Contamination

Contamination with extraneous DNA can be a problem for any PCR-based method. However, in the case of procedures such as PGD, the large number of PCR cycles necessary to amplify the DNA from a single cell to a detectable level exacerbates this problem and necessitates the use of particularly stringent precautions against contamination. An additional source of contamination is that derived from previously amplified DNA fragments. This problem, known as 'carry-over' contamination, tends to become progressively worse with passing time, as amplified DNA fragments accumulate in the laboratory. (For tips on decreasing these effects, refer to section 3.)

Single-cell PCR approaches often employ a nested PCR strategy, involving two sequential rounds of amplification. The first round provides a modest amplification of the fragment of interest. The amount of DNA produced is usually below the threshold of detection and does not present a major contamination risk for subsequent PCR amplifications. The second round takes DNA produced during the first round and amplifies it further using primers that anneal within the fragment. These internally situated (nested) primers give rise to a smaller fragment. Large quantities of DNA are produced during the second round, providing a detectable product that would normally be a significant risk for carry-over contamination. However, subsequent first-round amplifications cannot amplify these potential contaminants, as the first-round primers have no annealing sites within the smaller second-round fragments. As well as reducing the risk of carry-over contamination, nested PCR has the added advantage of increasing the specificity of the amplification, which can be a problem when amplifying small amounts of DNA. In the case of nested PCR, a product can only be generated if two pairs of primers can anneal efficiently, rather than the usual single pair.

Decontamination of work areas should be routine for laboratories amplifying minute quantities of DNA, such as single cells. Work surfaces can be decontaminated using commercially available reagents (e.g. DNAZap; Ambion) or by irradiation with UV light. DNA can be eliminated from PCR ingredients (e.g. nucleotides, polymerase buffer, water, etc.) by digestion using DNA-degrading enzymes such as DNase I, although primers cannot be treated in this way. An alternative that can permit the decontamination of primers is the use of restriction enzymes, carefully selected so that they cleave the intended amplicon but do not digest the primers. Of course, these strategies are no substitute for avoiding DNA contamination in the first place.

For single-cell diagnostics, it is advisable to perform a multiplex PCR (see *Protocol 3*), amplifying one or more microsatellites in addition to the diagnostic locus. Microsatellite loci are highly polymorphic and their analysis provides a simple form of DNA fingerprint. For PGD, embryos produced by a given couple can only have a limited, predictable combination of alleles at each microsatellite locus (one allele inherited from the father and one from the mother). If analysis of a single cell from an embryo reveals the presence of unexpected alleles, not present in either parent, then contamination is the most likely explanation. Additionally, the detection of more than two parental alleles may indicate contamination with parental cells/DNA or the presence of aneuploidy affecting the chromosome upon which the microsatellite is situated (e.g. the presence of two maternal alleles and one paternal allele may indicate a trisomy of maternal origin). The use of microsatellites reveals the incidence of contamination within each PCR tube and is particularly useful for detecting low-level or sporadic contamination that might escape detection in a standard PCR negative control.

1.3 Amplification efficiency

Single-cell PCR for the purposes of PGD has proved to be highly successful, leading to the diagnosis of more than 100 different monogenic diseases in human embryos and the birth of several thousand unaffected children as a result. However, the amplification of DNA from a single cell remains challenging and it is unlikely that amplification success rates will ever reach 100%. Even in experienced PGD laboratories, approximately 10% of cells do not yield a diagnostic result, due to failure of the locus containing the mutation to amplify.

There are several factors believed to influence amplification efficiency. For diagnosis based on analysis of a single cell, a seemingly obvious, yet important consideration, is the efficient transfer of the cell into the PCR tube. If the cell is lost during transfer (e.g. if it sticks in the pipette), then no amplification will be obtained. This is a less significant problem for well-developed, laser-capture microdissection protocols, but for PGD methods, the successful transfer of the cell remains reliant on the embryologist or scientist and their micromanipulation skills (see *Protocol 1*).

Once the cell has been successfully placed within a microcentrifuge tube, the accessibility and integrity of its DNA are the major factors affecting amplification efficiency. PCR amplification is only possible if the phosphodiester backbone of the target DNA fragment is intact. A double-strand DNA break at any site between the two primers creates a gap that the DNA polymerase is unable to amplify across. Similarly, single-strand breaks situated between the primers can also be refractory to PCR, provided that damage occurs on both of the DNA strands.

Cells derived from frozen and thawed samples, necrotic samples or from tissues with high levels of apoptosis usually display raised levels of amplification failure, due to increased amounts of DNA damage (2). In the case of PGD, lower amplification rates are frequently seen in embryos of poor morphology. Such embryos often display features such as cellular fragmentation and micronuclei, which may be associated with DNA damage (3).

Premature lysis of the cell during biopsy or washing can lead to rapid DNA degradation and a reduced probability of successful amplification. However, ultimately, the cell must be lysed in order for the DNA to be made accessible to PCR reagents. Lysis should only occur after the cell has been transferred to the PCR tube and enzymes that could potentially degrade the DNA have been inactivated. This is accomplished by lysing the cell in the presence of a strong denaturant such as potassium hydroxide, or using proteinase K (see *Protocol 2*) to digest cellular structures and enzymes.

Amplification success rates increase rapidly as the number of cells assessed is increased. If three or more cells are simultaneously amplified, it is possible to obtain amplification efficiencies that approach 100% (2).

1.4 Allele dropout

Many of the difficulties encountered when amplifying DNA from single cells are similar to those encountered in other PCR applications, only more severe. However, one problem unique to single-cell PCR is the phenomenon known as allele dropout (ADO). ADO is defined as amplification failure affecting one of the two alleles in a diploid cell. In most cases, each allele is equally likely to suffer ADO, although extreme differences in the length, DNA sequence, or secondary structure of alleles may cause one to have a greater tendency to display ADO than the other. As only one of the two alleles is amplified to a detectable level, a heterozygous cell with ADO appears to be homozygous. In the literature, the reported incidence of ADO varies widely, affecting anywhere from 0 to almost 40% of single cells. However, in the hands of an experienced laboratory, ADO rates of 5–15% are typical.

Not surprisingly, the high frequency of ADO represents a major problem for single-cell testing and has resulted in several PGD misdiagnoses. The problem is particularly pronounced in the diagnosis of dominant disorders, as amplification failure affecting the mutant allele can lead to an embryo being wrongly diagnosed as unaffected. As with total amplification failure (affecting both alleles), the incidence of ADO is higher in cells containing degraded DNA and a major cause may be DNA strand breakage. The probability of a particular allele suffering PCR-refractory DNA damage is related to the length of the fragment amplified: the longer the amplicon, the greater the chance of a strand break and the higher the ADO rate (2).

For single-cell PCR, the first few cycles are the most critical. At these early stages, poor replication fidelity can lead to problems in the subsequent detection of mutations and polymorphisms. Additionally, incomplete amplification of DNA fragments during early cycles can lead to skewed amplification of one allele relative to another (and in the most extreme cases results in ADO). Indeed, skewed amplification is very common in single-cell PCR, precluding the use of certain quantitative PCR strategies (e.g. quantitative fluorescent PCR for the purposes of aneuploidy detection).

It has been suggested that high denaturing temperatures during the first few PCR cycles reduce ADO by improving accessibility of the template to PCR reagents (4). Some studies have also suggested a link between ADO and the method used for cell lysis, although the published data is somewhat contradictory (5, 6). Currently, no method has been able to eliminate ADO entirely. At present, the most effective strategy for improving diagnostic accuracy is to perform multiplex PCR, amplifying informative linked polymorphisms in addition to the mutation site. This provides more than one opportunity to detect the chromosome bearing the mutation. ADO is independent for amplified fragments, even those that are closely linked.

1.5 Whole genome amplification

The minute amount of DNA available in single cells or small tissue samples limits the number of analyses that can be carried out before the sample is expended. In the case of single-cell PCR, only a single amplification reaction can be performed, as the cell is destroyed during the procedure and its DNA cannot be recovered and re-used. The most straightforward method for maximizing the genetic information obtained from a small sample is to perform multiplex PCR, simultaneously amplifying several DNA fragments. However, multiplex PCR has some significant technical limitations:

• The amplification of multiple loci in a single reaction can sometimes lead to problems in distinguishing the different DNA fragments (amplicons) produced.
• Some sets of primers are incompatible and cannot be combined in the same reaction. Incompatibility can occur if the optimal reaction conditions for two sets of primers are too dissimilar.
• Primers for different loci may anneal to each other or interact with the products that they produce in such a way that amplification is inhibited.
• There can be competition between amplified loci for PCR reagents.

These factors make it difficult to amplify more than a handful of distinct loci efficiently in a single reaction.

Whole genome amplification (WGA) represents an alternative to multiplex PCR and provides a source of DNA that can be used to set up multiple reactions. Most WGA techniques utilize the principles of PCR to replicate enzymatically the entire genome of a sample multiple times. Unlike most PCR strategies, the amplification aims to provide an increase in the copy number of all sequences in the genome, rather than a single defined fragment. Following WGA, aliquots of the reaction mixture can be taken and used to set up multiple separate PCR amplifications for specific loci, thus avoiding issues of primer incompatibility. Furthermore, the generation of WGA products creates a resource of DNA that can be stored or subjected to additional genetic testing at some point in the future.

One of the most widely applied forms of WGA is degenerate-oligonucleotide-primed PCR (DOP-PCR) (7, 8) (see Chapter 18). This method utilizes a heterogeneous mixture of oligonucleotide primers, consisting of specified 5′ and 3′ ends separated by a short stretch of random bases. A number of PCR cycles, conducted at low annealing temperatures, allows the semi-degenerate primers to anneal at numerous sites throughout the genome. Wherever primers anneal, DNA synthesis is initiated, providing a modest amplification of the genome. In subsequent PCR cycles, a change in reaction conditions (raised annealing temperature) causes the primers to cease annealing at random throughout the genome and instead anneal specifically to the 5′ ends of fragments generated during earlier cycles. This gives rise to an exponential amplification.

It has been shown that DOP-PCR performed on a single cell provides enough DNA for more than 90 subsequent PCR amplifications, although amplification of microsatellite loci may occur with low fidelity (8). Additionally, sufficient DNA is produced to permit the use of genetic methods that require large quantities of DNA. One example is comparative genomic hybridization (CGH), a DNA-based method that provides information on chromosome copy number (9). The CGH procedure requires at least 100 ng of DNA, and consequently WGA is necessary for application to small DNA samples (8, 10).

Alternative PCR-based WGA methods exist, such as primer-extension pre-amplification (8, 11) and linker-adaptor-mediated amplification strategies (12). Some of these methods have now been put into kit form and commercialized (e.g. GenomePlex; Sigma). Furthermore, nonPCR methods, based on the concept of multiple strand displacement using highly processive bacteriophage polymerases, have also been introduced (13).

2. METHODS AND APPROACHES

A flow diagram outlining the steps involved is shown in *Fig. 2.*

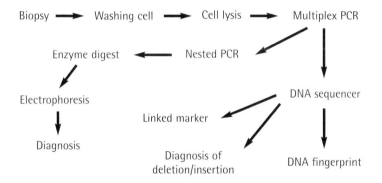

Figure 2. Flow diagram of the processes involved in PGD.
A similar strategy is also employed for other diagnostics involving single cells. Following cell washing and lysis, genomic DNA is amplified and the resultant products analyzed. Alleles or mutations that cause altered DNA fragment sizes are most easily assessed using fluorescent PCR, followed by electrophoresis using a DNA sequencer. Single-base changes can be detected using a wide range of mutation detection methods (e.g. enzyme digestion, single-strand conformation polymorphism, etc.). However, in this case, a further round of amplification using nested primers is usually required, as nonfluorescent methods of visualizing PCR products are less sensitive and therefore require the generation of more DNA.

Protocol 1

Cell washing and transfer to microcentrifuge tubes

Equipment and reagents
- Binocular dissecting microscope
- Phosphate-buffered saline (PBS; Gibco-BRL) (without magnesium or calcium) containing 0.1% (w/v) polyvinyl alcohol (PVA; Sigma)
- Aerosol-resistant pipette tips
- Sterile Petri dish
- 0.2 ml Microcentrifuge tubes
- Pulled Pasteur pipettes or Drummond micropipettes

Methods
1. Pipette several microdrops (5–10 µl each) of PBS/0.1% PVA onto the surface of the Petri dish.

2. Pipette 5–10 µl of cell suspension onto the surface of the Petri dish but away from the PBS/PVA[a].

3. Wash the micropipette/capillary by filling and expelling fresh PBS/PVA several times.

4. Pre-load the pipette with ~5 µl of fresh PBS/PVA and insert into the droplet containing cells. Under high magnification, attempt to draw a single cell into the micropipette/capillary[a].

5. Transfer the cell(s) to one of the droplets of fresh PBS/PVA. Attempt to minimize the volume of fluid expelled from the pipette[b]. Discard any PBS/PVA remaining in the micropipette/capillary and wash (as described in step 3). Reload the micropipette/capillary with PBS/PVA.

6. Insert the micropipette/capillary into the droplet containing the cell(s) (from step 5). A few microliters of the PBS/PVA should be expelled over one of the cells. This usually causes the cell to roll and move through the drop. Thus, the clean PBS/PVA expelled from the pipette washes the cell, whilst pushing it into areas of the drop containing more fresh PBS/PVA.

7. Repeat the washing steps 5 and 6 at least three more times, transferring the cell through several clean droplets of PBS/PVA.

8. After the final wash is complete, transfer the cell to a microcentrifuge tube in no more than 2 µl of clean PBS/PVA[c,d]. The tube should be capped as soon as the cell has been transferred.

9. Centrifuge briefly to deposit the cell at the bottom of the tube. Cells should be placed on ice immediately, or frozen for long-term storage.

10. A few microliters of PBS/PVA from the final wash drop to have contained the cell can be taken and used as a negative control. If contaminants have been sufficiently washed away, no amplification should be observed in this sample.

Notes
[a]Any cells in suspension may be separated in this way. Cells should be collected in sterile medium or PBS. Cells that grow attached to the substrate, such as fibroblasts, must be treated with trypsin before isolation of single cells. The concentration of the cell sample is not critical. Initially, cell suspensions are usually at high concentration and consequently it may be impossible to isolate a single cell to begin with. If this is the case, serially dilute the cell suspension to decrease the cell number.

[b]When moving a cell from one droplet to the next, it is advisable to transfer it in the smallest volume of fluid possible. This precaution minimizes the risk that contaminants in the vicinity of the cell will be transferred to the new droplet.

^cIf the cell is transferred in more than 3 µl of fluid, the reagents used for lysis and DNA amplification will be diluted to an unacceptable level.

^dBe careful not to expel any air after the cell has left the pipette, as the resultant bubbles can burst, propelling the cell further up the tube or causing it to lyse.

Protocol 2

Cell lysis using proteinase K

Equipment and reagents
- Proteinase K (PCR grade) (250 µg/ml; Roche)
- 17 µM Sodium dodecyl sulfate (SDS; Sigma)
- Thermal cycler
- Mineral oil (Sigma)

Methods
1. Add 2 µl of proteinase K solution and 1 µl of SDS solution to the microcentrifuge tube containing the cell[a].

2. Pulse centrifuge the tube to ensure that the proteinase K, SDS, and cell are taken to the bottom of the tube.

3. Overlay the fluid with ~10 µl of mineral oil (optional for thermal cyclers with heated lids).

4. Place the tubes in a thermal cycler and incubate at 37°C for 1 h (enzyme digestion), followed by 95°C for 10 min (enzyme inactivation).

5. DNA released during cell lysis may be amplified immediately or frozen for later use[b].

Notes
[a]Some people find it easier to add the cell to a microcentrifuge tube that already contains proteinase K and SDS.

[b]The DNA may be stored at –20°C for several weeks or at –70°C for several months.

Protocol 3

Single-cell multiplex PCR for cystic fibrosis

Equipment and reagents

- Lysed single cell (in <5 µl; see *Protocol 2*)
- HotMaster *Taq* DNA polymerase (5 units/µl), plus accompanying 10× PCR buffer (Eppendorf)
- 2.5 mM dNTPs (Invitrogen)
- Nuclease-free water (Promega)
- Oligonucleotides (all 100 µM):
  - ☐ ΔF508-F (forward) 5′-FAM-GTTTTCCTGGATTATGCCTGGCAC-3′
  - ☐ ΔF508-R (reverse) 5′-GTTGGCATGCTTTGATGACGCTTC-3′
  - ☐ D7S1799-F (forward) 5′-FAM-ATGGTATTAGGAGATGGGGC-3′
  - ☐ D7S1799-R (reverse) 5′-TTGCATAAGCCAATTTCCAT-3′
  - ☐ D7S1817-F (forward) 5′-PET-CAAATTAATGGCAAAAACTGC-3′
  - ☐ D7S1817-R (reverse) 5′-CCCCCCATTGAGGTTATTAC-3′
- Thermal cycler
- Fluorescent sequencer

Methods

1. Taking all precautions to maintain a DNA-free environment (see section 3), set up the following reaction mixture on ice:

Water	10.5 µl
10× Hotmaster reaction buffer	2.5 µl
dNTPs (2.5 mM)	2 µl
ΔF508-F primer (100 µM)	0.6 µl
ΔF508-R primer (100 µM)	0.6 µl
D7S1799-F primer (100 µM)	1.2 µl
D7S1799-R primer (100 µM)	1.2 µl
D7S1817-F primer (100 µM)	0.6 µl
D7S1817-R primer (100 µM)	0.6 µl
Hotmaster *Taq* polymerase (5 units/µl)	0.2 µl

2. Transfer this 20 µl of reaction mixture to the tube containing the isolated cell (in ~5 µl of PBS/PVA and lysis reagents).

3. Centrifuge briefly and then incubate according to the following PCR program:
   - 94°C for 2 min
   - 45 cycles of 94°C for 20 s, 57°C for 45 s, and 65°C for 30 s
   - Final incubation at 65°C for 5 min

4. Analyze the samples using a fluorescent sequencing apparatus, such as an ABI 310 Genetic Analyzer (Applied Biosystems).

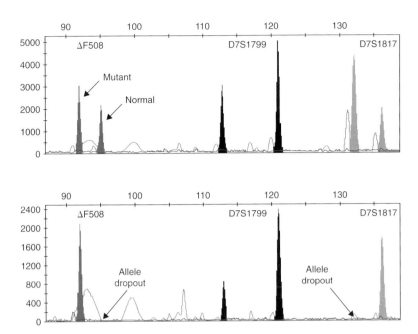

Figure 3. Results of single-cell multiplex PCR.
The upper image shows PCR-generated fragments from three loci, simultaneously amplified from a single cell. The cell is derived from a heterozygous carrier of the cystic fibrosis ΔF508 mutation, as revealed by the presence of normal and mutant alleles. The lower image shows PCR results for a second cell derived from the same individual. In this case, two of the loci have been affected by ADO. However, an accurate diagnosis is still possible, as the third locus is unaffected by ADO.

2.1 Expected results from multiplex PCR

The expected fragment sizes are 91 bp (ΔF508; the most common cystic fibrosis-causing mutation) and 94 bp (normal) for the cystic fibrosis gene, *CFTR* (see *Fig. 3*). The D7S1799 locus is situated ~2 Mb proximal from the *CFTR* gene, whilst D7S1817 is located 3.2 Mb distal. These microsatellite polymorphisms display a range of allele sizes from 150 to 300 bp. The amplification of an informative hypervariable-linked polymorphism provides a back-up for direct mutation detection and also has the potential to reveal DNA contaminants (e.g. contaminants not expected to be observed in the sample).

Protocol 4

Whole genome amplification via DOP-PCR

Equipment and reagents

- Lysed single cell or DNA sample (in <5 µl)
- SuperTaq Plus DNA polymerase (5 units/µl), plus accompanying 10× PCR buffer (Ambion)
- 2.5 mM dNTPs (Ambion)
- Nuclease-free water (Promega)
- 25 mM MgSO$_4$ (Ambion)
- 100 µM Primer (5'-CCGACTCGAGNNNNNNATGTGG-3')
- Thermal cycler
- 1% Agarose gel containing 10 ng/ml ethidium bromide
- 6× Orange loading dye solution (Fermentas)
- Equipment and reagents for agarose gel electrophoresis including 1× TBE agarose gel running buffer (10.8 g/l Tris base; 5.5 g/l boric acid; 4 ml/l 0.5 M EDTA, pH 8.0; diluted from a 10× stock; Sigma)
- DNA size marker (100 bp ladder; Invitrogen)
- UV light source

Methods

1. Taking all precautions to maintain a DNA-free environment (see section 3), set up the following reaction mixture on ice:

Water	31 µl
10× reaction buffer	5 µl
dNTPs (2.5 mM)	4 µl
Primer (100 µM)	1 µl
MgSO$_4$ (25 mM)	3.5 µl
Supertaq Plus DNA polymerase (5 units/µl)	0.5 µl

2. Transfer this 45 µl of reaction mixture to the tube containing the isolated cell or DNA sample (~5 µl).

3. Centrifuge briefly and then apply the following thermal cycling conditions:
   - 94°C for 4 min
   - 8 cycles of 94°C for 30 s, 30°C for 1 min, and 1°C/s ramp to 68°C; hold at 68°C for 3 min
   - 40 cycles of 94°C for 1 min, 62°C for 1 min, and 68°C for 1.5 min
   - 68°C for 8 min

4. Analyze the PCR products by mixing 5 µl of the reaction mix with 1 µl of 6× orange loading dye solution and resolving the sample by agarose gel electrophoresis alongside a DNA size marker (see *Fig. 4*).

3. TROUBLESHOOTING

- **How is the risk of contamination avoided?**
  Gloves and other clothing to block cellular contamination should be worn throughout single-cell PCR procedures and changed regularly. It is also

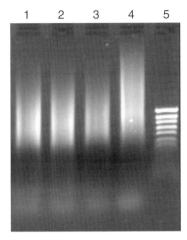

Figure 4. Analysis of DOP-PCR products generated from single cells.
The characteristic smear (lanes 2–4) indicates that many fragments of varied size are present. The mean fragment size is usually around 500 bp (determined by comparison with a molecular size marker, shown in lane 5). However, fragments of over 10 kb can also be detected.

advisable to set up reactions in a laminar flow hood. Single cells should be washed thoroughly in medium or PBS that has been demonstrated to be free of contaminating DNA (14).

- **Should pre- and post-PCR methods be performed in separate laboratories?**
Ideally, a separate room should be designated for PCR set-up. No amplified DNA should ever be brought into the room and access to personnel should be restricted. The room should contain dedicated equipment, reagents, gloves, and laboratory coats. Equipment should not be permitted to pass between this room and any laboratory where tubes containing amplified DNA are opened and PCR products analyzed. This restriction should even apply to laboratory items such as microcentrifuge tube racks, ice buckets, and marker pens, as all of these have been reported as sources of contamination.

4. REFERENCES

1. Wells D & Delhanty JD (2001) *Trends Mol. Med.* **7**, 23–30.
2. Piyamongkol W, Bermudez MG, Harper JC & Wells D (2003) *Mol. Hum. Reprod.* **9**, 411–420.
3. Wells D, Bermudez MG, Steuerwald N, *et al.* (2005) *Hum. Reprod.* **20**, 1339–1348.
4. Ray PF & Handyside AH (1996) *Mol. Hum. Reprod.* **2**, 213–218.
5. Thornhill AR, McGrath JA, Eady RA, Braude PR & Handyside AH (2001) *Prenat. Diagn.* **21**, 90–97.
6. El-Hashemite N & Delhanty JD (1997) *Mol. Hum. Reprod.* **3**, 975–978.
7. Telenius H, Carter NP, Bebb CE, Nordenskjold M, Ponder BA & Tunnacliffe A (1992) *Genomics*, **13**, 718–725.

8. Wells D, Sherlock JK, Handyside AH & Delhanty JD (1999) *Nucleic Acids Res.* **27**, 1214–1218.
9. Kallioniemi A, Kallioniemi OP, Sudar D, *et al.* (1992) *Science*, **258**, 818–821.
10. Voullaire L, Wilton L, Slater H & Williamson R (1999) *Prenat. Diagn.* **19**, 846–851.
11. Zhang L, Cui X, Schmitt K, Hubert R, Navidi W & Arnheim M (1992) *Proc. Natl. Acad. Sci. U.S.A.* **89**, 5847–5851.
12. Klein CA, Schmidt-Kittler O, Schardt JA, Pantel K, Speicher MR & Riethmüller G (1999) *Proc. Natl. Acad. Sci. U.S.A.* **96**, 4494–4499.
13. Dean FB, Hosono S, Fang L, *et al.* (2002) *Proc. Natl. Acad. Sci. U.S.A.* **99**, 5261–5266.
14. Wells D & Sherlock JK (1998) *Prenat. Diagn.* **18**, 1389–1401.

CHAPTER 11

Rapid generation of gene-targeting constructs

Trevor J. Wilson, Dirk Truman, Antonietta Giudice, and Paul Hertzog

1. INTRODUCTION

The completion of the sequencing of the human genome has initiated a new era of biomedical research. New genes and their predicted functions are now more likely to be identified through bioinformatics than basic research. However, elucidating the true *in vivo* function of these genes requires analysis in an intact organism. Due to size, cost, ease of manipulation, and similarity to the human, much of this research will occur in mouse models by the generation of mice with a targeted mutation in the desired gene (1, 2). Not only can gene modification demonstrate their function, but by disrupting cell development or survival, it can also demonstrate the function in specific cell types. Some mutant mice are murine models of human genetic diseases and demonstrate that a single gene defect is capable of causing disease. In addition, the crossing of mice that are heterozygous or homozygous for specific gene mutations provides important information on human diseases that are caused by multiple genetic defects. Thus, the challenge is to develop means of rapidly introducing mutations into the murine genome and identifying genes and alleles important in biological processes and human disease.

To generate mutant mice, there are two broad categories of approaches, the function-driven approach and the gene-driven approach.

1.1 The function (or phenotype)-driven approach

The function-driven approach includes the introduction of a reporter gene (gene trap) and selection of cells or animals for further analysis based on an expression pattern of interest. Usually these insertions also disrupt gene function and thus analysis of the homozygous animals (or cells) can identify gene function, as with a targeted knockout. Recently, various laboratories using this approach have formed the International Gene Trap Consortium, which aims to generate gene-

PCR: *Methods Express* (S. Hughes and A. Moody, eds.)
© Scion Publishing Limited, 2007

trap embryonic stem (ES) cell libraries comprising mutants of every gene in the genome (3, 4). The other commonly used function-driven approach is the use of *N*-ethyl-*N*-nitrosourea to generate large numbers of mice (or ES cells) harboring genetic mutations that may disrupt gene function. This chemical mutagen is being used by a number of groups to generate high point mutation rates in pre-meiotic spermatogonial stem cells (5, 6). These mice are subsequently mated for the rapid generation of a vast array of mutant mouse lines that carry point mutations throughout the genome. These lines are bred to identify and characterize interesting phenotypes that result from dominant or recessive mutations, and the mutation is subsequently identified. More recently, similar *N*-ethyl-*N*-nitrosourea mutagenesis has been performed on ES cells to generate libraries of mutated ES cells for analysis (7). These function-driven approaches have already identified and characterized many genes, but the success of this approach relies on the ability to screen large numbers of cells and/or mice to identify a phenotype of interest. Furthermore, it is likely that there is a bias in the genes susceptible to mutation, which means that the generation of some mutant genes will never occur using this methodology.

1.2 The gene-driven approach

In contrast, the gene-driven approach, which is the focus of this chapter, relies on the selection of a gene of interest for further analysis followed by the generation of cells and/or animals with a targeted mutation in that gene. Unlike the function-driven approach, the gene-driven approach allows the introduction of a specific mutation into the gene of interest. In some cases, this is the disruption of the entire protein, but in many cases the planned targeting of specific domains or important amino acid residues of a protein has become a valuable, arguably the preferred, means of investigating the function and structure/function role of a protein under physiological conditions (2). It is also possible to generate the targeted mutation in a conditional manner, such that the disruption can be regulated in a temporal and/or tissue-specific manner.

Generation of targeted mutations is performed by the generation of a targeting construct (see *Fig. 1*), which, when introduced into ES cells, directs the creation of a modified allele by replacement of endogenous sequences through homologous recombination. Homologous recombination occurs at a high frequency in ES cells, perhaps due to their unique embryonic disposition, very rapid growth, or relatively long S phase, allowing relatively efficient generation of targeted ES cell clones. The efficiency of targeting is reliant on a number of factors:

1. It is important to use isogenic (i.e. truly homologous) DNA (8). Whilst this is now relatively simple for targeting of C57BL/6 ES cell lines (and to a lesser extent 129 cells), as libraries of bacterial artificial chromosome (BAC) clones are easily accessible and annotated onto murine genome databases, targeting of ES cells derived from other strains still requires the isolation of isogenic DNA.

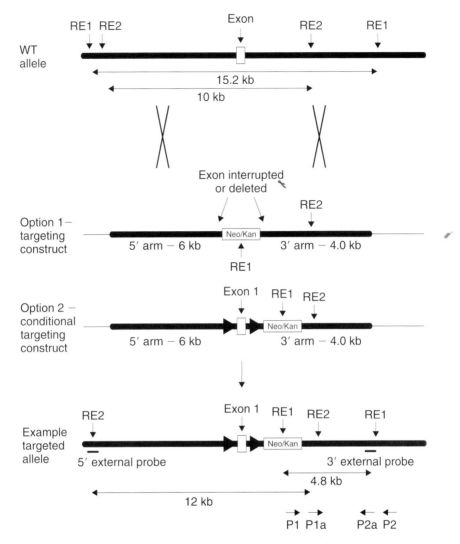

Figure 1. Example of a targeting and screening strategy.
Schematic of targeting strategy of a gene. Option 1 is a conventional targeting construct where an exon is interrupted or replaced by an antibiotic resistance gene. Option 2 is a conditional targeting construct where the endogenous exon is to be replaced with a floxed exon and the resistance gene. Homologous recombination at the genomic locus will result in the replacement of the exon with a neomycin/kanamycin cassette with or without a floxed exon to generate a conditional knockout using Cre recombinase. We usually include Flp recognition target (FRT) sites flanking the antibiotic resistance gene to allow its removal by Flp recombinase. Identification of putative correctly targeted clones can be performed by PCR screening with a 3′ external oligonucleotide and one that resides in the neomycin gene. Putative positive clones identified via this approach can be confirmed using Southern blot analysis with a 3′ external probe on restriction endonuclease (RE)1-digested DNA. Using this approach, correctly targeted ES cells will generate a wild-type band of 15.2 kb and a targeted band of 4.8 kb in this model targeting event. Southern blot analysis at the 5′ end on RE2-digested DNA from targeted clones should yield a wild-type band of 10 kb and a targeted band of 12 kb. ▶ represent *loxP* sites used in conditional targeting constructs.

2. The length of the homologous 'arms' flanking the desired insertion or mutation site. Although the optimal length of these arms is not clear and will probably depend on the particular locus, it is clear that longer homologous arms increase targeting efficiency (9).

3. The presence of repetitive sequences of homology in the arms can influence correct homologous recombination. If these are present, recombination appears to be more likely to occur at other sites or result in partial or duplicate insertions. Thus, the preferred construct design is large, homologous arms containing minimal repetitive elements, to allow efficient homologous recombination.

In addition, in designing the construct to maximize targeting efficiency, careful consideration must be given to the strategy used to identify targeted clones correctly, as large numbers of clones often need to be screened (see *Fig. 1*). Whilst PCR is a rapid and efficient means of screening clones, large homologous arms used to increase targeting efficiency may preclude identification of correctly targeted clones by PCR due to the size of the predicted amplified fragment. Southern blot screening of clones with a probe external to the targeting construct can often accommodate larger homologous arms, but this is dependent on the presence and location of suitable restriction sites. Therefore, if screening using a DNA methodology, a balance between the size of the arms and the ability to identify targeted clones rapidly must be achieved. One alternative strategy to overcome this issue is to screen mRNA/cDNA transcripts from putatively targeted ES cell clones; however, this is only possible where changes in exons are introduced and when the gene of interest is expressed in ES cells.

2. METHODS AND APPROACHES

The purpose of the methodology described herein is the rapid generation of constructs for gene targeting. The first step in the generation of the construct is to obtain sequence information (outlined in section 2.1) and design a suitable targeting strategy: *Fig. 1* shows two example constructs. In the first example (option 1), an exon is interrupted by the insertion of a neomycin/kanamycin antibiotic-resistance cassette to prevent protein translation or induce premature protein termination. In the second example (option 2), constructed *loxP* sites are included to flank or 'flox' the exon for conditional targeting, such that subsequent recombination by Cre results in deletion of the exon, preventing protein initiation or inducing a frame shift. Further information on the design of such constructs can be found elsewhere (10).

Once the design of the construct has been completed, suitable isogenic DNA is required for generation of the construct. Where suitable BAC clones are not available, DNA can be isolated from the cells to be targeted (see *Protocol 1*) and 5' and 3' homologous arms generated using high-fidelity, long-range PCR utilizing primers designed to contain suitable restriction enzyme sites for subsequent cloning (see *Protocol 3*).

Our optimal approach utilizes BAC clones where suitable isogenic DNA can be obtained from BAC libraries (see section 2.1). In this case, excision and incision fragments are generated by PCR (see *Protocol 4*) and ET cloning (homologous recombination using RecE and RecT) (11, 12) is used for efficient generation of the construct (see *Fig. 2* and *Protocols 5–10*). The methodology is identical for conditional targeting constructs; however, the floxed exon is generated by PCR and cloned into the construct as the final step (see section 2.3). During the production of the targeting elements, it is also important to develop and optimize a PCR screening strategy for screening putative targeted ES cell clones (see section 2.5 and *Protocol 11*). Other protocols such as for the transfection, selection, and culture of ES cells are outside this scope of this chapter, but are described in detail elsewhere (10, 13).

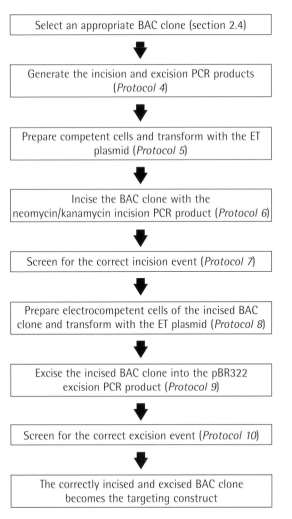

Figure 2. Overview of Red/ET recombination.

2.1 Obtaining gene sequences and identification of BACs

To target an allele of interest requires the isolation of fragments of DNA surrounding the region to be targeted and the design of a suitable targeting strategy using the principles outline above. The sequences of the gene of interest, the gene structure, and the genomic sequence can now be obtained rapidly by the on-line interrogation of many public domain databases. Furthermore, BAC clones for some mouse strains are also annotated on these databases and can be obtained readily for use in the generation of the targeting constructs. For strains where BACs are not available, the sequence data is still invaluable for the design and generation of the targeting construct. The approach we use to obtain such gene and sequence information is outlined below:

1. Access the NCBI data base (http://www.ncbi.nlm.nih.gov/) and search 'Gene' by entering the gene name or accession number and locate the correct Entrez Gene entry.

2. Within the appropriate Entrez Gene entry, click on the NC number under the heading **Genomic regions, transcripts, and products** and select the FASTA option. This will display the genomic sequence corresponding to the transcript (i.e. all introns and exons). You can change the size and region of the sequence displayed by changing the numbers in the range boxes. The fragment of chromosomal sequence retrieved should be used to create a map of the region to design suitable targeting and screening strategies and appropriate primers.

3. The Ensemble Genome Browser (http://www.ensembl.org/index.html) is useful for obtaining information about the locus in the mouse genome and/or identifying BAC clones to be used for building the targeting vector. Select the *Mus musculus* (mouse) genome and then under **Use Ensembl to...** select **Search Ensembl**. This can be used to search for a gene name or GenBank accession number (or other descriptor). The search will return one or more related matches depending on the search terms used. Select the Ensembl Gene Report for the gene of interest from the number of matches displayed. The Gene Report will include the Genomic Location and a number of other features including links to Exon Information. Clicking on **Exon Information** will show the exon boundaries that should be included in the map and design of the construct. To identify suitable BAC clones, go to the page entitled **Ensembl Mouse Contig View** by clicking on one of the links under the heading **Genomic Location** on the Ensemble Gene Report page. The initial genomic region displayed will depend on the link selected, and navigation buttons enable the display of regions further 5′ or 3′, or zooming in or out as required. To display suitable BAC clones, select **Decorations** under the Detailed view and enable **BAC map**, then click on **Close menu**. Select at least two suitable BAC clones that contain the region for targeting and order them. They can be obtained from resources such as the Roswell Park Cancer Institute (http://www.roswellpark.org/Research/Shared_Resources/Microarray_and_Genomics_Resource/RPCIBACandPACclones) and other local resources, e.g.

Australian Genome Research Facility (http://www.agrf.org.au). The RP23 series is readily available and is suitable for targeting C57BL/6 ES cells.

The UCSC Genome Bioinformatics Site (http://genome.ucsc.edu/) can also be used to obtain similar information. Under the heading **Mouse (*Mus musculus*) Genome Browser Gateway,** select **Genomes** and in the box for **position or search term** enter the gene name, GeneID from Entrez Gene, NM number from NCBI Entrez nucleotide, or DQ number from NCBI Entrez nucleotide. To show BACs, find **Mapping and Sequence Tracks**, go to **BAC End Pairs**, and select **full** and then hit the refresh button. It may be necessary to **zoom in** or **zoom out** or to **move start** or **move end**.

4. Design a suitable targeting construct and screening strategy specific for the gene of interest. We recommend analyzing the sequence to identify regions that contain repetitive sequences. The REPEATMASKER program is suitable (http://www.repeatmasker.org/cgi-bin/WEBRepeatMasker). Paste the sequence identified above into the sequence window, select mouse as the species of origin, and run the program. This will identify regions to be avoided for Southern blotting probes and suboptimal regions for homology arms. We recommend homology arms of approximately 10 kb, usually divided into one 4 kb and one 6 kb homology arm. Design of the construct should also consider the presence of restriction sites for putting the construct together and unique sites to release the completed targeting construct from the plasmid sequences.

Protocol 1

Isolation of isogenic DNA: DNA extraction from ES cells

Equipment and Reagents[a,b]
- ES cell cultures
- Lysis buffer (100 mM Tris/HCl, pH 8.0, 5 mM EDTA, 0.2% SDS, 200 mM NaCl, 100 µg/ml proteinase K (added fresh))
- Unifilter 800 (Whatman)
- Vacuum manifold for 96-well plates
- 1× PB buffer (Qiagen)
- 1× PE buffer (Qiagen)
- 0.25× TE (2.5 mM Tris/HCl, pH 7.5, 0.25 mM EDTA)

Method
1. Culture ES cells to confluency in 48-well tissue culture plate.

2. Add 100 µl of lysis buffer to each well and incubate for 2–3 h at 55°C (or overnight at 37°C).

3. Add 500 µl of 1× PB buffer, mix well, and add to a well of the Unifilter plate in a vacuum manifold. Set up the manifold according to the manufacturer's instructions.

4. Apply vacuum (–300 mmHg) until all of the mixture has passed through the filter.

5. Add 400 µl of PE buffer and apply the vacuum.

6. Allow the filter to dry thoroughly under vacuum.

7. Place the collection plate under the Unifilter plate and add 70 µl of 0.25× TE (warmed to 70°C).

8. After 1 min, apply the vacuum as above and elute the DNA into the plate.

Notes
[a]This protocol is suitable for isolation of DNA for the generation of large fragments (for construct generation) and for the screening of putative target clones.
[b]Larger amounts of DNA can be recovered by isopropanol precipitation following ES cell lysis (step 2). However, the purity of the DNA will be lower and it is essential that the DNA recovered from this protocol is high quality for robust long-range PCR.

Protocol 2

Amplification of genomic DNA fragments for the generation of constructs using cloning technology

Equipment and Reagents
- Thermal cycler with a gradient function[a]
- Platinum *Taq* DNA Polymerase High Fidelity, accompanying High Fidelity buffer, and 50 mM MgSO$_4$ (Invitrogen)
- 10 mM dNTPs (Invitrogen)
- Sterile Milli-Q water
- Oligonucleotide primers
- ES cell DNA extracted from cells to be targeted (see *Protocol 1*)
- Equipment and reagents for agarose gel electrophoresis

Method
1. Prepare a PCR master mix with the following reagents (per 50 µl reaction):
   - 5 µl of 10× High Fidelity buffer
   - 2 µl of 50 mM MgSO$_4$
   - 1 µl of 10 mM dNTPs
   - 1 µl of each oligonucleotide primer (10 µM)
   - 0.2 µl Platinum *Taq* DNA Polymerase High Fidelity
   - 37.8 µl H$_2$O

2. Aliquot 48 µl of the master mix into PCR tubes.

3. Add 2 µl of ES cell-derived DNA to each tube[b] (see *Protocol 1*).

4. Run in a thermal cycler programmed as follows:
   - 1 cycle of 94°C for 3 min
   - 35 cycles of 94°C for 30 s, annealing temperature gradient ranging from 52 to 60°C for 30 s, and 68°C for 46 min (1 min/kb of product)
   - 1 cycle of 68°C for 10 min

5. Analyze the PCR by running 25 µl of the amplified product on a 0.8% (w/v) agarose gel.

6. Identify conditions (annealing temperature) for optimal production of the correct-sized fragment.

Notes
[a]If a thermal cycler with a gradient function is not available, it is equally acceptable to perform separate PCRs at varying temperatures.
[b]Where BAC clones are unavailable, high-fidelity, long-range PCR and cloning can be used to generate constructs. Using sequence data obtained from online databases, the construct can be carefully designed and unique restriction sites included in all oligonucleotides to facilitate efficient cloning. Unique restriction sites in the most 5′ and 3′ oligonucleotides are also required to release the targeting construct from the plasmid backbone. The 5′ and 3′ homology arms are amplified separately and cloned.

Protocol 3

Cloning of PCR products

Equipment and Reagents
- QIAquick PCR purification kit (Qiagen)
- pGEM-T Easy vector kit (Promega)
- Plasmid mini kit (Qiagen)
- Electrocompetent DH10B cells (Invitrogen)
- Electroporator and cuvettes (Biorad Gene Pulsar)
- Luria–Bertani (LB) amp medium (1% tryptone, 0.5% yeast extract, 0.5% NaCl, 50 µg/ml ampicillin)
- LB amp plates (100 mm diameter plates containing ~10 ml of 1.5% agar in LB media plus 10 mM IPTG, 40 µg/ml X-gal, and 50 µg/ml ampicillin)

Method
1. Amplify the genomic fragments using *Protocol 2*.

2. Purify the PCR product using a QIAquick PCR purification kit[a], according to manufacturer's instructions.

3. Clone into pGEM-T Easy using the vector kit[a].

4. Add 5 µl of the ligation to 40 µl of DH10B cells and electroporate (Biorad Gene Pulser, 0.1 cm cuvette, 1.8 kV, 200 Ω, 25 µF).

5. Add 1 ml of LB amp medium and plate 200 µl of cells on an LB amp plate.

6. Incubate at 37°C overnight.

7. Select six to ten white colonies[b] for further growth (in 2 ml LB amp medium cultures) and prepare plasmid DNA using a Qiagen plasmid mini kit.

8. Determine that the cloned fragment is of the correct size by restriction digestion and sequencing.

Notes

[a]If restriction sites have been engineered into the PCR primers, the purified PCR products can be digested with the appropriate enzymes (based on the sites introduced by the oligonucleotides), purified, and then cloned directly into the linearized targeting vector. We have found, however, that the cloning efficiency is often low as many restriction enzymes cut poorly near fragment ends. Thus, initially cloning into the pGEM-T vector (TA-cloning) and then subcloning fragments into the targeting vector is more reliable.
[b]The pGEM-T system contains the *lacZ* gene. White colonies usually contain the insert for easy identification.

2.2 ET cloning procedure

Compared with standard cloning, the Red/ET recombination (Gene Bridges) approach is more rapid and enables cloning of larger fragments that are error-free (see *Fig. 2*). It works using homologous recombination in *Escherichia coli*, which is mediated by the phage-derived recombinases, either Redα/Redβ/Redγ or

RecE/RecT/Redγ. Essentially, there are two main steps in building a targeting construct via this method:

1. *Incision.* This involves introducing a kanamycin/neomycin (kan/neo) selection marker at a specific site into a BAC harboring the gene of interest (see *Protocol 6*).
2. *Excision.* Here, a fragment containing the kan/neo cassette flanked by the homology arms of the targeting vector is excised from the recombinant BAC into a bacterial vector (see *Protocol 9*).

2.2.1 Design of oligonucleotide primers for ET cloning and screening

Oligonucleotides for production of the incision fragment

Amplification of a kan/neo cassette flanked by short gene-specific sequences homologous to sequences surrounding the region to be deleted enables ET-mediated recombination. The PCR product generated in this step is referred to as the incision fragment as it is used to replace specific sequences in the BAC with the kan/neo cassette.

For the incision fragment design, primers need to contain 50–60 bp of sequence homologous to the region where recombination in the BAC is to occur and a short sequence of the kan/neo gene (sequences shown below). The 5′ primer will include the BAC homology sequences in the forward orientation followed by the pEGFP sequence 5′-AATTCGATTAGACG-3′ (see *Fig. 3a*, P1). The 3′ primer will include the BAC homology sequence in the reverse/complementary orientation followed by the pEGFP sequence 5′-GCGAATTCACTAGTG-3′ (see *Fig. 3a*, P2). It should be noted that, for the generation of conditional knockout targeting constructs, the oligonucleotides are designed to remove the region to be floxed during the incision and a unique restriction site is included in one of these oligonucleotides to allow subsequent insertion of the floxed exon(s).

For screening of clones that have undergone the correct incision event, design primers at the 3′ end of what will become the 5′ homology arm (see *Fig. 3d*, primer a) and at the 5′ end of what will become the 3′ homology arm (see *Fig. 3d*, primer d). These are used with primers in the kan/neo cassette (see *Fig. 3d*, primers b and c, respectively). Primer b (5′-CTGGTTCTTTCCGCCTCAGG-3′) lies 231 bp from the 5′ end of the kan/neo cassette and primer c (5′-CCTGCCATAGCCTCAGGTTACTC-3′) lies 239 bp from the 3′ end of the kan/neo cassette. It is convenient to design primers a and d so that a product of between 400 and 800 bp is amplified with the respective primer in the kan/neo cassette.

For the generation of incision fragments with the Flp recognition target (FRT) or *loxP* sites flanking this cassette, we use constructs containing these sites, so that, even though the sequences of the primers will be different to those mentioned above, the procedure remains unchanged.

Oligonucleotides for production of the excision fragment

The excision fragment is derived by amplification of a pBR322 backbone flanked by sequences homologous to the extreme boundaries of the homology arms of the

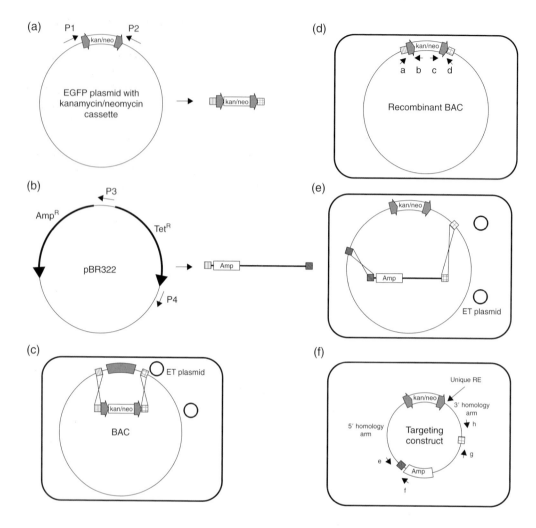

Figure 3. Basic strategy for generating a gene-targeting vector using Red/ET recombination.
The first steps (a, b; see *Protocol 4*) are the generation of the incision and excision PCR fragments. The incision fragment is amplified using primers P1 and P2 (a) and consists of the kan/neo cassette flanked by 50–60 bp of homology to the region of BAC to be deleted. The arrows either side of the cassette represent *loxP* sites, which can be used to remove the cassette following targeting. The excision PCR fragment is amplified using primers P3 and P4 (b) and consists of a pBR322 backbone including the ampicillin-resistance (Amp^R) gene flanked by 50–60 bp of homology to the region of BAC to be excised. The BAC clone harboring the ET plasmid is transformed with the incision PCR fragment (c). The sequence to be deleted is shown with downward diagonal lines and the regions of homology for the homologous recombination reaction are shown on either side. Recombinant BAC clones with the sequence to be deleted replaced by the kan/neo cassette are selected in the presence of kanamycin, and primers a, b, c, and d are used to screen for correctly incised clones (d; see *Protocol 7*). To generate the excision product, the correctly incised recombinant BAC clone harboring the ET plasmid is transformed with the excision PCR fragment (e; see *Protocol 9*). The regions on either side of the kan/neo cassette will form the homology arms of the targeting construct. Homologous recombination between the regions of homology in the BAC and the excision fragment will excise the region of interest into the PBR322 plasmid, which is selected by kanamycin/ampicillin to become the targeting construct (f). Primers e, f, g, and h are used to screen for correctly excised clones (f; see *Protocol 10*).

targeting construct and is used to excise the required fragment of BAC DNA into pBR322.

For the excision fragment design, primers need to contain short sequences homologous to the region where recombination in the BAC is to occur and a short sequence of the pBR322 vector (sequences shown below). The BAC-specific sequences should be designed to enable the excision of a ≥10 kb fragment of BAC DNA. It is desirable for one of the homology arms of the targeting vector to be shorter (≤4 kb) than the other, so that it is possible to perform the preliminary screen of G418-positive ES cells for putative correctly targeted clones via PCR. The pBR322 component of the primers will facilitate the amplification of the vector backbone minus the tetracycline resistance gene, but still including the ampicillin resistance marker. The 5′ primer will contain 50–60 bp of BAC homology in reverse/complement orientation, a unique restriction enzyme site (to release the plasmid sequences from the homology arms and positive selection marker prior to electroporation into ES cells) and the pBR322 sequence 5′-CACCTCGACCTGAATGGAAGCCG-3′ (see *Fig. 3b*, P3). The 3′ primer will include 50–60 bp of BAC homology in the forward orientation, the same unique restriction enzyme site used in the 5′ primer and the pBR322 sequence 5′-GATTTCATACACGGTGCCTGAC-3′ (see *Fig. 3b*, P4).

For screening of clones that have undergone the correct excision event, design primers at the 5′ end of the 5′ homology arm (see *Fig. 3f*, primer e) and at the 3′ end of the 3′ homology arm (see *Fig. 3f*, primer h). These are used with primers in pBR322 (see *Fig. 3f*, primers f and g, respectively). Primer f (5′-CATCGATAAGCTTTAATGCG-3′) is 200 bp from the integration site and primer g (5′-GTCCTCAACGACAGGAGCAC-3′) is 71 bp from the integration site. It is convenient to design primers e and h so that a product of between 400 and 800 bp is amplified with the respective primer in pBR322.

Correctly incised and excised BAC products can be verified by sequence analysis using primers b and c (see *Fig. 3d*) and f and g (see *Fig. 3f*).

2.3 Cloning of the floxed exon

A conventional knockout targeting construct will be generated from either the PCR and cloning strategy (see *Protocols 2* and *3*) or by ET cloning (see *Protocol 10*) as shown in *Fig. 3cf*). For the generation of a conditional targeting construct, a further step is required to insert back the targeted region, flanked by *loxP* sites. This is achieved by generating a PCR product using either the BAC or ES cell DNA as template using the method detailed in *Protocol 2*. However, the 5′ oligonucleotides will contain the unique restriction enzyme site (see *Fig. 3f*) followed by the *loxP* site (5′-ATAACTTCGTATAGCATACATTATACGAAGTTAT-3′) and then ~20 bases homologous to the genomic region. Similarly, the 3′ oligonucleotides will contain the same restriction enzyme site, followed by the *loxP* site (with the directional spacer reverse complementary sequence: 5′-ATAACTTCGTATAATGTATGCTATACGAAGTTAT-3′) and then ~20 bases homologous to the genomic region. Once produced, this fragment is cloned by conventional methods into the vector and verified by sequencing.

2.4 Preparation and electroporation of DNA

DNA for electroporation is prepared by standard methods, e.g. Qiagen maxi kit. Following preparation of the DNA, it is important to release the targeting sequences from the plasmid sequences and to repurify the DNA prior to electroporation. These procedures are covered in more detail elsewhere (10, 13).

Protocol 4

Generating incision and excision PCR products[a]

Equipment and Reagents
- Thermal cycler
- Spectrophotometer
- 37°C Incubator
- Platinum *Taq* DNA Polymerase High Fidelity (5 units/µl), accompanying High Fidelity buffer, and 50 mM MgSO$_4$ (Invitrogen)
- 10 mM dNTPs (Invitrogen)
- Sterile Milli-Q water
- Oligonucleotide primers (10 µM) (for incision, use primers P1 and P2, and for excision, use primers P3 and P4, as described above)
- pBR322 plasmid (50 ng/µl; template for excision PCR) (NEB)
- pEGFP plasmid (50 ng/µl; template for incision PCR) (Clontech) or suitable FRT- or *loxP*-flanked kan/neo cassette
- Restriction enzyme *Dpn*I and accompanying digestion buffer (NEB)
- MinElute Reaction Cleanup kit (Qiagen)
- Equipment and reagents for agarose gel electrophoresis

Method
1. Prepare a PCR master mix with the following reagents:
   - 5 µl of 10× High Fidelity buffer
   - 2 µl of 50 mM MgSO$_4$
   - 1 µl of 10 mM dNTPs
   - 1 µl of each oligonucleotide primer for incision *or* excision
   - 1 µl of pEGFP for incision product *or* pBR322 for excision product
   - 0.2 µl of Platinum *Taq* DNA Polymerase High Fidelity
   - 38.8 µl of H$_2$O

2. Aliquot the mixture into a suitable PCR tube and place in a thermal cycler programmed with the following conditions:
   - 1 cycle of 93°C for 3 min
   - 35 cycles of 93°C 30 s, 55°C for 30 s, and 68°C for 3 min
   - 1 cycle of 68°C for 10 min

3. Analyze the reactions by running 5 µl of the amplification products on a 0.8% (w/v) agarose gel. A product of ~2.0 kb should be amplified from pEGFP (the kan/neo cassette) and a product of ~3.0 kb should be amplified from pBR322.

4. Purify the PCR products using a Qiagen MinElute Reaction Cleanup kit, according to manufacturer's instructions.

5. Treat the purified products with the restriction enzyme *Dpn*I to digest the pEGFP or pBR322 plasmid used as template. Combine the 30 μl of eluted/purified PCR product with 20 units of *Dpn*I, 5 μl of reaction buffer, and sterile water up to a final volume of 50 μl. Digest samples overnight at 37°C[b].

6. Gel purify the PCR products using the Qiagen MinElute Reaction Cleanup kit, following the manufacturer's instructions.

7. Determine the concentration of the PCR products using a spectrophotometer.

Notes

[a]The incision product will be a kan/neo cassette flanked by sequences homologous to the BAC integration site. The excision product will be the pBR322 plasmid flanked by sequences homologous to the ends of the BAC region to be cloned.

[b]*Dpn*I will only cleave when its recognition site is methylated. Thus, it will digest away the template, leaving the PCR product intact.

Protocol 5

Preparation of electrocompetent cells harboring the BAC clone and transformation with the ET plasmid

Equipment and Reagents

- Centrifuge able to accommodate 50 ml tubes
- Spectrophotometer
- Electroporator and electroporation cuvettes (Biorad Gene Pulser)
- 30°C Shaking incubator
- 30°C Stationary incubator
- 37°C Shaking incubator
- BAC clone containing the locus of interest with sufficient flanking sequence to allow for 5′ and 3′ homology arms, approximately 10 kb (see *Protocol 1* for identification of suitable BAC clones)
- ET plasmid, tetracycline resistant (pSC101-BAD-gbaA^{tet}; Gene Bridges)
- LB broth (10 g/l tryptone, 5 g/l yeast extract, 10 g/l NaCl)
- 10% (v/v) Glycerol (chilled on ice; Sigma)
- Chloramphenicol (30 mg/ml stock; Sigma)
- Tetracycline (12 mg/ml stock; Sigma)

Method

1. Set up a 5 ml culture of the BAC clone in LB broth with 2.5 μl of chloramphenicol stock (final concentration 15 μg/ml). Incubate overnight at 37°C with agitation in a shaking incubator.

2. Add 400 μl of the overnight culture to 40 ml of LB broth with 20 μl of chloramphenicol stock (final concentration 15 μg/ml). Mix and divide equally among four 50 ml tubes.

3. Shake at 37°C until the culture reaches an OD$_{600}$ of ~0.4. This usually takes 2.5–3.5 h.

4. Combine the cultures from the four tubes into a single tube and centrifuge at 5000 *g* for 5 min (4°C).

5. Discard the supernatant and resuspend the cells in an equal volume (40 ml) of ice-cold 10% glycerol. Centrifuge at 2°C at 5000 *g* for 5 min.

6. Discard the supernatant and repeat the glycerol wash.

7. Resuspend the cells in a final volume of ~200 μl of ice-cold 10% glycerol and aliquot 50 μl into separate 1.5 ml microfuge tubes[a].

8. Add 50 ng of the ET plasmid to 50 μl of competent cells (containing a BAC clone) and electroporate (0.1 cm cuvette, 1.8 kV, 200 Ω, 25 μF).

9. Resuspend the bacteria in 1 ml of LB broth and shake at 30°C[b] for 75–90 min. Spread appropriate dilutions (100 μl of the 1 ml culture, then pellet the cells in the remaining 900 μl, resuspend in 110 μl of LB broth and plate 100 μl of neat cells and 100 μl of a 10^{-1} dilution) of the culture onto LB agar plates with chloramphenicol (15 μg/ml) and tetracycline (3 μg/ml) and incubate overnight at 30°C.

Notes

[a]Store unused cells at −80°C.
[b]The ET plasmid is temperature sensitive and will be lost if incubated at 37°C.

Protocol 6

Incision of BAC DNA with the neo/kan cassette

Equipment and Reagents
- Centrifuge able to accommodate 50 ml tubes
- Electroporator and electroporation cuvettes (Biorad)
- 30°C Shaking incubator
- 37°C Shaking incubator
- 37°C Stationary incubator
- Electrocompetent BAC clone transformed with the ET plasmid
- Incision PCR product (see *Protocol 4*)
- LB broth
- Kanamycin (30 mg/ml stock; Sigma)
- Chloramphenicol (30 mg/ml stock; Sigma)
- Tetracyline (12 mg/ml stock; Sigma)
- LB agar plates with kanamycin (15 µg/ml) and chloramphenicol (15 µg/ml)
- 10% (w/v) L-Arabinose in water (filter-sterilized)

Method

1. Set up a 5 ml overnight culture of the BAC clone harboring the ET plasmid (from *Protocol 5*) in LB broth with 2.5 µl of chloramphenicol stock (final concentration 15 µg/ml) and 1.2 µl tetracycline (final concentration 3 µg/ml). Incubate at 30°C[a].

2. Add 400 µl of the overnight culture to 40 ml of LB broth with 20 µl of chloramphenicol (15 µg/ml) and 9.6 µl of tetracycline (3 µg/ml). Mix and divide equally among four 50 ml tubes.

3. Shake at 30°C[a] until the culture reaches an OD_{600} of ~0.2 (2.5–3.5 h).

4. Induce expression of the ET recombinase by adding filter-sterilized L-arabinose to a final concentration of 0.1–0.2% and shake at 37°C[b] until the OD_{600} reaches 0.4 (30–45 min). Combine the cultures from the four tubes into a single tube and centrifuge at 2°C for 5 min at 5000 ***g***.

5. Discard the supernatant and resuspend the cells in an equal volume of ice-cold 10% glycerol. Centrifuge at 2°C for 5 min at 5000 ***g***.

6. Discard the supernatant and repeat the glycerol wash.

7. Resuspend the cells in a final volume of ~200 µl of ice-cold 10% glycerol and aliquot 50 µl into separate microfuge tubes. Store unused cells at –80°C.

8. Add 200–300 ng of the incision PCR product to 50 µl of competent cells and electroporate (0.1 cm cuvette, 1.8 kV, 200 Ω, 25 µF).

9. Resuspend the bacteria in 1 ml of LB broth and shake at 37°C for 75–90 min. Pellet the cells, resuspend in 100 µl of LB broth, and spread appropriate dilutions (try undiluted, 10^{-1}, and 10^{-2}) onto LB agar plates with chloramphenicol (15 µg/µl) and kanamycin (15 µg/ml) and incubate overnight at 37°C[c].

Notes

[a]The ET plasmid is temperature sensitive and will be lost if incubated at 37°C.
[b]To minimize the possibility of inappropriate intramolecular recombination, it is important for the recombination proteins encoded by the ET plasmid to be expressed transiently. This is achieved

through the arabinose-inducible promoter and incubation at 37°C to reduce the copy number of the plasmid.

ᶜIt may take more than an overnight incubation for colonies to appear. Allow the plates to incubate at 37°C for an additional 6–8 h and check for colonies.

Protocol 7

Screening of incision clones

Equipment and Reagents
- Thermal cycler
- Multichannel pipette
- Two 96-well PCR plates
- 200 μl Pipette tips
- Sterile Milli-Q water
- 10 mM dNTPs (Invitrogen)
- *Taq* DNA polymerase (5 units/μl), accompanying PCR buffer, and 25 mM MgCl$_2$ (Promega)
- Oligonucleotide primers (a: own design; b: 5′-CTGGTTCTTTCCGCCTCAGG-3′ from section 2.2.1).
- Equipment and reagents for agarose gel electrophoresis

Method
1. Add 40 μl of sterile water to the wells of a 96-well PCR plate (plate a).

2. Using sterile 200 μl pipette tips, pick 48–96 colonies from the agar plate and place into separate wells of the previously prepared PCR plate a (step 1), leaving the tip in the well.

3. Prepare a PCR master mix with the following reagents:
   - 5 μl of 10× *Taq* DNA polymerase buffer
   - 3 μl of 25 mM MgCl$_2$
   - 1 μl of 10 mM dNTPs
   - 1 μl of each oligonucleotide primer (10 μM a and b)
   - 0.4 μl of *Taq* DNA polymerase
   - 36.6 μl of H$_2$O

4. Aliquot 48 μl of the master mix into the wells of the second 96-well PCR plate (plate b).

5. Using a multichannel pipette and the tips used to pick colonies, resuspend the cells in plate a and add 2 μl to each corresponding well of the 96-well PCR plate b.

6. Put the plate into a thermal cycler programmed as follows:
   - 1 cycle of 94°C for 3 min
   - 35 cycles of 94°C for 30 s, 55°C for 30 s, and 72°C for 1 min
   - 1 cycle of 72°C for 5 min

7. Analyze the PCR by running 25 μl of the amplification products on a 0.8% (w/v) agarose gel.

8. Clones amplifying a product of the expected size should be confirmed by PCR analysis using the other set of primers (c: 5′-CCTGCCATAGCCTCAGGTTACTC-3′; d: own design (see section 2.2.1 and *Fig. 3d*)). A clone amplifying products of the expected size from both sets of primers should be selected for excision into pBR322.

Protocol 8

Preparation of electrocompetent cells harboring the incised BAC clone and transformation with the ET plasmid

Equipment and Reagents

- Centrifuge able to accommodate 50 ml tubes
- Spectrophotometer
- Electroporator and electroporation cuvettes (Biorad Gene Pulser)
- 30°C Shaking incubator
- 30°C Stationary incubator
- 37°C Shaking incubator
- BAC clone containing the locus of interest with sufficient flanking sequence to allow for 5′ and 3′ homology arms, approximately 10 kb
- ET plasmid, tetracycline resistant (pSC101-BAD-gbaA^{tet}; Gene Bridges)
- LB broth (10 g/l tryptone, 5 g/l yeast extract, 10 g/l NaCl)
- 10% (v/v) Glycerol (chilled on ice; Sigma)
- Chloramphenicol (30 mg/ml stock; Sigma)
- Tetracycline (12 mg/ml stock; Sigma)
- Kanamycin (30 mg/ml stock, Sigma)
- LB agar plates with kanamycin (15 µg/ml), chloramphenicol (15 µg/ml), tetracyline (3 µg/ml)

Method

1. Set up a 5 ml culture of the incised BAC clone (from *Protocol 7*) in LB broth with 2.5 µl of chloramphenicol stock (final concentration 15 µg/ml) and 2.5 µl of kanamycin (final concentration 15 µg/ml). Incubate overnight at 37°C with agitation in a shaking incubator.

2. Add 400 µl of the overnight culture to 40 ml of LB broth with 20 µl of chloramphenicol stock (final concentration 15 µg/ml) and 20 µl of kanamycin (final concentration 15 µg/ml). Mix and divide equally among four 50 ml tubes.

3. Shake at 37°C until the culture reaches an OD$_{600}$ of ~0.4. This usually takes 2.5–3.5 h.

4. Combine the cultures from the four tubes into one and centrifuge at 5000 **g** for 5 min (4°C).

5. Discard the supernatant and resuspend the cells in an equal volume (40 ml) of ice-cold 10% glycerol. Centrifuge at 2°C at 5000 **g** for 5 min.

6. Discard the supernatant and repeat the glycerol wash.

7. Resuspend the cells in a final volume of ~200 µl of ice-cold 10% glycerol and aliquot 50 µl into separate 1.5 ml microfuge tubes[a].

8. Add 50 ng of the ET plasmid to 50 µl of competent cells (containing the incised BAC clone) and electroporate (0.1 cm cuvette, 1.8 kV, 200 Ω, 25 µF).

9. Resuspend the bacteria in 1 ml of LB broth and shake at 30°C[b] for 75–90 min. Spread appropriate dilutions (100 µl of the 1 ml culture, then pellet the cells in the remaining 900 µl, resuspend in 110 µl of LB broth and plate 100 µl of neat cells and 100 µl of a 10^{-1} dilution) of the culture onto LB agar plates with chloramphenicol (15 µg/ml), kanamycin (15 µg/ml), and tetracycline (3 µg/ml), and incubate overnight at 30°C.

Notes

[a]Store unused cells at −80°C.
[b]The ET plasmid is temperature sensitive and will be lost if incubated at 37°C.

Protocol 9

Excision of incised BAC DNA into pBR322

Equipment and reagents

- Centrifuge able to accommodate 50 ml tubes
- Electroporator and electroporation cuvettes (Biorad)
- 30°C Shaking incubator
- 37°C Shaking incubator
- 37°C Incubator
- Electrocompetent incised BAC clone transformed with the ET plasmid
- Excision PCR product
- 10% (w/v) L-Arabinose in water (filter-sterilized)
- LB broth
- Tetracyline (12 mg/ml stock; Sigma)
- Ampicillin (50 mg/ml stock; Sigma)
- Chloramphenicol (30 mg/ml stock; Sigma)
- Kanamycin (30 mg/ml stock; Sigma)
- LB agar plates with ampicillin (50 µg/ml) and kanamycin (15 µg/ml)

Method

1. Set up a 5 ml overnight culture of the incised BAC clone (from *Protocol 8*) in LB broth with 2.5 µl of chloramphenicol (final concentration 15 µg/ml), 2.5 µl of kanamycin (final concentration 15 µg/ml), and 1.2 µl of tetracycline (final concentration 3 µg/ml). Shake at 30°C[a].

2. Add 400 µl of the overnight culture to 40 ml of LB broth with 20 µl of chloramphenicol (15 µg/ml), 20 µl of kanamycin (15 µg/ml), and 9.6 µl of tetracycline (3 µg/ml). Mix and divide equally among four 50 ml tubes.

3. Shake at 30°C until the culture reaches an OD_{600} of ~0.2 (2.5–3.5 h).

4. Induce expression of the ET recombinase by adding filter-sterilized L-arabinose to a final concentration of 0.1–0.2% and shake at 37°C[b] until the OD_{600} reaches 0.4 (30–45 min). Combine the cultures from the four tubes into a single tube and centrifuge at 2°C for 5 min at 5000 *g*.

5. Combine the cultures from the four tubes into a single tube and centrifuge at 5000 *g* for 5 min (4°C).

6. Discard the supernatant and resuspend the cells in an equal volume (40 ml) of ice-cold 10% glycerol. Centrifuge at 2°C at 5000 *g* for 5 min.

7. Discard the supernatant and repeat the glycerol wash.

8. Resuspend the cells in a final volume of ~200 µl of ice-cold 10% glycerol and aliquot 50 µl into separate 1.5 ml microfuge tubes[a].

9. Add 200–300 ng of the excision PCR product to 50 µl of competent cells and electroporate (0.1 cm cuvette, 1.8 kV, 200 Ω, 25 µF).

10. Resuspend the bacteria in 1 ml of LB broth and shake at 37°C for 75–90 min. Pellet the cells and resuspend in 100 µl of LB broth. Plate appropriate dilutions (try undiluted, 10^{-1}, and 10^{-2}) of the culture onto LB agar plates with kanamycin (15 µg/ml) and tetracycline (3 µg/ml) and incubate overnight at 37°C[c].

Notes
[a]The ET plasmid is temperature sensitive and will be lost if incubated at 37°C.
[b]To minimize the possibility of inappropriate intramolecular recombination, it is important for the recombination proteins encoded by the ET plasmid to be expressed transiently. This is achieved through the arabinose-inducible promoter and incubation at 37°C to reduce the copy number of the plasmid.
[c]It may take more than an overnight incubation for colonies to appear. Allow the plates to incubate at 37°C for an additional 6–8 h and check for colonies.

Protocol 10

Screening of excision clones

Equipment and Reagents
- Thermocycler
- Multichannel pipette
- Two 96-well PCR plates
- 200 µl Pipette tips
- Sterile Milli-Q water
- 10 mM dNTPs (Invitrogen)
- *Taq* DNA polymerase (5 units/µl), accompanying PCR buffer, and 25 mM MgCl$_2$ (Promega)
- Oligonucleotide primers (e: own design; f: 5′-CATCGATAAGCTTTAATGCG-3′ from section 2.2.1)
- Equipment and reagents for agarose gel electrophoresis

Method
1. Add 40 µl of sterile water to the wells of a 96-well PCR plate (plate a).

2. Using sterile 200 µl pipette tips, pick 48–96 colonies from the agar plate and place into separate wells of the previously prepared PCR plate a (step 1), leaving the tip in the well.

3. Prepare a PCR master mix with the following reagents:
   - 5 µl of 10× *Taq* DNA polymerase buffer
   - 3 µl of 25 mM MgCl$_2$
   - 1 µl of 10 mM dNTPs
   - 1 µl of each oligonucleotide primer (10 µM e and f)
   - 0.4 µl of *Taq* DNA polymerase
   - 36.6 µl of H$_2$O

4. Aliquot 48 µl of the master mix into the wells of the second 96-well PCR plate (plate b).

5. Using a multichannel pipette and the tips used to pick colonies, resuspend the cells in plate a and add 2 µl to each corresponding well of the 96-well PCR plate b.

6. Put the plate into a thermocycler programmed with the following conditions:
   - 1 cycle of 94°C for 3 min
   - 35 cycles of 94°C for 30 s, 55°C for 30 s, and 72°C for 1 min
   - 1 cycle of 72°C for 5 min

7. Analyze the PCR by running 25 µl of the amplification products on a 0.8% (w/v) agarose gel.

8. Clones amplifying a product of the expected size should be confirmed by PCR analysis using the other set of primers (g: 5′-GTCCTCAACGACAGGAGCAC-3′; h: own design; see section 2.2.1 and *Fig. 3f*). Clones amplifying a product of the expected size with both sets of primers should be incised correctly with the neo/kan cassette and excised into pBR322. This can be verified by sequence analysis using primers b, c, f, and g on miniprep DNA from the positive clone. A clone that has been incised and excised correctly forms the targeting construct and will be electroporated into embryonic stem cells.

Protocol 11

Screening of ES cell clones

Equipment and Reagents
- Thermocycler
- Multichannel pipette
- 10 mM dNTPs (Invitrogen)
- Platinum *Taq* DNA Polymerase High Fidelity (5 units/μl), accompanying High Fidelity buffer, and 50 mM MgSO$_4$ (Invitrogen)
- Sterile Milli-Q water
- Oligonucleotide primers (10 μM)
- ES cell DNA extracted from neomycin-resistant clones (see *Protocol 1*)
- Equipment and reagents for agarose gel electrophoresis

Method
1. Prepare a PCR master mix with the following reagents (per 50 μl reaction):
   - 5 μl of 10× High Fidelity buffer
   - 2 μl of 50 mM MgSO$_4$
   - 1 μl of 10 mM dNTPs
   - 1 μl of each oligonucleotide primer
   - 0.2 μl of Platinum *Taq* DNA Polymerase High Fidelity (5 units/μl)
   - 37.8 μl of H$_2$O

2. Aliquot 48 μl of the master mix into the wells of a 96-well PCR plate.

3. Using a multichannel pipette, add 2 μl of embryonic stem cell DNA to each well of the 96-well PCR plate.

4. Put the plate into a thermocycler programmed as follows:
   - 1 cycle of 94°C for 3 min
   - 35 cycles of 94°C for 30 s, 55°C for 30 s, and 68°C for 4 min
   - 1 cycle of 68°C for 10 min

5. Analyze the PCR by running 25 μl of the amplification product on a 0.8% (w/v) agarose gel.

2.5 Screening of ES cell clones

Following electroporation of the targeting construct into ES cells, those clones that survive G418 selection can be analyzed further. This is necessary to discriminate between antibiotic-resistant ES cell clones that have randomly integrated the construct and those that contain the correctly targeted allele generated by homologous recombination. A PCR assay can be used that will amplify a specific product from the DNA of correctly targeted cells. By specifically using oligonucleotides such as P1 (location in the resistance gene) and P2 (located external to the targeting construct) (see *Fig. 1*), a band is only expected in correctly targeted cells.

- Design a primer external to one of the homology arms (this may be the 5′ or 3′ arm, depending on which is shorter) to be used in a PCR with a primer in the neo/kan cassette (see *Fig. 1*, P2 and P1, respectively).
- When available, a correctly incised BAC clone (see *Protocol 7*) should be used as the template in a test PCR to determine the optimal conditions for the screening assay.
- A 2 µl aliquot of overnight culture is sufficient to act as template and if available it is useful to use the gradient function of the thermal cycler to determine the optimal annealing temperature.
- To establish a robust PCR when the BAC is not available, add <1 ng of the plasmid construct to ES cell DNA and use the P1 oligonucleotide with an internal test oligonucleotide (P2a) to establish optimal conditions. In this case, the P1 oligonucleotide binds to the neo/kan antibiotic resistance gene and the P2a oligonucleotides binds to the 3′ arm to generate a product. The presence of ES cell DNA mimics the conditions for screening from putatively targeted ES cells. Similarly, the P1a oligonucleotide is present in endogenous DNA and can be used with the P2 oligonucleotide to amplify product from ES cell DNA using the same conditions.

ES cell clones that are capable of producing a product of the expected size by PCR amplification are usually correctly targeted. To ensure that they are not false positives, it is recommended that targeting is confirmed by analysis of the 5′ and 3′ recombination sites using a Southern blot approach (see *Fig. 1*).

3. TROUBLESHOOTING

The biggest issue with all of the protocols described here is the occasional inability to produce a PCR product of the expected size or to identify correctly incised or excised ET cloning products. These can be due to a number of reasons:

- The wrong BAC, or incorrectly annotated BAC. Our approach of obtaining and using two BACs in parallel normally overcomes this. However, validation of the BAC clones prior to performing ET recombination will demonstrate whether this is an issue.

- The presence of repetitive sequences or other tertiary structures that prevent PCR/cloning. Analysis of the genomic region with REPEATMASKER (see section 2.1) will identify potentially problematic areas. The construct should be designed to avoid such areas.

- Incision or excision problems. Homologous recombination will only occur if the DNA molecules share regions of exact sequence identity. If you find that incision or excision is not occurring as expected, there may be a problem obtaining recombination with the area of sequence used in the oligonucleotide primers. Remember that the sequence information obtained from databases is not always correct and any errors in your primers will prevent recombination. This may be overcome by designing oligonucleotide primers to target a different region.

- Oligonucleotide problems. These protocols use long oligonucleotides, which may have production issues. Ensure that quality control is very high and resynthesize or redesign oligonucleotides to a different target sequence.

- If no colonies appear after transformation with the incision or excision products, check that the correct antibiotics have been used at the appropriate concentration. In some cases, it may be necessary to reduce the concentration of the antibiotic being used to select for the recombination event. This is usually only seen for the incision event, and lower concentrations of kanamycin will result in the growth of colonies.

- If no products are amplified from colonies resulting from transformation with the incision or excision PCR products (see *Protocols 7* and *10*), check that you are using the correct primer pairs for screening.

4. REFERENCES

★★ 1. van der Weyden L, Adams DJ & Bradley A (2002) *Physiol. Genomics*, **11**, 133–164. – *Review highlighting some of the approaches used in mouse genetics.*

2. Wilson TJ, Lazner F, Kola I & Hertzog PJ (2000) In: *Comparative Genomics*, pp. 97–121. Edited by M Clark. Kluwer Academic Publishers, Boston, USA.

3. Auwerx J, Avner P, Baldock R, *et al.* (2004) *Nat. Genet.* **36**, 925–927.

4. Austin CP, Battey JF, Bradley A, *et al.* (2004) *Nat. Genet.* **36**, 921–924.

5. Herron BJ, Lu W, Rao C, *et al.* (2002) *Nat. Genet.* **30**, 185–189.

6. Beier DR & Herron BJ (2004) *Genetica*, **122**, 65–69.

7. Chen Y, Yee D, Dains K, *et al.* (2000) *Nat. Genet.* **24**, 314–317.

★★★ 8. te Riele H, Maandag ER & Berns A (1992) *Proc. Natl. Acad. Sci. U. S. A.* **89**, 5128–5132. – *Paper highlighting the use of isogenic DNA to prepare the targeting vectors.*

★★★ 9. Yang Y & Seed B (2003) *Nat. Biotechnol.* **21**, 447–451. – *Paper describing a method for preparing targeted gene disruptions in ES cells using BACs.*

10. Muyrers JP, Zhang Y, Testa G & Stewart AF (1999) *Nucleic Acids Res.* **27**, 1555–1557.

11. Muyrers JP, Zhang Y, Buchholz F & Stewart AF (2000) *Genes Dev.* **14**, 1971–1982.

12. Turksen K (2006) *Embryonic Stem Cell Protocols*, 2nd edn, vols I and II. Humana Press, Totowa, New Jersey.

13. Tymms MJ & Kola I (2001) *Gene Knockout Protocols*. Humana Press, Totowa, New Jersey.

CHAPTER 12

Construction of long DNA molecules from multiple fragments using PCR

Nikolai A. Shevchuk and Anton V. Bryksin

1. INTRODUCTION

When a DNA molecule has to be assembled from three or more unrelated fragments, it can often be more convenient to use a PCR-based method called long multiple fusion (1) instead of the traditional restriction enzyme/ligation cloning. If an error rate of about 1 per 7000 bp is acceptable, long multiple fusion can be employed to assemble a linear recombinant DNA molecule of up to 10 kb from five fragments, or of up to 20 kb long from three fragments, precisely and quickly. This linear recombinant DNA molecule can then be cloned into a plasmid vector, if necessary.

The method described in this chapter could prove useful for such applications as the development of multi-domain vaccines and the construction of plasmid vectors and gene-targeting vectors, as well as for the assembly of viral genomes for basic and vaccine research.

The technique offers more flexibility and is less time- and labor-consuming than traditional cloning, which can often lead to introduction of unwanted sequence at fusion points. The long multiple fusion method has proved itself over a wide range of DNA sources, sizes of fused fragments, and final application of the obtained recombinant molecules (2–8).

2. METHODS AND APPROACHES

2.1 Principles of long multiple fusion

The method consists of three major steps (see *Fig. 1*).

- Step 1. During the first step, the fragments are amplified from the original source using chimeric primers carrying overlapping sequence at their 5′ ends.

PCR: *Methods Express* (S. Hughes and A. Moody, eds.)
© Scion Publishing Limited, 2007

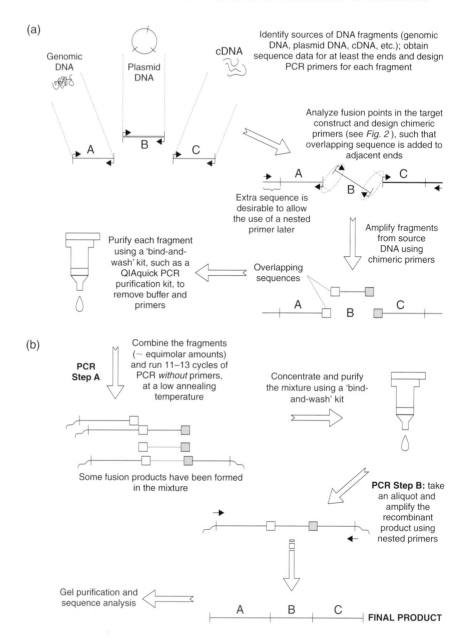

Figure 1. Outline of long triple fusion.
(*a*) Fragments are amplified form the original source using chimeric primers carrying overlapping sequence at their 5′ ends. (*b*) Fragments are allowed to anneal and form intermediary products (PCR Step A) and, finally, the recombinant end product is amplified (PCR Step B). The correctness of the assembly is then verified using agarose gel electrophoresis and/or sequencing of critical regions.

- Step 2. Fragments are allowed to anneal and form intermediary products (PCR Step A in *Fig. 1*).
- Step 3. The recombinant end product is amplified (PCR Step B in *Fig. 1*)

The accuracy of the assembly can then be verified using agarose gel electrophoresis and/or sequencing of critical regions. The method has several pitfalls that need to be avoided in order to assemble the recombinant product successfully. All of the key points that are often overlooked by researchers who follow the method described in our original paper (1) are discussed in the section below and in section 3.

The long multiple fusion technique was born from the synthesis of long PCR with overlap-extension PCR (9, 10). Existing protocols of overlap-extension PCR are limited to regular (short) PCR, i.e. with a limit of about 3–4 kb (11, 12), and have been limited to the fusion of only two DNA fragments (13). Recently, others have described methods for the assembly of up to ten short fragments using modifications of overlap-extension PCR (14), but the length of the end product was limited to about 5.5 kb. Our long multiple fusion method allows the creation of recombinant products as long as 20 kb from three fragments (1), and has been used successfully to assemble 10 kb products from four (1) or five (A.V. Bryksin, unpublished data) fragments. Long multiple fusion can facilitate the construction of highly complex recombinant DNA molecules for various applications (2–8, 14). For example, potentially it can allow a vaccine researcher to create multiple custom-made viral genomes within a short time frame or develop complex multi-domain vaccines and other recombinant proteins. It can also be used for the assembly of sophisticated gene-targeting constructs (7, 12).

2.2 Limitations of long multiple fusion

PCR, even when performed using a proofreading enzyme, will lead to a small number of base changes (PCR errors). In a typical long multiple fusion set-up, we found the error rate to be below 1 base change per 6.6 kb (1). To ensure fidelity of the assembly and the absence of PCR errors in critical regions of a recombinant construct, sequencing of critical regions prior to assembly is advisable. Resequencing of critical regions in the final product is recommended. The expected error occurrence in the final product can be calculated from the total number of PCR cycles used throughout the whole procedure and the error rates of the PCR kits used. If the error rate (expressed in errors per megabase (Mb) per cycle) is multiplied by the total number of PCR cycles, this will yield the expected number of errors in the final product per Mb, e.g. 150 errors/Mb. If this value is then divided by 1000, this will give the error rate per kilobase (kb), i.e. 0.15 errors/kb. Finally, if this value is then multiplied by the length of the final product, it will give the number of expected errors in the final fusion product. A typical error rate of a long triple fusion procedure is approximately one error per 6.6 kb (1). If this error rate is unacceptable, then traditional restriction enzyme/ligation cloning should be used. The planning of a complex cloning project using the traditional approach can be simplified with the software application VECTOR NTI ADVANCE (Invitrogen), available free of charge to academic users at the time of writing.

It should also be noted that a PCR product, even when gel purified, can contain a very small amount of nonspecific PCR products. To obtain an extremely pure product (which may not be necessary for many applications), the PCR fusion product may have to be cloned into a plasmid vector and verified by partial sequencing. For this purpose, TOPO vectors (Invitrogen) or T-vectors can be used.

2.3 Factors critical for successful long multiple fusion

- The component fragments shown at the top of *Fig. 1* have to be amplified with a polymerase that has a 3′→5′ proofreading activity, for example *Pfu* polymerase, rather than *Taq* polymerase. A combination of *Taq* with *Pfu*, as found in long-PCR kits, is also suitable. The problem with *Taq* polymerase is that it leaves single A nucleotide overhangs at the 3′ ends, which will disrupt priming of the overlapping regions during the overlap extension reaction (15), shown in *Fig. 1* (PCR Step A). An enzyme with proofreading activity will generate PCR products that have predominantly blunt ends (without extra A nucleotides added at the end) allowing successful overlap extension among DNA fragments (see *Protocol 2*). The following is a list of PCR kits that we have used successfully for amplification of component fragments:
  - ○ TripleMaster polymerase (Eppendorf)
  - ○ Advantage-HF 2 PCR kit (Clontech)
  - ○ Herculase HotStart DNA polymerase (Stratagene)
  - ○ EXL polymerase PCR kit (Stratagene)
  - ○ Long Template Expand polymerase (Roche)

  Avoid using the DeepVent polymerase kit, even though it is mentioned in the original publication (1); this kit requires extensive optimization and can cause many problems.

- Exposure of the PCR products to UV light should be avoided and all purification steps should be limited to 'bind-and-wash' or desalting methods, such as a QIAquick PCR purification kit (Qiagen). We found that UV light and handling of agarose slices containing DNA bands will make the DNA template unusable for long multiple fusion, most likely due to damage and nicking of the DNA. An exception can be made for gel purification when *both* of the following conditions are true:
  1. The length of the final product is below 10 kb.
  2. Crystal violet (Sigma) is used in the agarose gel instead of ethidium bromide.

  Crystal violet allows visualization of DNA bands without UV light, thus minimizing damage to the DNA template. A 10 kb final product was successfully assembled from five fragments that were gel purified using agarose gels containing crystal violet (A.V. Bryksin, unpublished data).

- No primers should be used in PCR Step A in *Fig. 1*. This will allow the overlaps to anneal and extend forming various fusion products, including the target product. Although omission of this step can sometimes successfully produce a fusion product (12, 14), in our experience, inclusion of this step produces reliable results. The 'bind-and-wash' purification steps before and after PCR Step A partially remove residual primers from the previous amplification reactions and thus are also critical for success. Gel purification with crystal violet visualization (instead of UV and ethidium bromide) can be used for purification of fragments before PCR Step A, but *only if* the length of the final recombinant product is less than or equal to 10 kb. We have not tested whether this works for 20 kb fusions.
- The length of overlapping sequence (see *Fig. 2*) for each fusion point should be at least 20–30 nt when the final product is under 3 kb, 35–40 nt when the final product is 3–10 kb, and 50–70 nt when the final recombinant product is greater than 10 kb. *Fig. 2* shows the design of two chimeric primers for one fusion point. The length of overlapping sequence equals the sum of the extra sequences included in the two chimeric primers shown in *Fig. 2*.
- It is possible and often desirable to design only one chimeric primer per fusion point and to use a regular primer on the other side. In our experience, chimeric primers do not work well when the fragment to be amplified is 7 kb or longer. Also, chimeric primers may not work when the fragment has to be amplified from genomic DNA. In this case, it is preferable to use regular (nonchimeric) primers to amplify this problem fragment and to use chimeric primers for amplification of the adjacent fragments. In this case, in *Fig. 2*, a chimeric primer would be used for amplification of fragment B and a regular primer would be used for amplification of fragment A. If only one chimeric primer will be used for a fusion point, it should contain twice the length of the extra sequence (overlap) that would normally be required for a fusion point assembled using two chimeric primers.
- It is strongly recommended that nested primers (20–50 nt away from the terminus) are used during amplification of the final product (PCR Step B in *Fig. 1*). This will improve the purity of the final product and also decrease the chance of failure of the project. Proofreading polymerases are believed to degrade PCR products at the ends due to their exonuclease activity. This can prevent binding of the original terminal primers (top of *Fig. 1*) to the final product (bottom of *Fig. 1*) because the homologous sequence at the ends has been degraded. This is why it is desirable to use nested primers in the final amplification step.
- Long PCR in general is much more sensitive to the quality of the template (16) than regular short PCR. Repeated freezing and thawing and/or storage in distilled water, rather than in a buffer with a pH above 7, will also render a template unusable for long PCR (16). It is recommended that DNA templates for long PCR are stored at 4°C in 5 mM Tris/HCl (pH 8 or 8.5). This rule is less critical for projects in which all fragments are shorter than 5 kb.

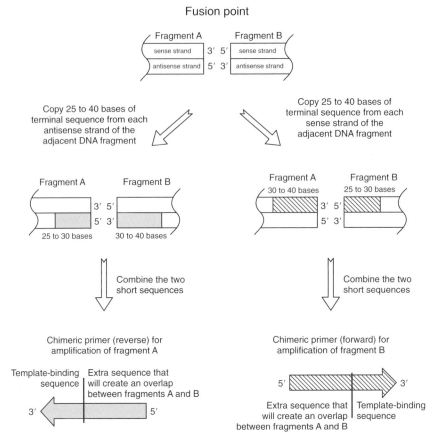

Figure 2. Designing chimeric primers.

Only adjacent ends of two DNA fragments are shown in the figure. One needs to know the terminal DNA sequence of the adjacent fragments in a prospective fusion point in order to design chimeric primers. Typically, two chimeric primers are designed for each fusion point. One chimeric primer will serve as a reverse primer for amplification of fragment A and this chimeric primer is created by combining short terminal sequences from *complementary* DNA strands of fragments A and B, as shown in the flowchart directed to the left. The forward primer for amplification of fragment A is not shown in the figure. The second chimeric primer will serve as a forward primer for amplification of fragment B and this chimeric primer is created by combining short terminal sequences from *direct* DNA strands of fragments A and B as shown in the flowchart directed to the right. The reverse primer for amplification of fragment B is not shown in the figure. A DNA sequence editor and/or primer design software would be helpful for this task. The length of overlapping sequence for each fusion point should be at least 40 nt when the final product is under 10 kb, and 50–70 nt when the final recombinant product is longer than 10 kb. The total length of overlapping sequence equals the sum of the lengths of extra sequences included in the two chimeric primers. For example, if the reverse chimeric primer has 40 nt of extra sequence in it, whilst the direct chimeric primer has 30 nt of extra sequence, then, after fragments A and B are amplified using those chimeric primers (and two other primers not shown in the figure), fragments A and B will have an overlap of 30 + 40 = 70 bp.

2.4 Recommended protocols

2.4.1 Long triple fusion

The protocols described below can be used for the creation of linear DNA molecules of 3–20 kb from three fragments.

Primer design

To design chimeric primers, it essential to know the sequence of each DNA fragment, or at least 40–50 bp of sequence at each end of every fragment if the complete sequence is unavailable.

For amplification of fragment A, identify the desired fusion point between the fragments of interest (fragments A and B in *Fig. 2*) and select 25–30 nt from the antisense strand of fragment A (as if you are designing a normal reverse primer for amplification of this DNA fragment). The exact number of terminal nucleotides will depend on the melting temperature of the resulting oligonucleotide, which should be between 62 and 68°C. Remember that the 5′ end of your oligonucleotide is fixed at the end of fragment A (fusion point) and you can only vary the 3′ terminus of the oligonucleotide.

Now add 30–40 nt of sequence from the terminus of fragment B (also from the complementary strand, as shown in the downward flow chart in *Fig. 2*) to the 5′ end of this reverse primer. The combined oligonucleotide is a reverse chimeric primer for amplification of fragment A. Design a forward primer for fragment A (not shown in *Fig. 2*). This should be a regular primer if fragment A is the first fragment in the construct. The extra sequence included in the reverse chimeric primer will ensure that after you have amplified fragment A from its source DNA, the amplicon will contain additional sequence that overlaps fragment B.

Use a similar approach to design a forward chimeric primer for amplification of fragments B and C. Make sure that the melting temperature of the template-binding moiety of each chimeric primer (see *Fig. 2*) is around 62–68°C. During amplification of the component fragments (see *Protocol 1* and top of *Fig. 1*), a high annealing temperature should be used (62–68°C). This will ensure a high specificity of PCR and a low level of side-products, as a high annealing temperature minimizes mispriming and the formation of secondary structures (15).

For each fusion point, make sure that the total length of extra 5′ sequence included in the chimeric primer(s) is at least:

- 20–30 nt when the final product is under 3 kb
- 35–40 nt when the final recombinant product is less than or equal to 10 kb
- 50–70 nt when the final recombinant product is between 10 and 20 kb long

When designing primers for generating amplification fragments, it is important to include an additional 20–50 nt at the extreme 5′ end of fragment A and at the 3′ end of fragment C (or the final fragment) in order to allow for nested primers at the final amplification step (see top and bottom of *Fig. 1*).

The steps outlined above should provide two primers for each DNA fragment and at least one chimeric primer for each fusion point. When ordering primers, request polyacrylamide gel electrophoresis (PAGE) purification for chimeric primers. If chimeric primers (long oligonucleotides) are not PAGE purified, they will contain erroneous oligonucleotides (resulting from the shortcomings of automated oligonucleotide synthesis), which can introduce base changes and deletions in the final product.

Protocol 1

Amplification of fragments from source DNA

Equipment and Reagents
- One of the following PCR kits:
  - ☐ TripleMaster polymerase (Eppendorf)
  - ☐ Long Template Expand polymerase (Roche)
  - ☐ Herculase HotStart polymerase (Stratagene)
  - ☐ EXL polymerase (Stratagene)
  - ☐ Advantage-HF 2 PCR kit (Clontech)[a]
- Nuclease-free water (Promega)
- DMSO (Sigma)
- Thermal cycler
- 10 mM Ultrapure dNTP mix (Sigma)
- GM3 synthase gene BAC clone DNA (50 ng/μl)
- GM3 synthase forward primer: 5′-TCTGAGAGTAACTGCCCTCTTGACATC-3′ (50 μM)
- GM3 synthase reverse primer: 5′-CATCTTGCTTTGAGCTCGGGTG-3′ (50 μM)
- p3XFLAG-CMV-9 vector (1 μg/μl; Sigma)
- p3XFLAG-CMV-9 vector forward primer (chimeric): 5′-GTGATTGCTCGAGGCCTTCCCTGCAATGGTACACCCGAGCTCAAAGCAAGATGATTGAACAAGATGGATTGCACGCAGGTTC-3′ (50 μM)
- p3XFLAG-CMV-9 vector reverse primer (chimeric): 5′-ATGCATTTTTTTCATGTCACATTCTTCAGTAGTATAATTTAACTTGAGGATATAAAGGATCCACACTCCAGGGAATTGATCCAGACATGATAAGATACA-3′ (50 μM)
- Human genomic DNA (0.1 μg/μl)
- Example forward primer: 5′-TGGAGTGTGGATCCTTTATATCC-3′ (50 μM)
- Example reverse primer: 5′-AGACCTTCTTCTGCCCATATACATC-3′ (50 μM)

Method
1. Amplify each of your three DNA fragments using a typical protocol recommended by the manufacturer of your PCR kit(s)[b,c].

2. For the generation of a 20 kb fusion, use the following protocols.

3. For fragment A (11.2 kb), combine:
   - 33 μl of nuclease-free water
   - 5 μl of reaction buffer (Herculase HotStart DNA polymerase kit)
   - 2.5 μl of dNTPs (10 mM)
   - 5 μl of human GM3 synthase gene BAC clone DNA
   - 1 μl of GM3 synthase forward primer
   - 1 μl of GM3 synthase reverse primer
   - 1 μl of Herculase HotStart polymerase
   - 1.5 μl of DMSO

4. Place the tube in a thermal cycler and run the following program:
   - Initial denaturation step at 92°C for 1 min
   - 27 cycles of denaturation at 92°C for 10 s, annealing at 65°C for 30 s, and extension at 68°C for 11 min 30 s (plus automatic extension of the extension time by 5 s per cycle)
   - Final additional extension step at 68°C for 13 min
   - Hold at 4°C

5. For fragment B (1.7 kb), combine:
   - 36.7 μl of nuclease-free water
   - 5 μl of HF buffer (Advantage-HF 2 PCR kit)
   - 5 μl of dNTPs (10 mM)
   - 0.3 μl of p3XFLAG-CMV-9 vector
   - 1 μl of p3XFLAG-CMV-9 vector forward primer
   - 1 μl of p3XFLAG-CMV-9 vector reverse primer
   - 1 μl of HF polymerase

6. Place the tube in a thermal cycler and run the following program:
   - Initial denaturation step at 94°C for 30 s
   - 26 cycles of denaturation at 94°C for 15 s, annealing at 65°C for 40 s, and extension at 68°C for 1 min 50 s
   - Final additional extension step at 68°C for 3 min
   - Hold at 4°C

7. For fragment C (7.5 kb), combine:
   - 35 μl of nuclease-free water
   - 5 μl of reaction buffer (Herculase HotStart DNA polymerase kit)
   - 2.5 μl of dNTPs (10 mM)
   - 3 μl of human genomic DNA
   - 1 μl of example forward primer
   - 1 μl of example reverse primer
   - 1 μl of Herculase HotStart polymerase
   - 1.5 μl of DMSO

8. Place the tube in a thermal cycler and run the following program:
   - Initial denaturation step at 92°C for 1 min
   - 26 cycles of denaturation at 92°C for 10 s, annealing at 60°C for 30 s, and extension at 68°C for 7 min 30 s (plus automatic extension of extension time by 4 s per cycle)
   - Final additional extension step at 68°C for 9 min
   - Hold at 4°C

Notes

[a]This kit will only amplify fragments shorter than 4 kb, but it offers the best PCR error rate.

[b]It is OK to use different kits for different fragments as long as they are from the list of recommended PCR kits above.

[c]Be sure to use a high annealing temperature (62–68°C) in your PCRs, especially the ones that include chimeric primers, in order to minimize the level of nonspecific PCR products.

Protocol 2 serves three purposes:

1. To remove primers and buffer from a PCR product.
2. To analyze the length of the PCR product.
3. To quantify the PCR product.

We usually use crystal violet instead of ethidium bromide for staining of agarose gels as it is safer and does not require UV exposure, but this protocol will also work if ethidium bromide is used throughout instead of crystal violet. If one of your fragments is shorter than 0.5 kb, which is rare for a long fusion project, then please see note e (*Protocol 2*).

Protocol 2

Analysis and purification of each fragment

Equipment and Reagents
- Crystal violet (2 mg/ml; Sigma)
- 1% (or 0.5% for long PCR products) agarose gel, containing 30 µl of the crystal violet stock solution per 150 ml of agarose.
- 5× Loading buffer (10% Ficoll 400, 0.1 M EDTA, pH 8.0, 120 µg/ml crystal violet)
- White light box and camera
- Apparatus required for electrophoresis
- 1× TAE running buffer (4.84 g/l Tris base, 1.142 ml/l glacial acetic acid, 0.372 g/l EDTA in distilled water)
- QIAquick PCR purification kit (Qiagen)
- QIAquick gel extraction kit (Qiagen) (optional)
- DNA molecular weight markers, such as a 1 kb DNA step ladder (Promega)
- Spectrophotometer

Method
1. Purify each PCR product using a 'bind-and-wash' DNA purification kit, such as a QIAquick PCR purification kit[a], following the manufacturer's instructions.

2. Elute the purified PCR product with 30 µl of 5 mM Tris/HCl (pH 8.5) (this is a twofold dilution of the Qiagen elution buffer supplied with the kit)[b].

3. Load and run 5 µl of each PCR product (see *Protocol 1*) on your agarose gel. Also include a lane with 0.5 µg of DNA marker in 10 µl of the same loading buffer as the PCR samples[c].

4. Use a white light table (instead of UV light)[d] to view the results of the electrophoresis[e].

5. Assess whether your PCR products are the correct length.

6. Quantify your PCR products either visually (by comparing the intensity of PCR bands with the intensity of DNA marker bands) or by using spectrophotometry to calculate approximate concentrations of the PCR products.

Notes
[a]This purification is necessary in order to remove (at least partially) the primers and buffer from the previous PCR.

[b]This is optional: if the length of your final recombinant product is less than or equal to 10 kb, you may use gel purification using a crystal violet-containing agarose gel and a gel extraction kit such as a QIAquick gel extraction kit (Qiagen). In this case you can skip the 'bind-and-wash' purification step in step 1, but may have to repeat electrophoresis in order to quantify your DNA samples. The gel purification procedure will completely remove previous primers and primer dimers from your PCR products.

[c]This is necessary for correct visual quantification of DNA amounts in PCR samples.

[d]You may use the regular ethidium bromide-stained agarose gels and UV light throughout this protocol for analysis of PCR results. However, if you decide to use gel purification of your PCR products (see note b), ethidium bromide and UV light should be avoided as they will damage the DNA and may lead to failure of the multiple fusion procedure.

[e]You may have a problem viewing fragments smaller than 0.9 kb using crystal violet. In this case, increase the amount loaded on the gel two- to threefold and use a dark room to view the results. It is almost impossible to see DNA shorter than 0.5 kb using crystal violet. If this is the case, use 0.75 µl of ethidium bromide (10 mg/ml stock) per 150 ml of agarose throughout this protocol instead of crystal violet.

Protocol 3

Overlap extension reaction (see *Fig. 1*, PCR Step A)

Equipment and Reagents
- One of the PCR kits mentioned in *Protocol 1*
- Nuclease-free water (Promega)
- DMSO
- Thermal cycler
- 10 mM Ultrapure dNTP mix (Sigma)

Method
1. Prepare a 50 µl PCR[a]. Combine equimolar[b] amounts of all three purified fragments (from *Protocols 1* and *2*) in a final volume of 10 µl, but *do not* add primers. An example for the generation of a 20 kb reaction is shown below. Combine:
   - 29 µl of water
   - 5 µl of reaction buffer (Herculase HotStart DNA polymerase kit)
   - 2.5 µl of dNTPs (10 mM)
   - 3 µl of 11 kb fragment A (total ~0.45 µg) in 5 mM Tris/HCl (pH 8.5)
   - 1.5 µl of 1.7 kb fragment B (total ~0.07 µg) in 5 mM Tris/HCl (pH 8.5)
   - 5.5 µl of 7.5 kb fragment C (total ~0.31 µg) in 5 mM Tris/HCl (pH 8.5)
   - 1 µl of Herculase HotStart polymerase
   - 2.5 µl of DMSO

2. Run the reaction for 11–15 cycles using an annealing temperature of 60°C[c] and the following program:
   - Initial denaturation step at 92°C for 1 min
   - 13 cycles of denaturation at 92°C for 10 s, annealing at 60°C for 1 min, and extension at 68°C for 21 min[d]
   - Final additional extension step at 68°C for 21 min
   - Hold at 4°C

Notes

[a]If a larger final volume is required, i.e. 200 μl, set up four separate 50 μl reactions and combine following cycling.

[b]Equimolar concentrations of fragments ensure that the overlapping sequence is as likely to anneal to an adjacent fragment as to its complementary strand in the same fragment. If one of the fragments has a higher molar concentration than the others, the overlapping sequence that it contains will be more likely to anneal with the complementary strand of the same fragment than with the other fragments. In the example, we have three purified DNA fragments of 11, 1.7, and 7.5 kb. The DNA concentration of fragments A, B, and C are 0.150, 0.047, and 0.056 μg/μl, respectively. The fragments should be in the following proportions by weight to achieve equimolar concentrations: 11 : 1.7 : 7.5 (molecular weight of DNA is roughly proportional to its length). From these weight proportions, we have to calculate the volumes of fragments given their known concentrations. We need to solve the equation:

$$V = V_A + V_B + V_C = (11 * M/C_A) + (1.7 * M/C_B) + (7.5 * M/C_C)$$

where V is the sum of volumes of all DNA fragments (in μl; 10 μl in our case); V_A, V_B, and V_C are the volume of fragments A, B, C in μl, respectively; C_A, C_B, and C_C are the concentrations of fragments A, B, and C in μg/μl, respectively; and M is the $1/P$ fraction of the total amount of DNA in the reaction (in μg).

The value of P is calculated by adding up the values in the weight proportions above (or just the lengths of the DNA fragments); therefore, $P = (11 + 1.7 + 7.5) = 20.2$. Hence, M is defined as approximately 1/20 of the total amount of DNA. It is not difficult to calculate M, which equals 0.041 μg. Now the formula for calculation of the final volume of each DNA fragment (corresponding to equimolar amounts of DNA fragments in the reaction) is:

$$V_x = L_x * M/C_x$$

where L_x is the length of a fragment (A, B, or C) in kilobases, M is the value we calculated above, and C_x is the concentration of the fragment (A, B, or C) in μg/μl.

For example, for fragment A:

$$V_A = L_A * M/C_A$$
$$V_A = 11 * 0.041/0.150$$
$$V_A = 3$$

Therefore, we have volumes of 3, 1.5, and 5.5 μl for fragments A, B, and C, respectively. The higher the total amount of DNA fragments that you use in this overlap-extension step, the better the outcome of the fusion; the recommended minimum total amount of DNA of all fragments in a 50 μl reaction is around 0.4 μg. Therefore, you should try to obtain the highest concentration of each fragment that is possible in *Protocol 2*. You could use one QIAquick column to process two 50 μl reactions of each fragment and elute the DNA with 30 μl of 5 mM Tris buffer.

[c]This should work for the annealing of 40 nt or longer overlaps. For troubleshooting, you can lower the annealing temperature to 55°C.

[d]Use an extension time that corresponds to the total length of the final recombinant product, i.e. 1 min per kb for most PCR kits. For example, if the length of your target product is 10 kb, use a 10 min extension time in your cycling program. Theoretically, overlap extensions cover only part of the total length of the final product, but we find that using an extra extension time leads to better results.

Protocol 4

Purification of intermediary product[a]

Equipment and Reagents
■ QIAquick PCR purification kit (Qiagen)

Method
1. Purify each PCR product using a 'bind-and-wash' DNA purification kit, such as a QIAquick PCR purification kit[a], following the manufacturer's instructions[b]. For the elution step, use 30 μl of 5 mM Tris/HCl (pH 8.5)[c].

Notes
[a]This purification step is optional if you used crystal violet gel purification in *Protocol 2* as gel purification removes all primers completely. You may still want to do this step to concentrate your sample.

[b]If a reaction volume larger than 50 μl was generated from *Protocol 3*, load the separate 50 μl reactions sequentially onto a single QIAquick column to achieve the highest concentration. It is not necessary to run agarose gel electrophoresis at this point, as you will not see your fusion product here due to its very low concentration.

[c]This is a twofold dilution of the Qiagen elution buffer supplied with the kit.

Protocol 5

Amplification of the final recombinant product[a]

Equipment and Reagents
- One of the PCR kits mentioned in *Protocol 1*
- Nuclease-free water
- Thermal cycler
- 10 mM Ultrapure dNTP mix (Sigma)
- Example forward primer: 5′-AAAGCAGGCAATTGAATGACAGTAATGATG-3′ (50 µM)
- Example reverse primer: 5′- GTGTAGCATTCAAGGCCTTTTGCTATCTGG-3′ (50 µM)

Method
1. Amplify the final product using a typical protocol recommended by the manufacturer of your PCR kit(s). For example, for a 50 µl reaction[b] (see *Fig. 1*, PCR Step B), combine[c]:
   - 28.2 µl of water
   - 5 µl of reaction buffer (Herculase HotStart DNA polymerase kit)
   - 2.5 µl of dNTPs (10 mM)
   - 10 µl of purified DNA from PCR Step A (*Protocol 4*)
   - 0.4 µl of example forward primer
   - 0.4 µl of example reverse primer
   - 1 µl of Herculase HotStart polymerase
   - 2.5 µl of DMSO

2. Place the tube in a thermal cycler and run the following program:
   - Initial denaturation step at 92°C for 1 min
   - 31 cycles of denaturation at 92°C for 10 s and combined annealing and extension at 68°C[d] for 20 min 40 s (plus automatic extension of extension time by 10 s per cycle)
   - Final additional extension step at 68°C for 30 min
   - Hold at 4°C

Notes
[a]This step is shown as PCR Step B in *Fig. 1*. The use of nested primers here (if you planned for them in *Protocol 1*) will improve purity and the chances of success of your fusion.
[b]A 50 µl reaction volume per tube offers fast enough exchange of heat between the thermal cycler and the reaction mixture and at the same time is convenient for mixing all of the components in the right amounts.
[c]An aliquot of your template (from *Protocol 5*) can comprise up to 1/5 of the volume of the PCR.
[d]Make sure that you use a high annealing temperature (65–68°C) to achieve the lowest level of nonspecific PCR products in your reaction. This assumes that you have designed primers that have a high melting temperature as described in section 2.4.1.

Protocol 6

Analysis and purification of final product

Equipment and Reagents
- Crystal violet (2 mg/ml; Sigma)
- 1% (or 0.5% for long PCR products) agarose gel, containing 30 µl of the crystal violet stock solution per 150 ml of agarose
- 5× Loading buffer (10% Ficoll 400, 0.1 M EDTA, pH 8.0, 120 µg/ml of crystal violet)
- White light box and camera
- Apparatus required for electrophoresis
- 1× TAE running buffer (4.84 g/l Tris base, 1.142 ml/l glacial acetic acid, 0.372 g/l EDTA in distilled water)
- QIAquick PCR purification kit or QIAquick gel extraction kit (Qiagen)
- DNA molecular weight markers, e.g. 1 kb DNA step ladder (Promega)

Method
1. Load and run 5 µl of each PCR on your agarose gel[a]. Also include a lane with 0.5 µg of DNA marker in 10 µl of the same loading buffer as the PCR samples[b].

2. Use a white light table (instead of UV light)[c] to view the results of the electrophoresis to determine whether the PCR products are of the correct length[d].

3. Purify each PCR product using a 'bind-and-wash' DNA purification kit, such as a QIAquick PCR purification kit, following the manufacturer's instructions.

4. Quantify your PCR products visually (by comparing the intensity of PCR bands with the intensity of DNA marker bands) or using spectrophotometry.

Notes
[a]Load 20–40 µl if gel purification is required.
[b]This is necessary for correct visual quantification of DNA amounts in PCR samples.
[c]You can use ethidium bromide-stained agarose gels and UV light throughout this protocol for both analysis and purification of PCR products. However, be advised that ethidium bromide and UV light can damage DNA and may decrease the quality of your final product.
[d]If you did not get the expected PCR product, refer to section 3 below.

Additional considerations

It is recommended that you sequence across the fusion points to verify whether the assembly has worked correctly. You can also sequence other critical regions of your construct to make sure they have no PCR errors.

Prepare a sufficient quantity of your final product for the downstream application. If the desired end product is linear, then it is not necessary to clone it into a plasmid vector to obtain a large quantity. Instead, one can set up a large-scale final amplification reaction (see *Protocol 5*), gel purify the PCR product and use it for a desired downstream application such as cell transfection. It is possible to obtain 5–10 µg of pure final product using this approach.

Fig. 3 shows a 20 kb product that we obtained successfully by fusing three fragments of 10.7, 1.7, and 7.5 kb (1). If the size of the final product is 12–20 kb, you may need to use pulsed-field electrophoresis to measure the size of your PCR product. It is possible to use regular electrophoresis with 0.5% agarose for proper assessment of DNA fragments up to 12 kb with appropriate DNA size markers.

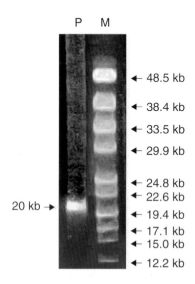

Figure 3. Pulsed-field gel electrophoresis of the 20 kb product.
Pulsed-field gel electrophoresis of the 20 kb product of long triple fusion of fragments of 10.7, 1.7, and 7.5 kb (adapted from 1). P, 20 kb product; M, DNA molecular weight markers.

2.4.2 Long multiple fusions

The process described below can be used for the assembly of four or five fragments into products up to 10 kb. The basic procedure is outlined briefly in *Fig. 4*. The detailed procedure is outlined in *Fig. 1*. The majority of the steps involved are identical to those outlined in the protocols above:

- *Analysis and purification of each fragment.* This protocol is identical to *Protocol 2*, except that, for a quintuple fusion, it is preferable to use gel purification using a crystal violet-based agarose gel.
- *Overlap extension reaction.* This step is depicted in *Fig. 1* as PCR Step A and in *Fig. 4* as PCR Steps A.1 and A.2. One should use two separate reactions for adjacent fragments: A+B and C+D as shown in *Fig. 4*. The protocol is largely identical to *Protocol 3*, except that extension times in PCRs should correspond to the length of products A+B or C+D, not the final product A+B+C+D (see *Fig. 4*).
- *Purification of intermediary product.* This protocol is identical to *Protocol 4* and is applied to products A+B and C+D separately.

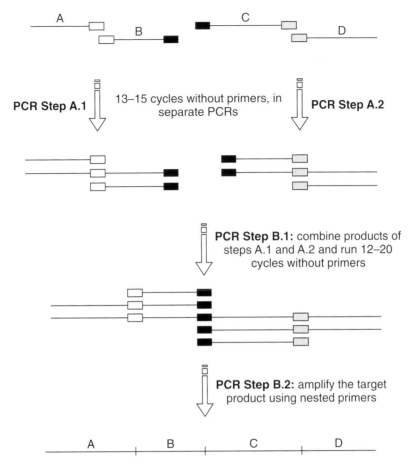

Figure 4. A brief outline of long quadruple fusion.
This technique was used successfully to assemble a 10 kb product. The procedure can also be used for the assembly of five fragments into products up to 10 kb. A more detailed procedure is outlined in *Fig. 1*, with the exception of the steps between PCR Step A and PCR Step B. For the quintuple we used pairwise fusions of A+B and D+E, and then mixed those two products with fragment C during PCR Step B.1.

- *Additional overlap extension reaction.* This step is depicted in *Fig. 4* as PCR Step B.1. The protocol is largely identical to *Protocol 3*, except that the calculation of eqimolar amounts is omitted, and equal aliquots of products A+B and C+D are used for a total of 1/5 of the final volume of the PCR. For the quintuple fusion, products A+B, D+E, and fragment C are mixed in equal aliquots and fused.
- *Amplification of final recombinant product.* This step is shown as PCR Step B/B.2 in *Figs 1* and *4*. The use of nested primers here will improve purity and the chances of success of your fusion. This protocol is identical to *Protocol 5*.

3. TROUBLESHOOTING

- **There are problems with amplifying a fragment with chimeric primers**
  Amplify the fragment with regular primers, purify it with a bind-and-wash kit, and then reamplify the fragment with chimeric primers. During the reamplification, use 25% or more of the total amount of DNA obtained in the first PCR and run the reamplification for only 10–12 cycles.
- **No product is obtained in the final amplification**
  Check the sequences of all chimeric primers and the method shown in *Fig. 2*. Ensure that you have adhered to the principles outlined in section 2.3.
- **An individual fragment cannot be amplified**
  - ○ Check whether the template contains extended GC-rich sequences.
  - ○ Make sure that you are using one of the five recommended kits.
  - ○ If you are using chimeric primers with genomic DNA or with long fragments (7 kb or more), try to redesign new primers such that chimeric primers are only used for shorter fragments or for fragments amplified from plasmid DNA.
  - ○ Try using the Herculase HotStart DNA polymerase kit (Stratagene) for the problem template.
  - ○ Some problems with amplification of individual fragments can arise from poor quality of the source DNA. If it is a genomic DNA, phenol extraction is recommended. A substantial amount of template is needed for long PCR, in addition to good quality of template. For a successful long PCR, 0.3–1 µg of DNA template per 50 µl reaction is often needed. This applies to fusion steps A and B, as well as to amplification of individual fragments.
- If you have read section 2.3 and followed the protocols but still cannot solve your problems with the method, feel free to contact me at shevchook@hotmail.com.

ACKNOWLEDGMENTS

This work was supported by NIH grants RO1 CA61010 (to Stephan Ladisch) and RO1 AI48856 (to Felipe C. Cabello).

4. REFERENCES

★★ 1. Shevchuk NA, Bryksin AV, Nusinovich YA, Cabello FC, Sutherland M & Ladisch S (2004) *Nucleic Acids Res.* **32**, e19. – *Original publication describing long multiple fusion.*
2. Ternes P, Sperling P, Albrecht S, *et al.* (2006) *J. Biol. Chem.* **281**, 5582–5592.
3. Dubytska L, Godfrey HP & Cabello FC (2006) *J. Bacteriol.* **188**, 1969–1978.
4. Strahilevitz J, Robicsek A & Hooper DC (2006) *Antimicrob. Agents Chemother.* **50**, 600–606.
5. Kato T, Muraski J, Chen Y, *et al.* (2005) *J. Clin. Invest.* **115**, 2716–2730.
6. Ojaimi C, Mulay V, Liveris D, Iyer R & Schwartz I (2005) *Infect. Immun.* **73**, 6791–6802.
7. Bugrysheva JV, Bryksin AV, Godfrey HP & Cabello FC (2005) *Infect. Immun.* **73**, 4972–4981.
8. Morozova OV, Dubytska LP, Ivanova LB, *et al.* (2005) *Gene,* **357**, 63–72.

★★★ 9. Yon J & Fried M (1989) *Nucleic Acids Res.* **17**, 4895. *– One of the first publications describing overlap-extension PCR.*

★★ 10. Yolov AA & Shabarova ZA (1990) *Nucleic Acids Res.* **18**, 3983–3986. *– One of the first publications describing overlap extension PCR.*

★ 11. Pont-Kingdon G (1997) *Methods Mol. Biol.* **67**, 167–172. *– A useful review of the application of PCR to the creation of recombinant molecules and to mutagenesis.*

12. Kuwayama H, Obara S, Morio T, Katoh M, Urushihara H & Tanaka Y (2002) *Nucleic Acids Res.* **30**, e2.

13. Horton RM (1995) *Mol. Biotechnol.* **3**, 93–99.

★★ 14. Xiong AS, Yao QH, Peng RH, *et al.* (2004) *Nucleic Acids Res.* **32**, e98. *– Successful fusion of ten DNA fragments using PCR.*

15. Sambrook J, Fritsch, E. & Maniatis, T. (2001) *Molecular Cloning: a Laboratory Manual.* Cold Spring Harbor Laboratory Press, New York.

16. *RedAccuTaq LA DNA polymerase mix* (1999) Technical bulletin no. MB-690. Sigma-Aldrich, Saint Louis, MO.

CHAPTER 13

Efficient PCR-based mutagenesis method applicable to diverse mutagenesis strategies using type IIs restriction enzymes

Jae-Kyun Ko and Jianjie Ma

1. INTRODUCTION

PCR-based *in vitro* mutagenesis is an important tool that allows defined mutations to be made *in vitro* to cloned DNA. Prior to its conception, approaches focused on the generation of random mutations, such as those introduced by radiation or chemicals. However, these methods did not allow mutations in a specific gene or chromosomal location to be investigated. PCR-based *in vitro* mutagenesis, which is now commonly used in many laboratories, can provide critical information on the regulation of gene expression, as well as protein structure and function (1). Numerous strategies have been developed to target defined regions of DNA, including;

- Base(s) substitution
- Deletion/insertion
- Chimeric gene generation
- Multiple-site mutagenesis

Many of these methods have been developed either commercially or noncommercially, for example, the overlap-extension method (2), the megaprimer method (3, 4), and the QuikChange method (Stratagene). However, none of these approaches can be applied to all of the diverse mutagenesis strategies mentioned above. The methods detailed in this chapter utilize type IIs restriction enzymes (see *Table 1*). These enzymes are made up of two distinct domains, one for DNA binding and one for DNA cleavage, and unlike other type II enzymes, which cut within the recognition sequence, type IIs enzymes cleave to one side of their recognition

PCR: *Methods Express* (S. Hughes and A. Moody, eds.)
© Scion Publishing Limited, 2007

Table 1. Type IIs restriction enzymes useful for mutagenesis primer design

Enzyme[a]	Recognition sequence[b]	Type of end	Isoschizomers
Bbsl	GAAGAC (2/6)	5′ overhang	Bpil, BpuAl, BstV2l
BfuAl	ACCTGC (4/8)	5′ overhang	Acc36l, BspMl
Bsal	GGTCTC (1/5)	5′ overhang	Bsol, Eco31l
Bsml	GAATGC (1/–1)	3′ overhang	BsaMl, Pctl
BsmBl	CGTCTC (1/5)	5′ overhang	Esp3l
BsrDl	GCAATG (2/0)	3′ overhang	Bse3Dl, BseMl
Btsl	GCAGTG (2/0)	3′ overhang	–
Earl	CTCTTC (1/4)	5′ overhang	Bst6l, Eam1104l, Ksp632l
Sapl	GCTCTTC (1/4)	5′ overhang	–
Aarl	CACCTGC (4/8)	5′ overhang	–

[a]Commercially available type IIs restriction endonuclease are listed.
[b]Numbers in parentheses indicate the cleavage point for type IIs enzymes. For example, GAAGAC (2/6) indicates cleavage two bases after the recognition sequence for the sense strand and six bases before the recognition sequence for the antisense strand:

5′-GAAGACNN/-3′
3′-CTTCTGNNNNNN/-5′

sequence. This chapter describes a rapid and highly efficient method, using digestion with type IIs restriction enzymes and *in vitro* ligation, to generate mutations. This technique is applicable to diverse mutagenesis purposes in molecular biology studies.

2. METHODS AND APPROACHES

2.1 Principles of mutagenesis

As outlined in *Fig. 1*, a target gene is amplified in two separate PCR fragments by four PCR primers. Each fragment is produced by one anchor primer and one mutagenic primer. The two anchor primers contain restriction sites (not type IIs) for cloning of a mutated gene and the two mutagenic primers contain the desired mutation near the recognition site of the type IIs restriction enzyme (see *Table 1*). After digestion of the two PCR fragments with a given type IIs restriction enzyme, the cohesive ends of each fragment remain complementary and can undergo ligation, resulting in the generation of specific mutations.

The generated mutant DNA fragment can be subcloned into a vector after digestion of the anchor region with appropriate restriction enzymes. Although similar approaches have been described (5–8), their applications are limited to mutation of only a single base or substitution of only a single codon. However, the method described here is very successful at generating mutations, not only for single-site substitutions but also for multiple-site substitutions, insertions, deletions, chimeragenesis, and random mutagenesis (9). This method is also rapid and highly efficient, and has been applied successfully to several different genes (9).

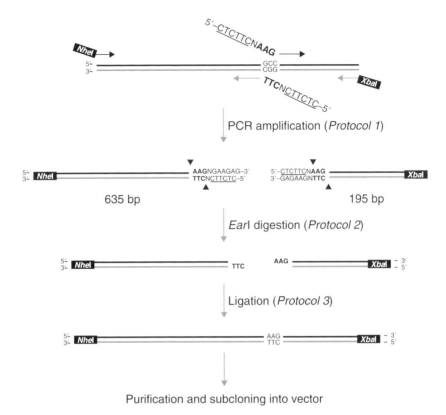

Figure 1. Schematic diagram of single codon mutagenesis (GCC→AAG) using *Ear*I type IIs restriction enzyme.
Two separate PCR fragments are produced from a template EGFP cDNA. Two mutagenic primers contain mutated codon sequences complementary to each other near the *Ear*I recognition site (5′-CTCTTC-3′). Each anchor primer contains a restriction site for downstream subcloning. After *Ear*I digestion, the two fragments are ligated together to generate mutated full-length EGFP cDNA. The ligation product is purified by agarose gel electrophoresis and subcloned into a cloning vector using *Nhe*I and *Xba*I cloning sites.

2.2 Design of mutagenic primers and choice of type IIs restriction enzyme for mutagenesis

To achieve successful mutagenesis, it is critical to choose an appropriate type IIs enzyme that cannot digest the gene of interest but can efficiently recognize the cleavage site generated by mutagenic primers at one end of the PCR product. Commercially available type IIs enzymes are listed in *Table 1*. When designing mutagenic primers, special caution is required, as type IIs enzyme recognize nonpalindromic sequences (where the sequence on one strand of the DNA does not read the same in the opposite direction on the complementary strand) and digests at only one side of the cleavage target sequence (see *Fig. 1*).

In this chapter, we describe protocols used to generate site-directed mutagenesis within the coding sequence of the enhanced green fluorescent protein (EGFP; 830 bp) present in the plasmid pEGFP-C1. For mutagenesis of Ala-206 (GCC) to Lys (AAG), two mutagenic primers containing the type IIs *Ear*I site and two anchor primers are used (*Fig. 1*).

The template EGFP gene is amplified into two separate PCR fragments using two pairs of anchor and mutagenic primers. Mutated sequences are located near the recognition site of *Ear*I type IIs restriction enzyme. After digestion of the two fragments with *Ear*I, the complementary sequences from the two fragments can then be ligated together to generate a mutated gene. The ligated DNA fragment containing the mutation can then be subcloned.

Protocol 1

PCR amplification and purification of PCR product

Equipment and Reagents
- TaKaRa Ex Taq (5 units/μl) (TaKaRa)[a]
- 10× TaKaRa Ex Taq reaction buffer
- 10 mM dNTP mix (2.5 mM each of dATP, dCTP, dGTP, and dTTP) (TaKaRa)
- Primers (each 100 μM stock in ddH$_2$O)[b]
- A206K-F: 5′-AA*CTCTTCT*AAGCTGAGCAAAGACCCCAACGAG-3′ (*Ear*I) and Anchor-R: 5′-TA*TCTAGA*TCCGGTGGATCCCG-3′ (*Xba*I)
- Anchor-F: 5′-CC*GCTAGC*GCTACCGGTCGC-3′ (*Nhe*I) and A206K-R: 5′-AA*CTCTTCT*C*TT*GTAGGACTGGGTGCTCAGGTAG-3′ (*Ear*I)
- Template pEGFP-C1 plasmid (100 ng/μl in ddH$_2$O; Clontech)[c]
- Thermal cycler
- QIAquick PCR purification kit (Qiagen)
- 2% Agarose gel containing 10 ng/ml ethidium bromide
- 6× Loading dye (Fermentas)
- Equipment and reagents for agarose gel electrophoresis including 1× TAE agarose gel running buffer (40 mM Tris/HCl, pH 8.0, 40 mM acetic acid, 1 mM EDTA)
- Spectrophotometer

Method
1. Prepare two PCR tubes and add primers as indicated below:
   - PCR1: 0.5 μl of Anchor-F and 0.5 μl of mutagenic primer A206K-R
   - PCR2: 0.5 μl of Anchor-R and 0.5 μl of mutagenic primer A206K-F

2. Prepare 98 μl of PCR master mix by adding the following reagents in the following order to a PCR tube:
   - 78 μl of ddH$_2$O
   - 10 μl of 10× reaction buffer
   - 8 μl of dNTP mix
   - 1 μl of plasmid template
   - 1 μl of TaKaRa Ex Taq[d]

3. Mix gently by pipetting, spin down briefly, and add 49 μl of the PCR master mix to each tube containing primers.

4. Place in a thermal cycler and perform the amplification reaction under the following cycling conditions:
    - 94°C for 4 min
    - 30 cycles of 94°C for 30 s, 55°C for 1 min, and 72°C for 1 min
    - 72°C for 7 min

5. Visualize the PCR products by electrophoresis of 5 μl of the reaction mix on a 2% agarose gel[e].

6. Purify each PCR product using the QIAquick PCR purification kit according to the manufacturer's instructions and elute the PCR products in 50 μl of ddH$_2$O. Quantify the DNA concentration using a spectrophotometer.

Notes

[a]For amplification of short DNA fragments less than 1 kb, normal *Taq* polymerase can be used for this protocol. However, high-fidelity polymerases are recommended for amplification of long DNA fragment (>1 kb).

[b]Purification of synthesized primers by polyacrylamide gel electrophoresis (PAGE) or high-performance liquid chromatography (HPLC) is not essential for short primers less than 45 bp but is recommended for long primers (>45 bp).

[c]Template pEGFP-C1 plasmid was prepared using a QIAprep Spin Miniprep kit (Qiagen).

[d]TaKaRa Ex Taq has proofreading activity, resulting in 4.5-fold lower mutation rates than standard *Taq* DNA polymerase.

[e]The expected sizes of the PCR products are 635 and 195 bp for PCR fragment 1 and PCR fragment 2, respectively (see *Fig. 2*, left panel).

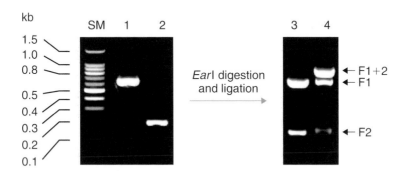

Figure 2. Agarose gel electrophoresis of PCR-amplified fragments and mutagenesis product.
Left panel: 2% agarose gel electrophoresis of two PCR-amplified products from an EGFP gene as a template. *Lane SM*, 100 bp ladder size marker (Promega); *lane 1*, 635 bp PCR fragment (F1) amplified using primers Anchor-F and A206K-R; *lane 2*, 195 bp PCR fragment (F2) amplified using primers Anchor-R and A206K-F. *Right panel*: electrophoretic analysis of the mutagenesis product on a 2% agarose gel. After *Ear*I digestion, PCR fragments F1 and F2 were ligated at room temperature for 1 h. *Lane 3*, ligation control reaction in the absence of T4 DNA ligase; *lane 4*, ligation reaction with T4 DNA ligase.

Protocol 2

*Ear*I digestion of PCR products

Equipment and Reagents
- *Ear*I type IIs restriction enzyme (10 units/µl; New England Biolabs)
- 10× NEBuffer 1 reaction buffer (New England Biolabs)
- Heat block or incubator
- QIAquick PCR purification kit (Qiagen)

Method

1. Prepare two 1.5 ml microtubes and add ~200–500 ng of each purified PCR product, PCR1 and PCR2, to separate tubes.

2. Add 5 µl of NEBuffer 1, 2 µl of *Ear*I, and sterile water to a final volume of 50 µl to each tube, mix well by pipetting up and down, and incubate at 37°C for 1 h.

3. Inactivate the *Ear*I by incubating the reaction at 65°C for 20 min.

4. Determine the number of picomoles of each DNA fragment using the following equation.

$$\text{pmol DNA ends} = \mu\text{g DNA} \times \left(\frac{\text{pmol}}{660\,\text{pg}}\right) \times \left(\frac{10^6\,\text{pg}}{1\,\mu\text{g}}\right)$$

This can be calculated automatically at http://www.promega.com/biomath, option dsDNA: Micrograms to Picomoles[a].

5. Combine equimolar amounts of the two digested fragments into one microtube and purify using a QIAquick PCR purification kit, following the manufacturer's instructions.

6. Elute the DNA fragments in 30 µl of ddH$_2$O.

7. Check the eluted DNA fragments by electrophoresis of 3 µl of sample on a 2% agarose gel.

Note

[a]In the example here, we have fragments of 635 and 195 bp for PCR1 and PCR2, respectively. The molarity of these fragments, assuming a DNA concentration of 200 ng, is 0.477 and 1.554 pmol of DNA. This is a ratio of 1:3.2; therefore, to have equimolar amounts, we would need 200 ng of PCR fragment 1 and 640 ng of PCR fragment 2.

Protocol 3

Ligation of two *Earl*-digested DNA fragments

Equipment and Reagents
- T4 DNA ligase (400 units/μl; New England Biolabs)[a]
- 10× Ligase reaction buffer (New England Biolabs)
- Heat block or incubator

Method

1. Prepare two 1.5 ml microtubes.

2. Combine 16 μl of elution sample from *Protocol 2* (containing the two *Earl*-digested DNA fragments), 2 μl of ligase reaction buffer, and 2 μl of T4 DNA ligase[a] in one microtube. Mix well by pipetting.

3. For a negative-control reaction, combine 16 μl of elution sample, 2 μl of reaction buffer, and 2 μl of ddH$_2$O into the second microtube.

4. Incubate the samples at room temperature for 1 h[b].

5. Inactivate the T4 DNA ligase by incubating the reaction at 65°C for 15 min.

Notes

[a]Using rapid ligation kits that are available from several suppliers is not recommended for this procedure. It has been observed in our laboratory that ligation samples containing rapid ligation solution do not separate adequately by agarose gel electrophoresis, which is required for the next procedure (see *Protocol 4*) .

[b]Room temperature or 37°C incubation is recommended for this ligation reaction, rather than a 16°C incubation, as this increases ligase activity and shortens the reaction time.

Protocol 4

Purification of the ligated DNA from the agarose gel

Equipment and Reagents
- 2% Agarose gel containing 10 ng/ml ethidium bromide
- 6× Loading dye (Fermentas)
- Single-edged razor blade
- QIAquick gel extraction kit (Qiagen)
- Equipment and reagents for agarose gel electrophoresis including 1× TAE agarose gel running buffer (40 mM Tris/HCl, pH 8.0, 40 mM acetic acid, 1 mM EDTA)
- Heat block or incubator
- Spectrophotometer

Method
1. Add 4 µl of loading dye to the ligation sample and the negative-control reaction, and mix well by pipetting up and down.

2. Load samples into the wells of a 2% agarose gel[a] and resolve by electrophoresis to assess fragment size and success of the ligation reaction (see *Fig. 2*, right panel).

3. Carefully cut out the DNA fragment with the expected size of EGFP cDNA (830 bp) and transfer to a clean microtube.

4. Purify the DNA fragment using the QIAquick gel extraction kit following the manufacturer's instructions and elute the ligation product in 30 µl of ddH$_2$O.

5. Quantify the product using a spectrophotometer[b].

Notes

[a]To achieve high resolution of DNA separation and to prevent sample loss by overloading, it is recommended that 8–12 µl of sample is loaded into two or three separate wells.
[b]Usually, ~200–500 ng of ligated DNA can be obtained. The purified DNA can be subcloned into a cloning vector after double digestion with *Nhe*I and *Xba*I by following general subcloning protocols (10).

To confirm the presence of the target mutation and the absence of undesired extra mutations, it is recommended to DNA sequence at least three clones containing the insert. In our laboratory, this method has shown mutagenesis efficiency approaching 100% with usually at least two correct clones obtained without erroneous mutations (9).

2.3 Application to diverse mutagenesis

2.3.1 General *in vitro* mutagenesis

The mutagenesis method described here is rapid and highly efficient for introducing specific mutations into any site of a target DNA sequence. There are several advantages to this method:

1. It is simple and cost-effective.
2. Mutated genes can be generated rapidly; typically, mutation at a target site can be produced in as few as 5 h.
3. The PCR-based mutagenesis strategy has very high efficiency and fidelity.
4. With specific mutagenic primers, this method can be used to create insertions, deletions, chimeric genes, and multiple-site directed mutagenesis as described in *Fig. 3*. For insertion of sequences longer than 50 bp into a target site, a pair of long mutagenic primers (>45 bp) may be required (9), and we recommend that these be purified using PAGE or HPLC.

2.3.2 Directed evolution

Perhaps the most unique aspect of this strategy is the ability to create random and multiple mutations in a given gene. As illustrated in *Fig. 4*, different mutations at a given site can be produced by the use of randomized PCR primers. This strategy will be useful for directed protein evolution studies. Directed evolution is a powerful approach that can be used to optimize properties of proteins or create new properties (11). This innovative method consists of two steps: the creation of a mutant library for the target gene of interest, and screening and selection of mutants possessing the desired properties.

2.4 Summary

Site-directed random mutagenesis using type IIs restriction enzymes can target specific residues to mutate based on prior structural and functional knowledge. By targeting specific residues that are important for the biological function of a protein, a hyperdiverse mutant library can be generated whilst minimizing structural alterations of the protein backbone. Therefore, the mutant library can contain a high proportion of functional mutants that are more likely to yield positive results (12).

3. TROUBLESHOOTING

- **More than one band is seen in the mutation PCR step**
  For optimal results, it is important to obtain specific PCR products of an expected size. For this, we recommend optimizing PCR conditions including $MgCl_2$ concentration, annealing temperature, extension time, the number of amplification cycles, and the amount of template DNA. If this is unsuccessful, it may be necessary to redesign one or both primers.
- **A low yield of expected product is obtained following ligation**
  The amount of ligation product generated is related directly to the activity of the type IIs enzyme used. We have observed that different type IIs restriction enzymes can demonstrate varying digestion activities, even when identical amounts of enzyme and incubation time are used. Therefore, we recommend performing control reactions using different type IIs enzymes in conjunction

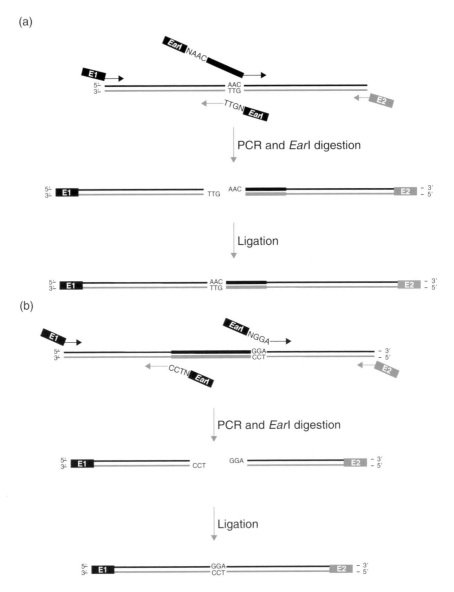

Figure 3. Various mutagenesis strategies using type IIs restriction enzymes.
Schematic diagrams of various mutagenesis strategies are shown including insertion (*a*),
deletion (*b*), and chimeric gene generation (*c*), E1 and E2 indicate restriction sites for
subcloning of mutated cDNA. The codons shown were chosen arbitrarily for illustrative
purposes.

with plasmids containing the restriction site in order to identify which
enzymes will work optimally.
- **DNA yield is to low to permit downstream cloning**
  The DNA can be reamplified by additional PCR using two anchor primers and
  the purified DNA as a template and then can be subcloned into a vector.

(c)

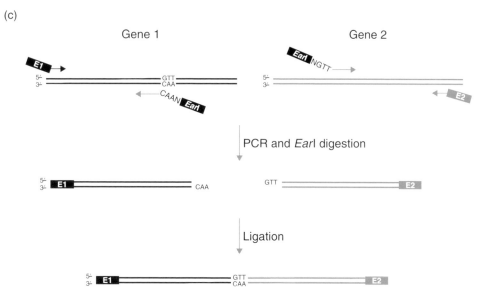

Figure 3 (cont'd). Various mutagenesis strategies using type IIs restriction enzymes.

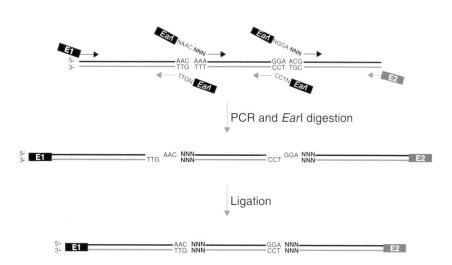

Figure 4. Experimental strategy for random mutagenesis at multiple sites.
Schematic diagram of random mutagenesis at two separate codons is shown. The mutagenic primers contain randomized codon sequences near the type IIs restriction enzyme cleavage site. Additional codons at different sites can be mutated by repeated mutagenesis reactions using the first-round randomized gene product as a template and additional random mutagenic primers *in vitro*. The codons shown here were chosen arbitrarily as an example.

4. REFERENCES

1. Ling MM & Robinson BH (1997) *Anal. Biochem.* **254**, 157–178.
2. Higuchi R, Krummel B & Saiki RK (1988) *Nucleic Acid Res.* **16**, 7351–7367.
3. Kammann M, Laufs J, Schell J & Gronenborn B (1989) *Nucleic Acid Res.* **17**, 5404.
4. Ke S-H & Madison EL (1997) *Nucleic Acid Res.* **25**, 3371–3372.
5. Urban A, Neukirchen S & Jaeger K-E (1997) *Nucleic Acid Res.* **25**, 2227–2228.
6. Zheng L, Baumann U & Reymond J-L (2004) *Nucleic Acid Res.* **32**, e115.
★ 7. Tomic M, Sunjevaric I, Savtchenko ES & Blumenberg M (1990) *Nucleic Acid Res.* **18**, 1656. – *Original publication describing the PCR-based mutagenesis method using type IIs restriction enzyme.*
8. Vilardaga J-P, Di Paolo E & Bollen A (1995) *BioTechniques*, **18**, 604–606.
★★ 9. Ko J-K & Ma J (2005) *Am. J. Physiol. Cell Physiol.* **288**, C1273–C1278. – *Application of the PCR-based mutagenesis method using type IIs restriction enzyme to the multiple-site directed mutagenesis and diverse in vitro mutagenesis strategies.*
10. Sambrook J, Fritsch E & Maniatis T (2001) *Molecular Cloning: a Laboratory Manual.* Cold Spring Harbor Laboratory Press, New York.
11. Yuan L, Kurek I, English J & Keenan R (2005) *Microbiol. Mol. Biol. Rev.* **69**, 373–392.
12. Parikh MR & Matsumura I (2005) *J. Mol. Biol.* **352**, 621–628.

CHAPTER 14

Inverse PCR-based restriction fragment length polymorphism for identifying low-level mutations in tumors

G. Mike Makrigiorgos

1. INTRODUCTION

Detecting the presence and diversity of low-level mutations in human tumors with genomic instability is desirable due to their potential prognostic value and their putative influence on the ability of tumors to resist drug treatment and/or metastasize. However, direct measurement of these genetic alterations in surgical samples has been elusive, as technical hurdles make mutation discovery impractical at low mutation frequency levels ($< 10^{-2}$) (1, 2). Mutation detection methods applied to the identification of low-level nucleotide changes in the hypoxanthine-guanine phosphoribosyltransferase and lipoprotein-associated coagulation inhibitor genes can, in fact, detect a very high number of low-level nucleotide changes in mismatch repair-deficient colon cancer human cell lines (3, 4) and in transgenic rat tumors (5). However, the inability to apply these mutation detection methods to nonclonal cells or to genes relevant to cancer has precluded direct examination of low-level mutations in surgical tumor samples, which potentially contain repair defects similar to those observed in the cell lines. Due in part to this technical hurdle, the influence of low-level mutations in tumor onset, progression, and response to treatment has been difficult to assess (6, 7).

PCR: *Methods Express* (S. Hughes and A. Moody, eds.)
© Scion Publishing Limited, 2007

2. METHODS AND APPROACHES

2.1 Principles of inverse PCR-based amplified restriction fragment length polymorphism

Inverse PCR-based amplified restriction fragment length polymorphism (RFLP) (iFLP) (8) is a new technology that combines inverse PCR, RFLP, and denaturing high-performance liquid chromatography (dHPLC) to allow scanning of the genome at several thousand positions per experiment for low-level point mutations. Using iFLP, widespread, low-level mutations at a mutation frequency of 10^{-2} to 10^{-4} can be detected in clinical tumor samples in genes located throughout the genome (8, 9).

We have developed a single-tube protocol that harnesses this simple principle on a genome-wide scale as follows (see *Fig. 1*):

1. Whole genome circularization: whole genomic DNA is first digested by *Msel* and the DNA sequences are then circularized (see *Protocol 1*).
2. *Taql* digestion: DNA circles are digested with a restriction enzyme (see *Protocol 2*), in this case *Taql* (see *Fig. 1*), although any enzyme can be used. DNA circles that do not normally contain a natural restriction site are converted to double-stranded linear DNA fragments only if they have acquired nucleotide changes leading to the formation of a *Taql* site(s).
3. Adaptor ligation: *Taql*-specific adaptors are then ligated onto the digested DNA fragments.
4. First (generic) PCR: a 'generic' first PCR step is performed using adaptor-specific primers. The PCR ensures that all linearized circles (with ligated adaptors) that contained *Taql* sites, including *Taql* sites in the original wild-type sequence and *de novo Taql* sites generated via single base changes (mutations), will be amplified (see *Protocol 2*).
5. High-throughput inverse PCR: the inverse PCR utilizes gene- or region-specific primers to screen the product of the first PCR for the presence of *de novo* (mutation-derived) fragments (see *Protocol 3*). Hence, PCR products are only generated if a mutation modifies the sequence to generate a *Taql* digestion site, thus permitting steps 2–5.

In addition to providing gene-specific information, the inverse PCR step ensures that only DNA fragments that have successfully circularized during the whole genome circularization step will be amplified. Thus, DNA fragments that remain linear despite the circularization step do not result in successful PCR, as the inverse primers point in the wrong direction.

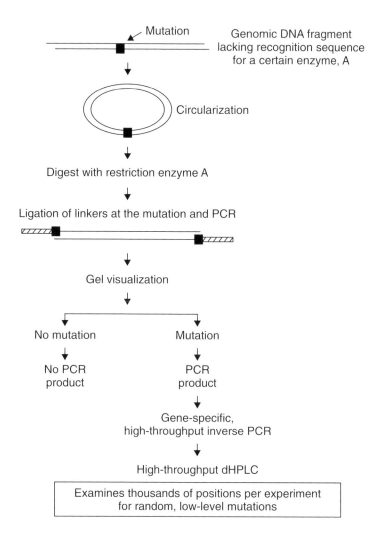

Figure 1. Outline of inverse PCR.
Whole genomic DNA is digested by *Mse*I and the DNA sequences are then circularized following a self-ligation reaction (see *Protocol 1*). The DNA circles are digested with a restriction enzyme (see *Protocol 2*), in this case *Taq*I. DNA circles that do not normally contain a natural restriction site are converted to double-stranded, linear DNA fragments only if they have acquired nucleotide changes leading to the formation of a *Taq*I site(s). Specific adaptors are ligated on to the DNA fragments. A 'generic' first PCR step is performed, which ensures that all linearized circles (with ligated adaptors) that contain *Taq*I sites (including both wild-type and *de novo Taq*I sites), will be amplified (see *Protocol 2*). The product of the 'generic' PCR is then used in an inverse PCR to screen on a gene-by-gene/region-by-region basis for the presence of *de novo* (mutation-derived) fragments.

Genomic regions selected for amplification using inverse PCR must belong to DNA fragments that do not contain wild-type *Taq*I regions, and therefore are not expected to yield a product unless a mutation has modified the sequence to generate a *Taq*I digestion site. In control experiments, use of an alternative enzyme whose recognition sequence is present in the wild-type sequences can be utilized to serve as a positive control for iFLP.

A 500 bp DNA fragment contains, on average, 80–100 mutable sites that can form a *Taq*I site following a single nucleotide substitution, deletion, or insertion. Therefore, each PCR from the high-throughput PCR set evaluates 80–100 potential sequence changes simultaneously. When using high-throughput dHPLC for mutation detection (at a selectivity of ~1 mutant per 10^4–10^5 wild-type sequences) and depending on the number of genes/regions examined, several thousand genomic positions can be examined in a single iFLP experiment.

Here, we present our genome-wide iFLP approach. This technique is presented in distinct steps to facilitate interpretation.

Protocol 1

Digestion of genomic DNA and circularization

Equipment and Reagents
- *Mse*I restriction enzyme (10 000 units/ml; New England Biolabs)
- T4 DNA ligase (2 000 000 units/ml; New England Biolabs)
- 10× T4 DNA ligase buffer (50 mM Tris/HCl, pH 7.5, 10 mM MgCl$_2$, 10 mM DTT, 1 mM ATP, 25 µg bovine serum albumin (BSA; New England Biolabs))
- *Escherichia coli* exonuclease I (20 000 units/ml; New England Biolabs)
- Lambda exonuclease (5 000 units/ml; New England Biolabs)

Method
1. Combine:
   - Genomic DNA (1–1000 ng[a])
   - 5 µl T4 DNA ligase buffer[b]
   - 0.5 µl *Mse*I
   - Sterile water up to a final volume of 50 µl

2. Incubate for 1.5 h at 37°C to digest the DNA and produce sticky ends for subsequent circularization.

3. Inactivate the *Mse*I by incubating at 70°C for 30 min.

4. Circularize the *Mse*I-digested DNA by the addition of 0.6 µl (1200 units) of T4 DNA ligase to the solution from step 1 and incubation at 14°C overnight.

5. Inactive the T4 ligase by incubation at 70°C for 30 min.

6. Add 0.6 µl (12 units) of *E. coli* exonuclease I and 2.4 µl (12 units) of lambda exonuclease to the circularized DNA and incubate at 37°C for 1.5 h to eliminate noncircularized DNA.

7. Inactive the exonucleases by incubation at 80°C for 30 min[c].

Notes
[a]The iFLP protocol will work at any concentration between 1 and 1000 ng. If a high selectivity of detecting the mutant within excess wild type is required, several genomic copies of starting material must be used, and therefore several hundred nanograms of DNA is required. On the other hand, if the starting DNA material is limited (e.g. tissue obtained via minute biopsies), then lower starting amounts of DNA are used, at the expense of the overall selectivity.

[b]The buffer used is compatible with the restriction enzyme.

[c]Optionally, iFLP can be performed on a genomic fraction rather than on the full genome in order to enrich target DNA sequences and allow higher selectivity in mutation detection. For this purpose, following *Mse*I digestion, separate the DNA on a 1% agarose gel, excise the DNA in a particular molecular weight range (e.g. 400–700 bp), and purify using a gel purification kit (e.g. Qiagen). The iFLP procedure can then be performed with the purified, enriched genomic DNA, which is estimated to contain about 5–10% of the original genome (8).

Protocol 2

*Taq*I digestion, adaptor ligation, and generic PCR

Equipment and Reagents
- Sterile water
- *Taq*I restriction endonuclease (20 000 units/ml; New England Biolabs) and accompanying reaction buffer (NEB buffer 1)
- Oligonucleotides (linkers):
    - Linker 1: 5′-AGGCAACTGTGCTATCCGAGGGAA-3′ (2 μg/μl)
    - Linker 2: 5′-CGTTCCCTCGGA-3′ (1 μg/μl)
- Linker-specific primer: 5′-AGGCAACTGTGCTATCCGAGGGAA-3′ (0.2 μM)
- 10× T4 DNA ligase (2 000 000 units/ml; New England Biolabs)
- 10× T4 DNA ligase buffer (50 mM Tris/HCl, pH 7.5, 10 mM $MgCl_2$, 10 mM DTT, 1 mM ATP, 25 μg BSA; New England Biolabs)
- TITANIUM *Taq* PCR kit (Clontech)
    - ☐ 10× TITANIUM *Taq* PCR buffer
    - ☐ 50× dNTP mix (10 mM each of dATP, dCTP, dGTP, and dTTP)
    - ☐ 50× TITANIUM *Taq* DNA polymerase

Method
1. Combine (per reaction):
   - Circularized DNA from *Protocol 1*
   - 0.5 μl of *Taq*I enzyme (10 units/μl)
   - 1 μl of NEB Buffer 1
   - Sterile water up to a final volume of 60 μl

2. Incubate for 90 min at 65°C to digest the DNA[a].

3. Inactivate *Taq*I by incubating at 80°C for 40 min.

4. For double-stranded oligonucleotide adaptor formation (per reaction), combine 2 μl of linker 1 and 2 μl of linker 2 and incubate the sample using a programmed temperature gradient in a thermal cycler, starting at 50°C and slowly cooling down to 10°C (ramping at 1°C/min).

5. For adaptor ligation (per reaction), combine:
   - 10 μl of *Taq*I-digested DNA
   - double-stranded oligonucleotide adapters (from step 4)
   - 2 μl of T4 DNA ligase buffer
   - 1 μl of T4 DNA ligase
   - Sterile water up to a final volume of 20 μl

6. Incubate the solution at room temperature for 30 min[b].

7. For PCR amplification (per reaction), combine:
   - 4–8 μl of the ligated DNA
   - 2 μl of linker-specific primer
   - 5 μl of 10× TITANIUM *Taq* PCR Buffer
   - 1 μl of 50× dNTP mix
   - 1 μl of 50× TITANIUM *Taq* DNA polymerase
   - Sterile water up to a final volume of 50 μl

8. Use the following PCR profile for amplification:
   - 72°C for 8 min
   - 22 cycles of 95°C for 30 s, 70°C for 1 min, and 68°C for 5 min
   - 4°C hold

Notes

[a] In certain experiments, TaqI can be substituted by the enzyme HpyCH4IV.
[b] The described procedure outlined in Protocols 1 and 2 takes about 30 h to perform, including the overnight circularization.

Protocol 3

Gene-specific inverse PCR and size-separation analysis of PCR products

Equipment and Reagents
- TITANIUM Taq PCR kit (Clontech)
- Sets of inverse PCR primers specific for genomic fragments of interest (0.2 µM primer concentration)[a]

Methods
1. For high-throughput PCRs (per reaction) combine:
   - 0.5 µl of the PCR product from Protocol 2
   - 0.5 µl of each gene-specific primer[a]
   - 1 µl of 10× TITANIUM Taq PCR buffer
   - 0.2 µl 50× dNTP mix
   - 0.2 µl 50× TITANIUM Taq DNA polymerase
   - Sterile water up to a final volume of 10 µl

2. The PCR program used is specific to the particular primer pair used. However, we have adjusted the melting temperature (T_m) of all of our primers to work with the following protocol:
   - 95°C for 1 min
   - 30 cycles of 95°C for 30s, 68°C for 1 min, and 68°C for 5 min
   - 4°C hold

3. Run 5–10 µl of the PCR products (from Protocol 2) including the negative control on a 1% agarose gel.

Notes

[a] Inverse primers should be designed to amplify a short segment (~50–200 bp) from the genomic fragment of interest. Genomic region-specific PCR using the following inverse PCR primers for p53, 5′-AATGCCGTTTTCTTCTTGACTG-3′ (forward) and 5′-TTCCGTCCCAGTAGATTACCAC-3′ (reverse), generates a 200 bp sequence. The presence of an amplicon following PCR indicates mutations leading to new restriction sites in a 612 bp circularized segment of p53 (nt 14361–14972), which was originally negative for restriction sites.

2.2 Size-separation analysis of PCR products and estimation of mutation frequencies

Inverse PCR products can be analyzed either by agarose gel electrophoresis or (preferably) via HPLC on a WAVE dHPLC system (Transgenomics). The WAVE system is equipped with UV and fluorescent detectors and two 96-well autosamplers that enable high-throughput analysis of PCR products. PCR products (8 μl) are injected and analyzed under nondenaturing conditions at 48°C. DNA elution profiles are recorded and displayed as chromatograms (see *Figs 2* and *3*).

To estimate the approximate mutation frequency, discovered by iFLP, the iFLP positive controls generated via cutting with an enzyme present within the wild-type sequence (for example *Hpy*CH41V can be used instead of *Taq*1 in sequences that have a single *Hpy*CH41V recognition sequence in wild type DNA; thereby one can expect 100% digestion of *Hpy*CH41V is used for a positive control) are diluted into mutation-negative samples in order to produce samples with known ratios of mutant to wild-type alleles. iFLP is carried out using samples of known mutation frequency in parallel with unknown samples and inverse PCR amplicons are quantitated (i.e. the mutation frequency determined) using the WAVE System's NAVIGATOR software (8).

2.3 Sample results

2.3.1 DNA from cell lines and human tumors

DLD1 and HCT116 cells are deficient in mismatch-repair activity and display microsatellite instability and a high spontaneous mutation rate at coding sequences (3, 10, 11). SW480 cells do not display mismatch repair deficiency (12) or microsatellite instability (10, 13–15).

2.3.2 iFLP

Genome-wide iFLP was combined with automated dHPLC detection and applied to analyze gene regions using genomic DNA from the colon cancer cell lines DLD1, HCT116, and SW480 (see *Fig. 2*), as well as ten colon cancer and corresponding normal tissue surgical samples (see *Fig. 3*). The chromatographs (see *Fig. 2a–c*) corresponding to regions of interest from 8-oxoguanine DNA glycosylase (*OGG1*), breast cancer 2 early onset (*BRCA2*), and mutS homolog 2 (*MSH2*) showed small peaks at the same retention time as the mutation-positive controls for DLD1 cells (repair deficient), but not SW480 cells (repair proficient). These results demonstrated the formation of new *Taq*I restriction sites for DLD1 but not for SW480.

To estimate mutation frequencies for the identified mutations, dHPLC was performed using dilutions of the iFLP positive control (diluted in mutation-negative sample) alongside DNA from DLD1 and SW480 cells. *Fig. 2(d)* depicts a typical dilution experiment carried out for the *BRCA2* region for HCT116 and

SW480 cells. Comparison of the results from the HCT116 cells with the results from the dilution series determined a mutation frequency of $\sim2 \times 10^{-4}$ for the HCT116 cells. This approach also indicated that the detection limit of the method

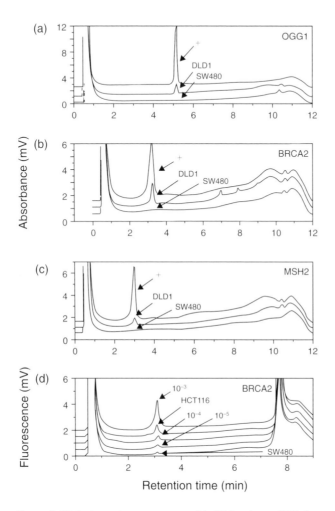

Figure 2. High-throughput, genome-wide iFLP using a dHPLC autosampler and the scheme presented in *Fig. 1.*

Chromatograms from gene segments indicating mutations in DLD1 cells and absence of mutations in SW480 cells are depicted. The chromatograms in (*a–c*) depict iFLP-screened fragments from the genes *OGG1*, *BRCA2*, and *MSH2*. Mutation-positive controls (+), obtained via digestion with *Hpy*CH4IV positive-control enzyme, indicate the signals and retention times expected if mutations are present in 100% of cells. The mismatch repair-deficient DLD1 cells demonstrate peaks that are absent from the mismatch repair-proficient SW480 cells in the genes *OGG1*, *BRCA2*, and *MSH2*. The chromatogram in (*d*) (high-sensitivity fluorescence detector) depicts profiles for a *BRCA2* region of the mismatch repair-deficient cells HCT116. Dilution of mutation-positive into negative (SW480) samples at mutation frequencies of 10^{-3}, 10^{-4}, and 10^{-5} were performed in parallel to estimate mutation frequency for the *BRCA2* region of HCT116 cells, and are depicted in (*d*).

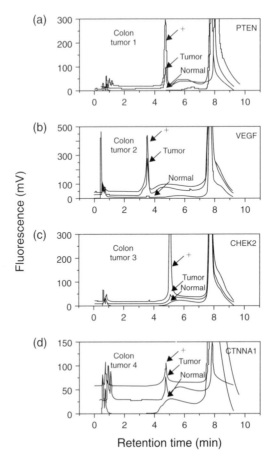

Figure 3. Colon cancer and normal tissue surgical samples examined via genome-wide iFLP.
Typical chromatograms are depicted from four gene segments indicating mutations in four colon cancer/normal tissue specimens: gene segments *PTEN* (*a*), *VEGF* (*b*), *CHEK2* (*c*), and *CTNNA1* (*d*). Mutation-positive controls (+) for each gene segment diluted 100-fold for (*a*), (*b*), and (*c*), and 1000-fold for (*d*), run under same conditions to indicate the retention times expected if mutations are present, are also depicted. Experiments were repeated three times.

was an approximately $1:10^5$ mutant-to-wild-type ratio (see *Fig. 2d*). Therefore, for starting genomic DNA material of 500 ng, which contains about 10^5 genomic copies, the lowest mutation frequency detectable is $\sim 10^{-4}$ to 10^{-5}. iFLP-detected mutations can be further verified by sequencing (16, 17).

2.3.3 iFLP-screening of clinical samples

Similar to the results obtained with repair-deficient colon cancer cell lines, iFLP screening of paired colon tumor and normal tissue with primers for phosphatase and tensin homolog (*PTEN*), vascular endothelial growth factor (*VEGF*), CHK2 checkpoint homolog (*CHEK2*), and catenin (*CTNNA1*) identified products in seven out of ten colon cancer samples that were absent in the corresponding normal tissues. This indicated the presence of *TaqI*-forming mutations in these tumors (8). These data indicate widespread genomic instability in sporadic colon tumors that manifest as numerous low-level mutations.

2.3.4 Potential applications of iFLP

Carcinomas are a generally heterogeneous populations of cells, generated by transformations including diverse mutations in individual genes and the loss or gain of entire chromosomes (aneuploidy) (18). To this end, diverse and widespread low-level mutations are expected to have profound implications in the ability of tumors to resist cytotoxic drug treatment (18). For example, specific mutations conferring resistance to treatment by the drug STI-571 in chronic myelogenous leukemia often pre-exist as low-level mutations in the untreated tumor (19–22). Following selection by the drug treatment, some of these low-level mutations become causative, prevalent genetic changes (19, 23–25). Accordingly, assessment of the extent and diversity of low-level mutations in tumors would be useful for estimation of the likelihood of drug resistance and tumor progression (18). iFLP provides a methodology for achieving this purpose in surgical specimens, and demonstrates the existence of low-level mutations in sporadic colon cancers at mutation frequencies of $\sim 10^{-2}$ to 10^{-4}.

iFLP enables the identification of unknown, random point mutations at low mutation frequencies. In contrast to methods such as denaturing gradient gel electrophoresis, dHPLC, and constant denaturant capillary electrophoresis, iFLP is not affected by polymerase errors as the genotypic selection step (*TaqI* digestion) occurs prior to PCR amplification, and it allows mutation detection at thousands of sequence positions per experiment. The number of sequences examined can be scaled at will by adjusting the number of inverse PCRs performed. It is estimated that about 3–5% of all possible mutations in each sequence can generate the *TaqI* recognition sequence via a single nucleotide change and therefore can be identified by iFLP.

By replacing *TaqI* with different restriction enzymes, more sequence changes can be detected (with an estimated upper limit of about 20% of all possible mutations). As the assay will work with nanogram amounts of input genomic DNA (8), it should be possible to perform high-throughput mutation scanning from minute biopsies obtained by needle aspiration or laser-capture microdissection.

3. TROUBLESHOOTING

- **The iFLP single-tube protocol is robust but time-consuming**
  To verify successful circularization of MseI-digested DNA prior to the high-throughput PCR step, which is the most 'expensive' step in terms of resources, the following approach can be used. Following the TaqI digestion step, and prior to adaptor ligation (see *Protocol 2*), the inverse PCR can be applied directly to the TaqI-digested product for ~45 cycles for two to three genes (this omits the generic PCR step). As long as the iFLP protocol has been applied correctly and circularization has occurred successfully, the anticipated inverse PCR products should produce gel bands with approximately equal intensities from both wild-type and other DNA, irrespective of whether or not the sample contains mutations.

- **Inverse PCR produces spurious dHPLC peaks that lie close to the position expected for the mutations**
  To control for this, each experiment is run in parallel with wild-type DNA expected to contain no mutations (8), and the inverse PCR step must be repeated at least twice. Only a consistent appearance of dHPLC peaks in the interrogated sample in the absence of such peaks in the wild-type control is indicative of true TaqI-forming mutations.

ACKNOWLEDGMENTS

The assistance of Mohamet Miri and Frank Haluska in obtaining tissue specimens from the Massachusetts General Hospital Tumor Bank (MA, USA) and of the Cooperative Human Tumor Network is acknowledged. This work was supported in part by training grant NIH 5 T32 CA09078 (to W. Liu) and by the JCRT Foundation.

4. REFERENCES

1. Sidransky D (1997). *Science*, **278**, 1054–1059.
2. Loeb LA (2001). *Cancer Res.* **61**, 3230–3239.
3. Parsons R, Li GM, Longley MJ, *et al.* (1993). *Cell*, **75**, 1227–1236.
4. Bhattacharyya NP, Skandalis A, Ganesh A, Groden J & Meuth M. (1994). *Proc. Natl. Acad. Sci. U. S. A.* **91**, 6319–6323.
5. Andrew SE, Reitmair AH, Fox J, *et al.* (1997). *Oncogene*, **15**, 123–129.
6. Loeb LA (1991). *Cancer Res.* **51**, 3075–3079.
7. Loeb LA & Christians FC (1996). *Mutat. Res.* **350**, 279–286.
★ 8. Liu W-H, Kaur M, Wang G, Zhu P, Zhang Y & Makrigiorgos GM (2004). *Cancer Res.* **64**, 2544–2551. – First description of the inverse PCR-based RFLP technique.
★★ 9. Wang F, Kaur M, Liu W-H, *et al.* (2005). *Clin. Chem. Lab. Med.* **43**, 810–816. – Application of the inverse PCR-based RFLP technique to clinical samples.
10. Umar A, Boyer JC, Thomas DC, *et al.* (1994). *J. Biol. Chem.* **269**, 14367–14370.
11. Bhattacharyya NP, Ganesh A, Phear G, Richards B, Skandalis A & Meuth M (1995). *Hum. Mol. Genet.* **4**, 2057–2064.
12. Glaab WE & Tindall KR (1997). *Carcinogenesis*, **18**, 1–8.
13. Parsons R, Li GM, Longley M, *et al.* (1995). *Science*, **268**, 738–740.
14. Shibata D, Peinado MA, Ionov Y, Malkhosyan S & Perucho M (1994). *Nat. Genet.* **6**, 273–281.

15. Malkhosyan S, McCarty A, Sawai H & Perucho M (1996). *Mutat. Res.* **316**, 249–259.
16. Kaur M, Zhang Y, Liu W-H, Tetradis S, Price BD & Makrigiorgos GM (2002). *Mutagenesis*, **17**, 365–374.
18. Liu W-H, Kaur M & Makrigiorgos GM (2003). *Hum. Mutat.* **21**, 535–541.
19. Kitano H (2003). *Nature*, **426**, 125.
20. Roche-Lestienne C, Soenen-Cornu V, Grardel-Duflos N, *et al.* (2002). *Blood*, **100**, 1014–1018.
21. Hofmann WK, Komor M, Wassmann B, *et al.* (2003). *Blood*, **102**, 659–661.
22. Kreuzer KA, Le Coutre P, Landt O, *et al.* (2003). *Ann. Hematol.* **82**, 284–289.
23. Liu W-H & Makrigiorgos GM (2003). *Leuk. Res.* **27**, 979–982.
24. Luzzatto L & Melo JV (2002). *Blood*, **100**, 1105.
25. Barthe C, Gharbi MJ, Lagarde V, *et al.* (2002). *Br. J. Haematol.* **119**, 109–111.
26. Gorre ME, Mohammed M, Ellwood K, *et al.* (2001). *Science*, **293**, 876–880.

CHAPTER 15

PCR methods for infectious disease diagnosis

Padmini Ramachandran, Andrew Hardick, Charlotte Gaydos, Samuel Yang, and Richard Rothman

1. INTRODUCTION

With the advent of PCR, methods for the molecular diagnostics for infectious diseases were forever changed (1). Significant technological advances from the basic method have occurred since the original introduction of PCR, increasing the ability of PCR methods to aid clinicians in diagnosing a range of infectious diseases (2, 3).

Conventional culture-based microbiological methods have been largely encumbered by the prolonged microbial growth requirement, which delays detection times (>24 h) and reduces sensitivity, particularly in the presence of antibiotics or in the detection of particular organisms (4). Conventional methods also increase biohazard risks in the hospital laboratory due to unnecessary propagation of highly contagious pathogens such as human immunodeficiency virus (HIV) and *Francisella tularensis* in culture-based systems (5). PCR assays permit the amplification of microbial genetic targets, thus allowing the detection of small amounts of pathogens including bacteria, viruses, fungal species, and parasites (e.g. *Trichomonas vaginalis*) in clinical specimens in a cultivation-independent manner. PCR has the potential to generate billions of copies of target DNA from a single copy in less than 2 h, significantly reducing the time to detection without compromising detection sensitivity (3, 4). In addition, detailed information stored in the genetic content of the detected pathogen can be uncovered for highly specific phylogenetic identification and pathogen characterization, making PCR an attractive alternative approach for infectious disease diagnosis in the clinical setting (3, 4).

PCR: *Methods Express* (S. Hughes and A. Moody, eds.)
© Scion Publishing Limited, 2007

1.1 PCR for infectious disease diagnosis in a clinical setting

PCR has potential use in a variety of practice settings, from rural or inaccessible field venues (i.e. military settings) to tertiary emergency and critical care settings (3). Multiple examples exist where PCR has influenced clinical practice and a few are provided below.

1.1.1 *Mycobacterium tuberculosis*

One of the earliest PCR applications for infectious disease diagnosis was detection of M. tuberculosis (3). Rapid detection of M. tuberculosis using PCR has allowed early recognition, isolation, and treatment of infected patients. This is in contrast to culture-based tests with week- to month-long delays, which can significantly impact on public health infection control measures (3).

1.1.2 Group B streptococci (GBS)

Other PCR tests have been demonstrated to have immediate life-saving applications. One well-established example is a real-time PCR assay for GBS, one of the commonest causes of neonatal bacterial infections in developed countries and a serious cause of neonatal morbidity and mortality. In 2002, the Centers for Disease Control and Prevention revised its guidelines for the prevention of pre-natal GBS disease to recommend that all pregnant women be screened for GBS carriage between 35 and 37 weeks of gestation to guide the use of intrapartum GBS antibiotic prophylaxis (6). In response, many institutions developed in-house PCR assays for rapid detection of GBS (7). Current assays have excellent performance characteristics (e.g. 99.6% sensitivity and 100% specificity), and the beneficial impact has been realized in the significant decline in the incidence of early-onset GBS infections in live births (6).

1.1.3 *Chlamydia trachomatis*

PCR-based methods have allowed significant progress to be made in recognition of asymptomatic and difficult-to-cultivate bacteria such as C. trachomatis, where practical limitations include:

- Difficulty in maintaining cell lines
- Transportation of samples at cold temperatures (cold chain)
- Nonviability of some chlamydial infectious elementary bodies (the stage in the chlamydia life cycle responsible for the transfer of infection from the host cell)
- Extended length of time for cultures to grow

These limitations render rapid and sensitive PCR the preferred method for routine screening and diagnosis (3, 8). Thus, PCR is now the preferred method for diagnosing C. trachomatis, Chlamydia pneumoniae and less frequently seen agents such as Chlamydia psittaci and Chlamydia pecorum (8). Cultivation of these agents would require a specialized laboratory with added expense and long delays in producing results using standard culture methods, in contrast to PCR, where equipment overheads and maintenance costs are relatively low (3).

1.1.4 Detection of other microbes

A variety of PCR-based assays are now commonly available in hospital-based laboratories. Examples of tests in typical tertiary care centers include those capable of detecting a range of common human viruses (e.g. Epstein–Barr virus, cytomegalovirus, varicella-zoster virus, adenovirus, enterovirus, herpes simplex virus types 1 and 2, human papillomavirus, respiratory syncytial virus, metapneumovirus, influenza A and B viruses, HIV types 1 and 2, hepatitis B and C viruses, parovirus B-19, BK virus, and JC virus) and bacteria (e.g. *Chlamydia* spp., *M. tuberculosis*, *Ehrlichia* spp., *Bordetella pertussis*, *Legionella* spp., and *Mycoplasma pneumoniae*) (4, 9).

1.2 Antimicrobial resistance profiling

PCR can also be useful in antimicrobial resistance profiling to aid clinician treatment decisions (3). A good example is its use in conjunction with sequencing for genotypic resistance testing in HIV; this method is now considered a standard of care in the management of HIV-infected individuals (3, 4). Another well-known example is its use in detecting methicillin-resistant *Staphylococcus aureus* (MRSA) where identification of the presence of resistance gene (*mec*A) by PCR has gained widespread acceptance and has significantly improved the ability of clinicians to target effective treatment (3, 10).

1.3 PCR and biodefense

The looming threat of a bioterrorism event also underscores the importance of a rapid, accurate, and high-throughput diagnostic method capable of large-scale screening of exposed victims. PCR-based assays for a number of category A bioterrorism agents have been developed, including variola major (smallpox), *Bacillus anthracis* (anthrax), *Yersinia pestis* (plague), and *Francisella tularensis* (tularemia) (11–13) The aforementioned characteristics of a PCR-based assay make it an ideal testing method in the event of a biothreat outbreak.

2. METHODS AND APPROACHES

2.1 Cost of PCR

The high cost associated with the Food and Drug Administration (FDA) approval process often limits commercialization of many PCR assays. In light of this restriction, a myriad of in-house PCR assays have been developed for a variety of infectious agents and have been published in many peer-reviewed scientific journals. Occasionally, biotechnology companies will offers analyte-specific reagents for which the diagnostic reagents are FDA cleared, but generally the entire kit is not cleared as a diagnostic method and there are no standing established protocols available. Instead, each laboratory must perform validation studies (3).

2.2 False-negative and -positive results

The balance between specificity and sensitivity can lead to the generation of false-negative and -positive results. Although neither is particularly desirable, no test is perfect, and the nature of PCR detection can lead to false-positive and -negative results. Both of these must be considered in the interpretation of PCR findings (3, 14).

The particular infectious agent usually dictates whether more false-positive or -negative results are acceptable in setting the cut-off point in the performance characteristics of the PCR assay under development. For example, a good screening assay will not miss any true positives but instead might produce some false positives. Although this may seem problematic, a second more-specific test can then be used to confirm the initial potential positive results, thus eliminating the false positives in the screening algorithm (3, 14).

False-positive results can be generated by contamination by previously amplified DNA copies, which is one of the most problematic aspects facing diagnostic PCR (3, 14). This problem can be dealt with in a variety of ways:

- In the laboratory, equipment such as pipettes, hoods, and centrifuges should be cleaned using 10% bleach.
- The use of commercial DNase and RNase solutions should be encouraged.
- The use of disposable plasticware helps to keep samples and equipment free of contamination.
- Pipette tips should always be aerosol barrier tips.
- Pre-PCR reagent preparation and sample processing should be performed in laboratory areas or rooms separate from post-amplification activities.
- Good laboratory practices should prevent background contamination as well as cross-contamination of samples.
- Monthly environmental sampling is recommended, where laboratory surface areas and equipment are sampled with sterile swabs moistened with water or buffer and then tested for amplicon contamination in the PCR of interest. Positive PCR results in the environment would then dictate extensive cleaning and decontamination of any area found to be contaminated.
- The use of enzymes such as uracil *N*-glycosylase in the PCR can further reduce contamination risk (3, 14).

False-negative results can be the result of inhibitors within the sample, improper PCR processing, or a low copy number of genetic material in a test sample (3, 14). The sample type, volume, and fraction (e.g. whole blood, plasma, buffy coat, etc.) should be considered carefully to maximize microbial yield. Sample processing methods based on the sample type chosen should be optimized (see below). Internal amplification controls (e.g. human β-globin gene) should always be incorporated into the PCR assay to monitor the presence of both purified sample DNA and potential PCR inhibitors (3, 14).

2.3 Sample processing for PCR

Sample processing is usually done with the purpose of extracting the DNA or RNA in a sample whilst simultaneously separating unwanted material such as cell-wall components and potential inhibitors of PCR from the sample of interest. Although processing is different for various sample types, they all follow the same basic steps:

- The sample is first collected in a medium that preserves the nucleic acid of an organism, although in some situations it may be collected without the addition of preservative medium. Sometimes, swabs transported in a dry state are acceptable for PCR (for example, for the detection of *Chlamydia*).
- The sample is then treated in order to lyse the infectious organism and free the DNA or RNA. In the case of bacteria, the cell wall is lysed, whereas viruses are treated to break down the protein coat. Treatment steps include:
  - Heating
  - Freezing
  - Chemical lysis
  - Sonication
- Once the organism has been lysed, DNA or RNA can be isolated and purified. The RNA or DNA can be extracted using phenol:chloroform, followed by precipitation in ethanol, column purification (e.g. Qiagen), or, in the case of robotic extraction equipment, subjection to magnetic bead isolation. Once the DNA or RNA is isolated, the rest of the sample is removed, and the DNA or RNA is eluted into a stable solution for the PCR (15).

Basic types of PCR used for infectious disease identification are:

- *Standard PCR*: the use of thermocycling to obtain large numbers of copies of specific target sequences.
- *Reverse transcriptase PCR (RT-PCR)*: the detection of RNA by converting the RNA into cDNA using reverse transcriptase, followed by PCR amplification.
- *Real-time PCR*: the use of fluorescent detection systems, capable of monitoring PCR product accumulation during the amplification process. This approach has improved reliability and speed, and allows quantitative analysis of PCR products.

2.4 Sample preparation from various human specimens

The use of PCR to diagnose infections operates on the principle that patient samples (such as urine, blood, cerebrospinal fluid, and urethral, vaginal, cervical, mucosal, sputum, and nasopharyngeal aspirates) from healthy individuals will not usually contain DNA from the infectious pathogen under consideration. Exceptions exist and include the presence of low copy numbers of organisms that can exist as part of the normal flora.

Sample processing is a critical and labor-intensive step that must be performed prior to PCR. The infectious agents/pathogens need to be isolated from the sample and concentrated to detectable levels in order to carry out amplification of the target DNA. In complex biological samples, there are known to be many inhibitory substances (9). This section discusses the processing of the samples to separate the nucleic acids from complex matrices and PCR inhibitors. The most common steps that can be carried out to avoid inhibitors and concentrate the target organism include culturing in enrichment medium, centrifugation, filtration, and two-phase separation methods.

2.4.1 Blood

Blood is one of the most complex sample matrices. Whole blood or its various fractions, including plasma, sera, and white cell buffy coat, should be chosen carefully for testing based on the fraction where the target organism sediments. For example, plasma or serum is preferred for detection of pathogens such as hepatitis B or C virus. Blood contains PCR inhibitors, including porphyrin from hemoglobin, heparin, and EDTA, which should be inactivated or eliminated prior to PCR. In general, thermal denaturation and phenol:chloroform:isoamylalcohol DNA extraction followed by ethanol precipitation is capable of inactivating and removing PCR inhibitors (9).

2.4.2 Urine

Urine can be processed in a number of different ways. For *C. trachomatis* for example, 10 ml of first-catch urine is centrifuged and the resulting pellet is treated further for extraction of DNA. Alternatively, the Roche Magna Pure LC robot can be used to perform direct DNA extraction from the urine sample, eliminating the need for centrifugation and DNA extraction steps. Urinary inhibitors vary widely among individuals and may vary from day to day in a single individual. Inhibitors in the urine may be removed by centrifugation, heating to 95°C for 10 min, freeze/thawing, or dilution (16).

2.4.3 Cerebrospinal fluid (CSF)

CSF has been used for the diagnosis of *Toxoplasma gondii* encephalitis, herpes simplex virus, enterovirus, cytomegalovirus, and meningitis-causing bacteria. Generally, small volumes can be used directly for phenol:chloroform extraction and ethanol precipitation. Alternatively, commercially available silica gel columns (e.g. QIAamp blood kit; Qiagen) can be used for processing CSF for extraction of DNA (9).

2.4.4 Sputum

Sputum samples have been evaluated extensively for the preparation of DNA from *M. tuberculosis*. Sputum is a complex matrix for DNA extraction because it is difficult to penetrate the viscous mucus and lyse the mycobacterial cell wall. Preliminary steps to liquefy sputum using mucolytic agents such as *N*-acetyl-ʟ-cysteine should be performed prior to nucleic acid extraction (17).

2.4.5 Stool samples

Stool samples are the most difficult and least studied with respect to extraction methods but generally require multiple steps. Samples should initially be resuspended in Tris/EDTA (TE) buffer with the addition of 5 µl of 100 mg/ml lysozyme. Following a 1 h incubation at 37°C, additional lysis is carried out by adding 50 µl of 5% SDS in TE buffer and 2.5 µl of 20 mg/ml proteinase K, followed by 10 min of boiling and centrifugation at maximum speed in a bench centrifuge to remove large particles. These steps are repeated until a clear solution is obtained, after which phenol:chloroform extraction and ethanol precipitation are performed (9).

2.4.6 Swabs

Vaginal and cervical swabs are generally tested for sexually transmitted diseases. The most common pathogens detected from the vaginal swab are *Chlamydia*, *Neisseria gonorrhoeae*, *Trichomonas*, and human papillomavirus. The swabs can be placed in TE buffer and the solution processed for the DNA extraction using a Roche MagNA Pure LC robot (8).

2.4.7 Tissue samples

Tissue samples are prepared according to the size and type of sample being evaluated. For example, for analysis of *Helicobacter pylori* from a gastric tissue biopsy specimen, a sample of approximately 20–30 mm^2 is required. In the case of kidney or lung biopsies, the sample should weigh approximately 10–200 mg (10). Prior to DNA extraction, the tissue should be minced and homogenized. Use of a Tekmar tissue homogenizer or alternative methods that disrupt the tissue are suitable. The tissue can then be processed for DNA or RNA extraction using conventional nucleic acid extraction methods (e.g. phenol:chloroform extraction or Qiagen column purification) (18).

2.5 Recommended protocols

The basic steps and protocols described in this section include: DNA/RNA extraction (*Protocols 1–4*), primer design, and conditions for standard PCR (*Protocol 5*) and RT-PCR from cDNA (*Protocols 6* and *7*).

Although individual microorganisms may require a unique extraction procedure, here we have included robust techniques for the preparation of nucleic acids from bacteria, fungi, and parasites that yield a suitable PCR template (19). Although pathogen DNA or RNA can be obtained directly from human specimens, it is often essential to prepare overnight cultures of pathogens in order to generate sufficient numbers of bacteria, fungi, parasites, or virus for nucleic acid extraction. Following extraction, it is essential to identify whether the process has been successful. A spectrophotometer can be used to identify whether or not DNA or RNA has been obtained.

2.5.1 DNA extraction from pathogens

The following protocols describe procedures for extraction of DNA from various pathogens.

Protocol 1

Bacterial DNA extraction (19)

Equipment and Reagents
- Overnight bacterial culture
- Lysozyme/RNase mixture (10 mg/ml lysozyme, 1 mg/ml RNase, 50 mM Tris/HCl, pH 8.0)[a] (all reagents from Sigma)
- STET (8% sucrose, 5% Triton X-100, 50 mM EDTA, 50 mM Tris/HCl, pH 8.0)[b] (all reagents from Sigma)
- STET-saturated phenol (Sigma)
- 4 M Lithium chloride (Sigma)
- Isopropanol
- 80% Ethanol
- TE buffer (10 mM Tris/HCl, pH 8.0, and 1 mM EDTA)
- 1.5 ml Microfuge tubes

Method
1. Pellet the bacteria from a 15 ml overnight culture into a 1.5 ml microfuge tube.

2. Resuspend the pellet with 300 µl of STET buffer and add 30 µl of lysozyme/RNase mixture.

3. Mix the solution and boil for 1 min 15 s.

4. Centrifuge at 15 000 *g* for 15 min.

5. Take the supernatant and phenol extract it with 150 µl of STET-saturated phenol.

6. Spin and take the supernatant. Add 0.1 vols 4 M lithium chloride (autoclaved). Incubate on ice for 5–10 min

7. Spin and take the supernatant. Add an equal volume of isopropanol, mix and incubate at room temperature for 5 min.

8. Centrifuge at 15 000 *g* for 15 min to pellet the DNA[c].

9. Wash the pellet with 500 µl of 80% ethanol[d].

10. Resuspend the pellet in 50–200 µl of TE buffer.

Notes
[a]Store at –20°C in small aliquots. Do not refreeze after thawing.
[b]Filter sterilize and store at 4°C.
[c]The pellet will not be visible.
[d]Use of 95% ethanol will cause any residual Triton X-100 to precipitate.

Protocol 2

Fungal DNA extraction (19)

Equipment and Reagents

- Hexadecyl trimethyl ammonium bromide (CTAB) extraction buffer (0.1 M Tris/HCl, pH 7.5, 1% CTAB, 0.7 M NaCl, 10 mM EDTA, 1% 2-mercaptoethanol). Add proteinase K to a final concentration of 0.3 mg/ml prior to use (all reagents from Sigma)
- Chloroform:isoamyl alcohol (24:1) (Sigma)
- Isopropanol
- 70% Ethanol
- TE buffer (10 mM Tris/HCl, pH 8.0, 1 mM EDTA) (Sigma)
- RNase A (Sigma)
- 2 ml Disposable centrifuge tubes
- Water bath

Method

1. Transfer the tissue sample[a] to a 2 ml disposable centrifuge tube.

2. Add 1.0 ml of CTAB extraction buffer.

3. Incubate in a 65°C water bath for 30 min.

4. Cool and add an equal volume of chloroform: isoamyl alcohol (24:1).

5. Mix and centrifuge at 2000 *g* for 10 min at room temperature.

6. Transfer the aqueous supernatant to a new tube and add an equal volume of isopropanol.

7. High-molecular-weight DNA should precipitate upon mixing and can be spooled out with a glass rod.

8. Rinse the DNA with 70% ethanol.

9. Centrifuge at 15 000 *g* for 2 min.

10. Air dry and add 1–5 ml of TE buffer containing 20 µg/ml RNase A. Resuspend the samples by placing the tubes in a 65°C water bath. Alternatively, allow pellets to resuspend overnight at 4°C.

Note

[a]Please refer to section 2.4.7 for the processing of tissue samples.

Protocol 3

Parasitic DNA extraction from blood (20)

Equipment and Reagents

- Extraction buffer: 1× SSC (0.15 M sodium chloride, 0.15 mM sodium citrate, pH 7.2) containing 0.01% saponin.
- Buffered sucrose solution (0.3 M sucrose, 5 mM $MgCl_2$, 10 mM Tris/HCl, pH 7.4)
- Mitochondrial lysis solution (10 mM Tris/HCl, 30 mM NaCl, 20 mM EDTA, 2% SDS, pH8.0)
- 4 M Sodium chloride (Sigma)
- Chloroform:isoamyl alcohol solution (24:1)
- Ultra-Turrax homogenizer (Rose Scientific Ltd)
- Proteinase K (100 µg/ml) (Sigma)

Method

1. Centrifuge the blood sample (200–400 µl) at 1000 g for 10–15 min, remove the supernatant, and resuspend the pellet in 1× SSC buffer (2.5 ml per 100 µl of original sample) and leave at 0°C for 15 min.

2. Wash the lysate with 2.5 ml per 100 µl of 1× SSC followed by centrifugation (3000 g) for 10 min. A small grey brown pellet is obtained[a].

3. For PCR diagnostic purposes, the mitochondrial DNA of the parasite is required. Resuspend the pellet in 8 ml of buffered sucrose solution.

4. Homogenize the suspension for 15 s using an Ultra-Turrax homogenizer at 0°C[b].

5. Centrifuge the homogenate at 2000 g for 5 min at 4°C.

6. Repeat steps 4 and 5 until no precipitate is detectable

7. Centrifuge the resulting supernatant at 12 000 g for 10 min at 4°C. Resuspend the precipitate in 4 ml of mitochondrial lysis solution. Incubate at 37°C for 20 min.

8. Add 100 µl of proteinase K and incubate at 37°C for 1 h. Add 0.5 ml of sodium chloride to give a final concentration of 0.5 M. Extract the crude mitochondrial DNA with an equal volume of chloroform:isoamyl alcohol (24:1).

9. Dialyze the resulting aqueous phase against a 1000-fold volume of 1× SSC.

Notes

[a]For total parasitic DNA extraction, resuspend the pellet in buffer A (0.1 M Tris/HCl, pH 8.0, 0.1 M EDTA containing 4% (w/v) sodium lauryl sarcosinate and 100 µg/ml proteinase K) and incubate the solution at 60°C for 60 min with occasional stirring. Fractionate the crude extract using a CsCl gradient (1.06 g CsCl per 1 ml extract) containing ethidium bromide in a Beckman TL-100. Remove the ethidium bromide and CsCl by standard procedures.
[b]Alternative homogenizers should yield similar results.

2.5.2 RNA extraction from viruses

For serum samples, collect the blood in sterile tubes with no anticoagulants, allow the blood to clot for 30 min, centrifuge, and remove the serum aseptically within 4 h. For plasma samples, collect the blood in sterile tubes using nonheparin anticoagulants, centrifuge, and remove the plasma aseptically within 4 h.

Protocol 4

RNA extraction from viruses

Equipment and Reagents
- Serum or plasma
- RNAzol B (Cinna/Biotecx)
- Chloroform:isoamyl alcohol (24:1)
- Isopropanol solution (0.3 ml of 3 M sodium acetate (Sigma), pH 4.0, 49.7 ml of isopropanol)
- 75% Ethanol
- DEPC-treated water

Method
1. Add 0.9 ml of RNAzol B to 0.1 ml of sample (serum or plasma) and vortex vigorously.

2. Add 0.1 ml of chloroform:isoamyl alcohol and vortex vigorously for 15 s.

3. Incubate on ice for 5 min. Centrifuge for 15 min at 12 000 g in a refrigerated microcentrifuge.

4. Transfer the aqueous layer (approximately 0.5 ml) to a fresh tube, add 0.5 ml of isopropanol solution and mix.

5. Store the samples on ice or at 4°C for 45 min. Centrifuge for 15 min at 12 000 g in a refrigerated microcentrifuge. Pour off the supernatant, resuspend the pellet in 25 μl of 75% ethanol, and vortex.

6. Centrifuge for 10 min at 12 000 g in a refrigerated microcentrifuge.

7. Pour off the supernatant and dry the pellet. Resuspend the pellet in 20 μl of DEPC-treated water.

2.5.3 Isolation of mRNA from total RNA

In some instances, isolation of mRNA from total RNA is required. Generally, mRNA is isolated from total RNA for studying gene expression through PCR. However, it is not necessary to isolate mRNA when creating cDNA or amplifying rRNA. Many protocols and commercial technologies are available for the extraction of mRNA from total RNA; however, they all operate under the same basic procedure (21). The system uses a biotinylated oligo(dT) primer to hybridize in solution at high efficiency to the 3′ poly(A) tail present on most eukaryotic mRNA molecules. The hybrids are captured using streptavidin coupled to paramagnetic particles and a magnetic separation stand (see *Fig. 1*) and washed at high stringency using RNase-free deionized water. The mRNA is eluted from the solid phase by the simple addition of RNase-free deionized water. This procedure yields an essentially pure fraction of mature mRNA after only a single round of magnetic separation. The isolated mRNA is suitable for all molecular biology applications, including *in vitro* translation and cDNA synthesis.

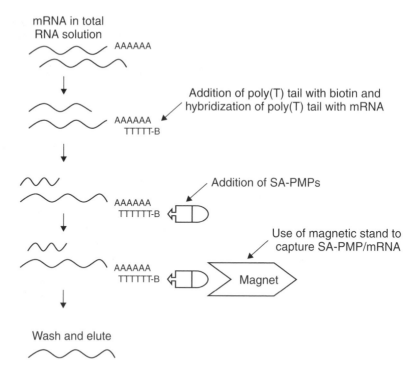

Figure 1. Schematic diagram of mRNA isolation from total RNA.
Total RNA solution is obtained from a patient specimen. mRNA is separated in a series of
steps starting with the addition of a poly(T) tail conjugated to biotin. The poly(T) tail
hybridizes only with the poly(A) tail of mRNAs. Streptavidin paramagnetic particles (SA-
PMPs) are added and bind the biotin, which is already hybridized to the mRNA. The SA-
PMP hybrids are then captured and washed with the aid of a magnetic stand. Finally, the
mRNA is eluted with RNAse-free deionized water yielding a pure fraction.

2.5.4 Primer design

The selection of primers for PCR can determine the efficiency and specificity of
the PCR. General recommendations for PCR primer selection and various aspects
to be considered when designing primers are as follow:

- *Location.* Primers are short, artificial DNA strands that are usually 18–25 bp
  and are complementary to the beginning or end of the DNA fragment to be
  amplified. This target is usually chosen to be a conserved region of the
  infectious agents' genome.
- *Amplicon size.* In general, amplicons range from 100 to 1000 bp. The lower
  limit is caused by the need to be able to visualize the amplicon on an agarose
  gel. The upper limit of 1000 bp is due to inherent difficulties in amplifying
  larger amplicons.

- *G/C content.* This is defined as the proportion of bases in the primer that are either G (guanine) or C (cytosine). Good PCR primers are generally selected to have a GC content of between 40 and 60%.
- *Considerations for optimal PCR.* The issues concerning PCR primer design can be divided into two categories: efficiency and specificity. Both of these are important to consider in most applications, but often factors that promote one of these will affect the other adversely.

There are many software programs available for primer design such as PRIMER PREMIER (www.premierbiosoft.com/primerpremier) and MACVECTOR (http://www.accelrys.com/products/macvector/). These programs are useful in designing primers with maximal efficiency for particular PCR conditions (19). Several example primers for common infectious agents are included here (see *Table 1*). Readers are advised to consult individual references for more detailed information regarding primer selection for a particular pathogen.

Multiplex PCR, in which several primer sets amplify several amplicons in the same reaction, adds a degree of complexity to the design of optimal primers (19). The additional issues to consider are those of possible heterodimer formation between the primers, as well as possible alternative hybridization sites within any of the target sequences.

Table 1. Primers for common infectious agents that can be diagnosed using PCR

Infectious agent	Target gene	Length of target sequence (bp)	Primer sequences (5′→3′)	Reference
Chlamydia pneumoniae	16S rRNA	197	GGTCTCAACCCCATCCGTGTCGG TGCGGAAAGCTGTATTTCTACAGTT	22
Chlamydia trachomatis	16S rRNA	315	GGCGTATTTGGGCATCCGAGTAACG TCAAATCCAGCGGGTATTAACCGCCT	9
Neisseria gonorrhoeae	CppB	241	CTTATCGTTTGGTTGATTC ACCAAGACCAAAGGTTTGACACTG	23
Trichomonas vaginalis	β-Tubulin	112	CATTGATAACGAAGCTCTTTACGAT GCATGTTGTGCCGGACATAACCAT	24
Streptococcus pneumoniae	16S rRNA	161	TGGAGCATGTGGTTTAATTCGA TGCGGGACTTAACCCAACA	2
Toxoplasma gondii	B1	98	TCCCCTCTGCTGGCGAAAAGT AGCGTTCGTGGTCAACTATCGATTG	25
Treponema pallidum	Bmp	617	CTCAGCACTGCTGAGCGTAG AACGCCTCCATCGTCAGACC	26
Mycobacterium tuberculosis	IS1081	306	CGACACCGAGCAGCTTCTGGCTG GTCGGCACCACGCTGGCTAGTG	27

2.5.5 Standard PCR

Following extraction of DNA from the pathogen and design of suitable primers, a standard PCR can be carried out.

Protocol 5

Standard PCR (28)

Equipment and Reagents
- Template DNA (5–10 ng/μl)
- 10× Amplification buffer (Invitrogen)[a]
- 50 mM dNTP solution (Invitrogen)
- 25 mM MgCl$_2$ (Invitrogen)
- *Taq* DNA polymerase (5 units/μl; Invitrogen)
- Forward primer (20 μM) in water (primer 1)
- Reverse primer (20 μM) in water- (primer 2)
- Nuclease-free water (Promega)
- Light mineral oil (Sigma)
- Thermal cycler

Method
1. Combine, per reaction[b]:
   - 5 μl of 10× amplification buffer
   - 1 μl of 50 mM dNTP solution
   - Primer-dependent MgCl$_2$ concentration[c]
   - 2.5 μl of 20 μM forward primer (primer 1)
   - 2.5 μl of 20 μM reverse primer (primer 2)
   - 1 μl of *Taq* DNA polymerase
   - 5 μl of template DNA
   - Water to a final volume of 50 μl

2. Centrifuge tubes or plates to consolidate the sample(s).

3. Place the tubes or the PCR plate into a thermal cycler.

4. Amplify the nucleic acids using the following cycling conditions[d]:
   - 30–50 cycles of 94°C for 30 s, 55°C for 30 s, and 72°C for 1 min
   - 1 cycle of 72°C for 7 min

Notes

[a]Alternatively, 'homemade' 10× amplification buffer can be prepared consisting of 500 mM KCl, 100 mM Tris/HCl (pH 8.3), and 15 mM MgCl$_2$. Autoclave the 10× buffer for 10 min at 15 p.s.i. on a liquid cycle. Allow the solution to cool to room temperature before making final adjustments to the pH. Divide the sterile buffer into aliquots and store at −20°C.
[b]If the thermal cycler is not fitted with a heated lid, overlay the reaction mixtures with 50 μl of light mineral oil.
[c]For details on how to optimize MgCl$_2$ concentrations, please refer to Chapter 1.
[d]Times and temperatures may need to be adapted to suit the particular reaction conditions. Extension should be carried out for 1 min for every 1000 bp of the target DNA.

2.5.6 RT-PCR

RT-PCR can be performed using either a one-step reaction, where a single master mixture containing both the reverse transcriptase and *Taq* DNA polymerase are included at the start of the reaction, or a two-step reaction, where RNA is initially reverse transcribed to cDNA and then the desired target cDNA is PCR amplified using specific primers. We use a two-step reaction as standard, as this can detect fewer than ten copies of target mRNA.

Protocol 6

cDNA synthesis from RNA (29)

Equipment and Reagents:
- mRNA (obtained from 3.6 µg of total RNA)
- dNTP mix (2.5 mM of each; Invitrogen)
- 0.1 M Dithiothreitol (Sigma)
- RNasin (40 units/µl; Invitrogen)
- Random primers (3 µg/µl; Invitrogen)
- 5× Reverse transcriptase buffer (Invitrogen)
- Reverse transcriptase (200 units/µl; Invitrogen)
- Nuclease-free water (Promega)

Method
1. For cDNA synthesis combine, per reaction (total volume of 18.8 µl):
   - 2.5 µl of dNTP mix
   - 2.0 µl of 0.1 M dithiothreitol
   - 0.4 µl of RNasin
   - 50–250 ng of random primers
   - 11.9 µl of RNA
   - nuclease-free water up to a total volume of 18.8 µl

2. Heat the reaction mixture for 5 min to 70°C and then cool quickly on ice for 5 min.

3. Centrifuge briefly to consolidate the sample(s).

4. To each sample add:
   - 5.0 µl of reverse transcriptase buffer
   - 1.2 µl of reverse transcriptase

5. Incubate the mixture at 37°C for 60 min followed by denaturation at 90°C for 5 min and cool quickly on ice for 5 min.

Protocol 7

PCR from cDNA (23)

Equipment and Reagents
- 10× Amplification buffer (Invitrogen)
- dNTP mix (2.5 mM of each; Invitrogen)
- *Taq* DNA polymerase (5 units/µl; Invitrogen)
- 25 mM MgCl$_2$ (Invitrogen)
- Forward primer (20 µM) in water (primer 1)
- Reverse primer (20 µM) in water (primer 2)
- Nuclease-free water (Promega)
- Light mineral oil (Sigma)
- Thermal cycler

Method
1. Combine, per reaction[a]:
   - 4.0 µl of dNTP mix
   - 5.0 µl of *Taq* DNA polymerase
   - 3.0 µl of MgCl$_2$
   - 2.0 µl of forward primer
   - 2.0 µl of reverse primer
   - 2.5 µl of cDNA from the reverse transcriptase reaction (*Protocol 6*)
   - Water up to 50 µl

2. Centrifuge the tubes or plates to consolidate the sample(s).

3. Place the tubes or PCR plate into the thermal cycler.

4. Amplify the nucleic acids using the following cycling conditions:
   - 30–50 cycles of 94°C for 45s, 53°C for 1 min, and 72°C for 2 min
   - 1 cycle of 72°C for 7 min

Note

[a]If the thermal cycler is not fitted with a heated lid, overlay the reaction mixtures with 50 µl of light mineral oil.

2.5.7 Nested RT-PCR

Sensitivity and specificity problems can arise when trying to detect low-copy-number transcripts. These can be overcome by using nested RT-PCR. The process is based on two consecutive rounds of amplification. The 3′-end external primer (see *Fig. 2*, P1) is used for the reverse transcriptase reaction and the 3′- and 5′-end primers (*Fig. 2*, P1 and P2) are used to perform the first PCR. The resulting amplification product is transferred to another Eppendorf tube containing a second pair of nested primers (nested PCR) that are internal to the initial pair (*Fig. 2*, P3 and P4). Alternatively, one of the external primers and a single nested primer (hemi-nested PCR) can also be used. The larger amplification fragment produced

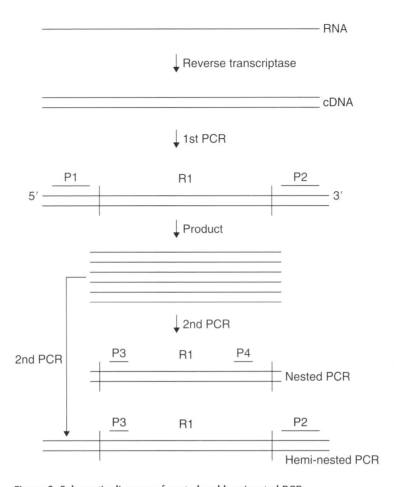

Figure 2. Schematic diagram of nested and heminested PCR.
cDNA is generated using reverse transcriptase from RNA using the P1 primer. The cDNA is then used as template for the first PCR using the P1 and P2 primer pair for amplification. Both of these primers are located outside the target of interest, R1. For nested PCR, a second PCR is performed using the product from the first PCR and two new primers, P3 and P4, that will amplify only R1. A hemi-nested PCR is similar to a nested PCR but uses one primer (P3) located inside R1 and one primer (P2) that is outside R1.

during the first reaction is used as the target for the second (nested or hemi-nested) PCR (19).

The final result of this nested or hemi-nested RT-PCR is a yield at least 100 times higher than a conventional RT-PCR. The main advantages of this new approach of nested RT-PCR are the high sensitivity afforded without risk of contamination and the possibility of using external primers with the lowest annealing temperature and internal primers with the highest, in contrast to previously described protocols.

2.5.8 Interpretation of PCR results in clinical settings

PCR results are used in a variety of settings for diagnosis of infectious diseases, but mainly they fall into two groups, clinical and research. Clinical results are used for patient treatment, whilst research results are generally not used for patient treatment but for pure research or surveillance. Regardless of use, most interpretations of PCR results are similar whether the application is for clinical or research use. For commercial assays, the manufacturer's package insert will dictate specific interpretation of the results. Generally, whether for commercial assays, clinical use, or research use, the run must include both positive and negative controls. This ensures that sample processing and extraction procedures are controlled:

- The negative processing control can inform the technician whether contamination has occurred during sample handle, processing, or amplification (3).
- The positive extraction control can indicate whether the sample extraction and PCR were performed properly.

These controls are handled identically to samples. The use of positive and negative amplification controls is also essential. Amplification controls determine whether the reagents used for the PCR are performing correctly. If these controls are acceptable, i.e. positive and negative extraction and amplification controls give expected results, then further analysis can be performed.

2.5.9 Conventional PCR versus real-time PCR

Most PCRs are qualitative, yielding a positive or negative result, but some are quantitative, i.e. real-time PCR, and can be used to determine relative amounts of infectious agent. The results of quantitative real-time PCR can be used to tailor treatment for patients.

The advantages of real-time PCR include: a very sensitive range of detection, (1–5 copies of target), smaller sample input, no post-PCR processing, and the ability to view amplification as it happens in real time within a reaction. The primary disadvantage is the necessity for synthesis of different probes for different target sequences (3).

One example of the utility of PCR in a clinical setting is exemplified in a study conducted by Yang *et al.* (2002) (2) at The Johns Hopkins University, where rapid diagnosis using a quantitative PCR assay was conducted on sputum samples for pneumococcal pneumonia in patients from an adult emergency department with suspected community-acquired pneumonia. Pneumococcal pneumonia is difficult to diagnose accurately in an acute-care setting due to inadequate sensitivity of conventional diagnostic tests. A further complicating factor is that sputum cultures, which are likely to have the highest diagnostic yield compared with other samples types, are considered to be unreliable due to the inability to differentiate colonization from infection (2). A rapid, quantitative, real-time PCR assay was compared with a composite reference standard comprising Gram staining of sputum samples and sputum/blood cultures. Calculated sensitivity and specificity were 90.0 and 80.0%, respectively, for the quantitative pneumolysin PCR assay (2). Although the authors envisaged this PCR being used in conjunction with conventional culture, its rapidity may be particular helpful in an acute-care setting for assistance in the selection of the most appropriate antibiotics to be used for treatment (2).

Although many PCRs for infectious agents show much promise, many have very large hurdles to overcome. For example, an article in the September 2006 issue of the *Journal of Clinical Microbiology* examined practical problems in implementing 16S rRNA gene sequencing in routine clinical microbiology laboratories. 16S rRNA gene sequencing can be used to identify difficult-to-identify or rarely encountered bacteria. Problems encountered include the cost of the test, cost of access to commercial databases for 16S rRNA gene sequences, incomplete sequences, and insufficient strain characterization in public databases such as GenBank. Mixed culture, mixed or ambiguous sequences, and the inability to provide identification for some samples past the genus level has also hampered the use of 16S rRNA gene sequencing. Although this method can be used successfully to identity infectious agents, it does highlight practical problems in using PCR methods in routine clinical microbiology laboratories.

In conclusion, PCR is a very basic procedure that has been tailored successfully for identification of a variety of infectious agents. Whilst this method holds the promise of identifying nearly every infectious agent in a range of clinical samples, optimization of sample processing methods and individual assay conditions are required before assays can be used in clinical situations.

3. TROUBLESHOOTING

Troubleshooting for infectious disease-orientated PCR is very similar to the problems that one would encounter with any other PCR. Problems are usually discovered only during analysis of PCR results. For standard or real-time PCR (see *Chapter 3*), problems are evident when no band (or no signal) or a weak band (or signal) is generated for controls. Standard PCR methods use end-point analysis methods such as agarose gels to allow detection of PCR amplification products, whilst real-time PCR has the advantage of the data being collected and displayed in real time. Improvements in imaging equipment have made gel resolution for PCR analysis easier and more accurate. No signal from amplification on a real-time instrument or no bands on an agarose gel (or specific matrix) can be indicative of several problems:

- Instrument trouble
    - ○ Proper instrument maintenance is essential for the successful generation of PCR results. Some real-time instruments require periodic calibration with regard to fluorescent dye absorption. The instrument user guide should be consulted to make sure that the instrument is serviced correctly.
    - ○ Contamination of the amplification block of real-time instruments can be detrimental to any PCR. Many real-time instruments have an automated

contamination check for wells of the amplification block. This check should be run periodically. Environmental swipe checks can be used to check for contaminating PCR amplicons. Decontamination procedures are instrument specific and should be consulted in the user's manual.

- Poor technique
  - Strict training and oversight is needed with novice technicians. Molecular biology techniques, specifically PCR testing, can be hampered by human factors. A few common operator errors include cross-contamination of samples, incomplete mixing of reagents, formation of air bubbles during pipetting, and imprecise pipetting. All of these problems can be addressed through practice and supervision.
- Poor extraction
  - Poor extraction is hard to identify but can have a huge impact on PCR results.
  - Use of positive and negative extraction controls is necessary for trustworthy PCR results. Results must confirm a negative and a positive extraction control. The controls can show cross-contamination of samples in the case of a positive result for a negative extraction control, or failed DNA/RNA extraction in the case of a negative result for a positive extraction control.
    - Positive and negative extraction controls should be made fresh with each run.
    - Re-extraction of samples with new controls is needed if extraction controls fail.
  - Low copy number in sample.
    - Retest, reprocess, or, if necessary and possible, recollect the sample.
- Poor release of DNA/RNA from the infectious agent
  - Use different lysis reagents.
    - Consult PubMed with regard to the best enzymes to use for specific infectious agents.
    - Commercial products usually have specifications for use, such as concentration and incubation temperature for specific lysis in different sample types.
    - Use different concentrations of lysis reagents.
  - New stock of lysis enzymes.
    - Carry out a temperature check of heaters used for lysis reactions.
    - Check control of the temperature and storage of lysis reagents.
    - Consult the expiration dates of lysis reagents.
    - Specific enzymes may be helpful for the release of DNA/RNA from specific types of infectious agents such as bacteria.
  - For optimized PCR, different concentrations of lysis reagents may be useful.
  - Try additional steps in the phenol:chloroform extraction.
  - Repeat the centrifugation steps until a clear supernatant is obtained.

- Poor removal of inhibitors
  - Carry out a dilution series of the sample for removal of inhibitors. Try a 10-fold, 100-fold, and 1000-fold serial dilution of the original sample.
  - Reprocessing/re-extraction of sample. Re-extraction of the original sample can correct possible errors in removal of inhibitors during the first extraction.
  - Recollection of the sample. If possible, some samples may have to be recollected if removal of inhibitors is not feasible.
  - The Internet can be a powerful tool to identify common inhibitors in specific sample types. Clinical samples can contain a myriad of inhibitors such as heme, microbicides, nitrate crystals in urine, etc.
- Poor storage of extracted samples
  - Samples can degrade if not stored at the proper temperature. Extracted samples can be stored in a refrigerator at 4°C overnight, at –20°C for short-term storage (generally 1 day to 3 months), or at –80°C or below for long-term storage.
  - Avoid excessive freeze/thawing of the samples. Alternatively, samples can be aliquotted out for single use, thereby reducing the number of total freeze/thaw cycles, which can degrade DNA/RNA.
- Poor primer/probe design
  - Poor primer or probe design can lead to poor binding affinity and a reduction in the utility of the PCR. Primer and probes may have to be redesigned.

For troubleshooting purposes, all of the above concerns should be addressed. The most useful points when troubleshooting PCRs specific for infectious agent diagnosis are extraction of genetic material from the clinical sample and removal of inhibitory substances within the sample (15).

ACKNOWLEDGMENTS

Funding for this project was from the Middle Atlantic Regional Center of Excellence (MARCE) for Biodefense and Emerging Infectious Diseases, Public Health Response Project 1, NIAID, NIH U54 AI57168-04. Under a licensing agreement with Cambrex Bioscience Walkersville Inc. and The Johns Hopkins University, Drs Rothman, Gaydos, and Yang are entitled to a share of the royalties received by the university on the sale of products related to the 16S rRNA gene assay described in this chapter. The terms of this agreement are being managed by The Johns Hopkins University in accordance with its conflict of interest policies.

4. REFERENCES

1. Erlich H, Gelfand D & Sninsky JJ (1991) Science, 252, 1643–1651.
2. Yang S, Lin S, Kelen G, et al. (2002) J. Clin. Microbiol. 40, 3449–3454.
3. Yang S & Rothman R (2004) Lancet Infect. Dis. 4, 337–348.
4. Church JD, Jones D, Flys T, et al. (2006) J. Mol. Diagn. 8, 430–432.
5. Farlow J, Wagner D, Dukerich M, et al. (2005) Emerg. Infect. Dis. 11, 1835–1841.

6. Rallu F, Barriga P, Scrivo C, Martel-Laferriere V & Laferriere C (2006) *J. Clin. Microbiol.* **3**, 725–728.
7. Bergeron MG, Ke D, Menard C, *et al.* (2000) *New Engl. J. Med.* **343**, 175–179.
8. Madico G, Quinn TA, Boman J & Gaydos CA (1999) *J. Clin. Microbiol.* **38**, 1085–1093.
9. Persing DH, Smith TF, Tenover FC & White TJ (eds) (1993) *Diagnostic Molecular Biology – Principles and Applications*. ASM Press, Washington, DC.
10. Healy CM, Hulten KG, Palazzi DL, Campbell JR & Baker CJ (2004) *Clin. Infect. Dis.* **39**, 1460–1466.
11. Fournier PE & Raoult D (2003) *J. Clin. Microbiol.* **41**, 5094–5098.
12. Yagupsky P & Baron E (2005) *Emerg. Infect. Dis.* **11**, 1180–1185.
13. Bode E, Hurtle W & Norwood D (2004) *J. Clin. Microbiol.* **42**, 5825–5831.
14. Hoorfar J, Wolffs P & Radstrom P (2004) *APMIS*, **112**, 808–814.
15. Lantz PG, Al-Soud WA, Knutsson R, Hahn-Hagerdal B & Radstrom P (2000) *Biotechnol. Ann. Rev.* **5**, 87–130.
16. Skov S, Miller P, Hateley W, Bastian IB, Davis J & Tait PW (1997) *Med. J. Aust.* **166**, 468–471.
17. Maher M, Glennon M, Martinazzo G, *et al.* (1996) *J. Clin. Microbiol.* **34**, 2307–2308.
18. Ross JA, Nelson GB & Holden KL (1991) *Nucleic Acids Res.* **19**, 6053.
★ 19. Barlett J & Stirling D (2003) *PCR protocols. Methods in Molecular Biology*, 2nd edn, vol. 226. Humana Press, New Jersey. – *Describes the DNA extraction protocols used in this chapter.*
20. Flores MV, Stewart TS & O'Sullivan WJ (1991) *Int. J. Parasitol.* **21**, 605–608.
21. Becker-Andre M & Hahlbrock K (1989) *Nucleic Acids Res.* **17**, 9437–9446.
22. Hardick J, Maldeis N, Theodore M, *et al.* (2004) *J. Mol. Diagn.* **6**, 132–136.
23. Herrmann B, Nystorm T & Wessel H (1996) *J. Clin. Microbiol.* **34**, 2548–2551.
24. Hardick J, Yang S, Lin S, Duncan D & Gaydos C (2003) *J. Clin. Microbiol.* **41**, 5619–5622.
25. Lin M-H, Chen T-C, Kuo T-T, *et al.* (2000) *J. Clin. Microbiol.* **38**, 4121–4125.
26. Bruisten SM, Cairo I, Fennema H, *et al.* (2001) *J. Clin. Microbiol.* **39**, 601–605.
27. Ahmed N, Mohanty AK, Mukhopadhyay U, *et al.* (1998) *J. Clin. Microbiol* **36**, 3094–3095.
28. Sambrook J, Fritsch E & Maniatis T (2001) *Molecular Cloning: a Laboratory Manual*. Cold Spring Harbor Laboratory Press, New York.
29. Raeymaekers L (2000) *Mol. Biotechnol.* **15**, 115–122.

CHAPTER 16

Use of PCR for DNA methylation analyses

Mario F. Fraga and Manel Esteller

1. INTRODUCTION

Genomic DNA methylation is one of the most important epigenetic modifications in eukaryotes and affects approximately 1% of DNA bases in humans. Most methylation occurs at the 5′ position of the cytosine pyrimidine ring with the resulting methylcytosine mainly being found in cytosine–guanine (CpG) dinucleotides (1). These dinucleotides are clustered into 'CpG islands' in the promoter regions of many genes. There are three main mechanisms that link methylation to cancer: transcriptional silencing via hypermethylation in the promoter regions of tumor suppressor genes (1), hypomethylation (demethylation) resulting in the failure to repress expression of tissue-restricted or proto-oncogenes, and genome-wide hypomethylation (neoplasia) leading to increased mutation rates and chromosome instability. The presence of methylcytosine in the promoter alters the binding of transcription factors and other proteins to DNA and recruits methyl-DNA-binding proteins and histone deacetylases that compact the chromatin around the gene transcription start site. Methylation of cytosine residues in genomic DNA plays a key role in the regulation of gene expression (2).

DNA methylation can be studied using a wide range of techniques. Global levels of methylcytosine in the genomic DNA can be measured by high-performance separation techniques or by enzymatic/chemical means. The latter are not as sensitive as the former and sometimes their resolution is restricted to endonuclease cleavage sites. Despite these drawbacks, enzymatic/chemical approaches are still commonly used, as they do not require expensive and complex equipment. When separation devices are available, high-performance capillary electrophoresis (HPCE) may be the best choice, as it is faster, cheaper, and more sensitive than high-performance liquid chromatography (3). Information on tissue-specific methylation patterns can be obtained using *in situ* hybridization-based methods using labeled anti-methylcytosine antibodies. DNA methylation

PCR: *Methods Express* (S. Hughes and A. Moody, eds.)
© Scion Publishing Limited, 2007

can be monitored in metaphase chromosomes and hetero/euchromatin within individual cells, which allows differences between normal and cancer cells to be assessed in the same sample. The use of methylation-specific microarrays is an emerging technology that allows the methylation status of many different genes to be determined in a single experiment.

2. METHODS AND APPROACHES

Genomic DNA methylation is a focus of scientific attention because it is a mechanism for control of gene expression. Most available methods for studying changes in the DNA methylation status of gene promoters containing a CpG island are based on bisulfite modification of DNA, which fixes the methylation pattern (4) (see *Protocol 1*), and subsequent PCR amplification, which allows methylated and unmethylated alleles to be distinguished. Without prior bisulfite treatment, methylation patterns would not be retained during subsequent PCR amplification. However, a disadvantage of the bisulfite treatment is that it degrades the DNA and reduces the yield.

2.1 PCR-based techniques

PCR-based techniques are used to study the methylation status within specific DNA regions, for example CpG islands within gene promoters. The two main approaches can be divided dependent upon whether the DNA is first modified by bisulfite (sodium bisulfite) treatment (see *Protocol 1*):

- *Bisulfite-treated approach.* All bisulfite-associated methods require subsequent PCR amplification of the bisulfite-modified DNA (discussed below), either by methylation-specific PCR (MSP), which can be used in conjunction with methylation-sensitive restriction endonucleases (combined bisulfite restriction analyses), genomic sequencing, pyrosequencing, or other approaches.
- *Nonbisulfite-treated approach.* For this approach, genomic DNA is used directly without any modification. The methodology relies on the use of methylation-sensitive restriction endonucleases (5) combined with Southern blotting or PCR detection, which may mean that results are limited to cleavage sites. This technique is described in more detail in Chapter 17.

2.1.1 PCR methods for studying DNA methylation following bisulfite treatment

There is an extensive range of methods for assessing the methylation status of cytosines located in specific DNA regions. Bisulfite modification converts unmethylated cytosine to uracil, while methylated cytosine does not react (see *Fig. 1*). This allows discrimination between methylated and unmethylated DNA. Bisulfite transformation of DNA can be followed by several methods, including MSP (see *Protocol 2*), bisulfite sequencing (BS; see *Protocol 3*), pyrosequencing (not detailed in this chapter; see http://www.pyrosequencing.com/ for further

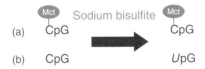

(a)

(b)

Figure 1. Sodium bisulfite treatment of CpG motifs.
Methylated cytosines are protected and remain unchanged (*a*), whilst unmethylated cytosines are deaminated to uracil after treatment with sodium bisulfite (*b*).

information), and combined bisulfite restriction analyses. All of these approaches are PCR-based and are thus suitable for analyzing paraffin-embedded tissues and scarce, purified DNA.

The study of methylation was greatly advanced by the use of sodium bisulfite to convert genomic DNA and stably fix the methylation pattern. However, as the subsequent PCR amplification step will amplify both genuine and artifactual observations exponentially, the success of the bisulfite treatment must be monitored using controls in all downstream assays to ensure correct interpretation of results. Possible issues include partial denaturation of DNA as only methylcytosines located in single-strands are susceptible to modification (4), incomplete modification, renaturation induced by high salt concentrations, and incomplete desulfonation. Although not described in this chapter, commercially available kits have been optimized to reduce the possibility of these problems.

2.1.2 Methylation-specific PCR

MSP (see *Protocol 2*) uses primers to distinguish methylated from unmethylated cytosines in bisulfite-modified DNA and is widely used for studying the methylation of CpG islands in gene promoters (6–8). The versatility of MSP has led to it being widely recommended as a rapid and cost-effective clinical tool for the initial evaluation of a wide range of patients with DNA-methylation-dependent diseases. MSP has also been employed successfully to assess the responsiveness of human cancer patients to alkylating agents. Using MSP, we have demonstrated hypermethylation of the DNA repair enzyme MGMT (8), a key factor in resistance to alkylating agents. MSP has also been useful in the detection of the presence of tumor DNA in the serum of cancer patients (9). However, there are a number of limitations to MSP that must be considered. MSP is not a quantitative technique, as it simply detects the presence or absence of methylation in a given region. In addition, experimental design and the use of appropriate controls are critical to ensure accurate interpretation of results, such as the use of wild-type control PCR primers to monitor and ensure complete conversion of nonmethylated cytosines to uracil and reduce the risk of false-positive results.

2.1.3 Experimental considerations

The differences between methylated and unmethylated alleles that arise from sodium bisulfite treatment are the basis of MSP (see *Fig. 1*) and are especially valuable in CpG islands due to the abundance of CpG sites. Primer design is a critical and complex component of the procedure and there are several computer programs that facilitate primer design (e.g. METHYL PRIMER EXPRESS, available at https://products.appliedbiosystems.com/ab/en/US/adirect/ab?cmd=catNavigate2&catID=602121). Bisulfite-converted DNA strands are no longer complementary, so primer design must be customized for each DNA chain. To optimize the PCR amplification step, the following critical requirements must be considered when designing the primers:

- The annealing temperature of both primers must be similar (±3°C) and always between 55 and 65°C.
- The PCR product should be small, between 80 and 175 bp, because the DNA will have been fragmented by the bisulfite treatment.
- Each primer should contain at least two CpG pairs. The number of CpG dinucleotides determines the methylation stringency of the primers.
- The sense primer should contain a CpG pair at its 3′ end in order to reduce the number of false positives.
- In order to avoid amplification of unmodified DNA, primers should contain nonCpG cytosines.
- The primers should avoid runs of three or more Ts.
- For newly designed primers, the assay should first be tested with unmethylated and *in vitro*-methylated control sequences.

We strongly recommend that primer design be performed using specialized software, as standard packages are not suitable due to the reduced complexity of sequence (i.e. three bases, not four).

The sensitivity of MSP enables the study of the methylation status of small samples of DNA, even those from paraffin-embedded or microdissected tissues. However, quantitative data and identification of cellular heterogeneity in populations are still not possible with this technique, whilst methylation status at the nucleotide level requires sequencing of the PCR products.

When performing methylation analysis using MSP, it important to consider that the extent of methylation can be greatly overestimated if:

- The PCR involves too many amplification cycles without ensuring that the reaction is in the linear response range with respect to the template concentration.
- The sequence can be amplified with both the methylation-specific primers and the primers for the unmethylated sequence. The appropriate control DNA sequences should first have been shown to yield no PCR product with the methylation-specific primers. Wild-type primers should also be used to show that the bisulfite conversion is complete.

2.1.4 Detection methods

The most straightforward way to visualize MSP reactions is to resolve the sample by agarose gel electrophoresis and UV exposure. Alternatively, combining MSP and denaturing high-performance liquid chromatography (MSP/dHPLC) can yield information on allele-dependent DNA methylation (10). Following PCR amplification, alleles can be resolved from the two populations of PCR products by dHPLC, as they differ at several positions within the amplified sequence. Distinct clonotypic epigenotypes can also be isolated and characterized by denaturing-gradient gel electrophoresis (11), which allows the detection of a small sequence change in every DNA region, including differentially methylated sequences.

2.1.5 Bisulfite sequencing

Sequencing bisulfite-altered DNA (see *Protocol 3*) is the simplest means of detecting cytosine methylation. After denaturation and bisulfite modification, dsDNA is obtained by primer extension and the fragment of interest is amplified by PCR. Methylcytosine may then be detected by standard DNA sequencing of the PCR products. Primers are designed so that they do not contain potentially methylated CpG sites, so that the target region is amplified in all samples. To optimize the PCR amplification step, the following critical requirements must be considered when designing the primers:

- The annealing temperature of both primers must be similar ($\pm\,3\,°$C) and always between 55 and 65°C.
- The PCR product should be between 200 and 400 bp. Bisulfite modification degrades the DNA and results in a bias towards shorter fragments. In addition, the treatment produces a biased base composition, making sequencing of long DNA fragments difficult.
- Each primer must not contain CpG dinucleotides to avoid methylation-slanted clone amplification.
- In order to avoid amplification of unmodified DNA, primers should contain nonCpG cytosines.

As with primer design for MSP, we strongly recommend that primer design for BS be performed using specialized software.

Cloning PCR products into plasmid vectors and then sequencing individual clones is an alternative method that, although slower, can provide methylation maps of single DNA molecules instead of the average values of the methylation status in the population of molecules provided by directly sequencing the PCR products. This method has been used to examine the detailed methylation status of many tumor-suppressor genes (12). Moreover, Radlinska *et al.* (13) described an alternative method by which methylcytosines can be localized directly in the primary product of the bisulfite treatment instead of in the PCR amplification product. They used a primer-extension mixture containing only three deoxynucleotides (dATP, dCTP, and dTTP), lacking dGTP, which produced an elongation stop at methylcytosine points. The positions of the methylcytosine are recognized by comparing the localization of run-off products and those of products of sequencing reactions performed on bisulfite-untreated template DNA.

2.1.6 Additional methods

- *Combined bisulfite restriction analysis* (14) is a highly specific approach based on the creation or modification of a target for restriction endonuclease after bisulfite treatment. The major advantage of the method is that it provides semi-quantitative data on the methylation status at specific regions in any DNA sample. Although it is specific, quantitative, and sensitive, this approach entails complete bisulfite modification and cannot be used for all DNA sequences as it is confined to restriction targets.

- *Methylation-sensitive single-nucleotide primer extension* employs bisulfite PCR combined with single-nucleotide primer extension to analyze DNA methylation status quantitatively at a particular DNA region without using restriction enzymes. Similar to other quantitative bisulfite PCR methods, single-nucleotide primer extension has a broad range of clinical applications due to its high sensitivity, specificity, and requirement for only small amounts of sample. Despite all of these advantages, PCR bias and analyses in CpG-rich regions can be a problem due to the difficulty of successfully designing primers lacking CpG dinucleotides.

- *Methylation-sensitive, single-strand conformation analysis.* Bisulfite modification of DNA generates accurate sequence disparities between methylated and unmethylated alleles, which can be resolved by single-strand conformation polymorphism. Methylcytosine detection via conventional single-strand conformation polymorphism requires PCR amplification of the DNA fragments of interest and denaturation of the double-stranded product, followed by either nondenaturing slab gel electrophoresis and detection via silver staining or HPCE techniques to detect PCR polymorphisms produced after bisulfite modification of the DNA (15). The HPCE approach has several advantages over the classical method:
  - ○ Simultaneous separation of alleles.
  - ○ Sequential analyses of a large number of samples.
  - ○ The relative percentage of methylation can be quantified accurately by using a suitable computer program to calculate the area under methylated and unmethylated peaks on the electropherogram.

- *'MethyLight'* (16). DNA is modified by the bisulfite treatment and amplified by fluorescence-based, real-time quantitative PCR using locus-specific PCR primers flanking an oligonucleotide probe with a 5′ fluorescence reporter dye and a 3′ quencher dye. The reporter dye is released enzymatically during the PCR, and fluorescence, which is proportional to the amount of PCR product and thus to the degree of DNA methylation, is detected using an automated nucleotide sequencer. Fluorescence detection greatly increases the sensitivity of the method, making it possible to detect a single methylated allele in 10^5 unmethylated alleles. The quantitative nature of the assay is based on real-time PCR, the inclusion of a methylated reference DNA, and the generation of standard curves using both methylated and unmethylated DNA. This assay allows the comparison of normal tissue samples with low levels of methylation with tumors with significantly more methylation. It also allows quantitative assessment of DNA methylation in specific sequences using high-throughput techniques, thereby enabling the rapid analysis of many DNA samples at many different sites.

Protocol 1

Treatment of genomic DNA with sodium bisulfite

Equipment and Reagents
- Genomic DNA (1 µg/sample)
- Sterile water
- 3 M NaOH (3 g of solid NaOH (Sigma) in 25 ml of water)
- 16 mM Hydroquinone[a] (27 mg of solid hydroquinone (Sigma) in 15 ml of water)
- 4 M Sodium bisulfite[b] (dissolve 3.8 g of solid sodium bisulfite[c] (Sigma) in 7 ml of water and adjust to pH 5.0 by adding drops of a concentrated solution of NaOH; add water to a final volume of 10 ml)
- Heating blocks or water baths set at 37 and 50°C
- Wizard DNA clean-up kit (Promega)
- Glycogen solution (10 mg/ml)
- 5 M Sodium acetate solution (dissolve 4.083 g of solid sodium acetate (Sigma) in 8 ml of water and adjust to pH 5.2 by adding drops of glacial acetic acid; add water to a final volume of 10 ml)
- Cold 70 and 100% ethanol

Method
1. Add 1 µg of DNA to a clean 1.5 ml Eppendorf tube and dilute to a final volume of 50 µl.
2. Add 5.7 µl of NaOH to the DNA and incubate for 10 min at 37°C.
3. Add 33 µl of the hydroquinone solution and 530 µl of $NaHSO_3$ solution.
4. Incubate overnight (16–18 h) at 50°C[d].
5. Purify the DNA using a Wizard DNA clean-up kit following the instructions of the supplier[e]. Elute the DNA in 50 µl of warmed (50°C) sterile water.
6. Add 5.7 µl of NaOH and incubate for 15–20 min at 37°C.
7. Precipitate the DNA by adding:
   - 1 µl of glycogen solution
   - 5.5 µl of sodium acetate solution
   - 125 µl of cold 100% ethanol
8. Mix by inverting the tube 10–20 times and then incubate overnight (16–18 h) at –80°C.
9. Centrifuge at 7000 r.p.m. at 4°C for 20 min. Discard the supernatant.
10. Wash the pellet with 400 µl of cold 70% ethanol, centrifuge again, and discard the supernatant.
11. Dry the pellet at room temperature.
12. Dissolve in 35–45 µl of water[f].

Notes
[a]Preparation of fresh hydroquinone is recommended for each experiment. *Hydroquinone is both light and air sensitive.* Once prepared, wrap the tube containing the hydroquinone in silver foil to avoid excessive exposure to light. In addition, to avoid excessive exposure to air, only open tubes containing hydroquinone just before they are required and quickly seal the tubes afterwards.
[b]Preparation of fresh sodium bisulfite is recommended for each experiment.
[c]As solid, it is sodium metabisulfite.
[d]Do not incubate for longer than 20 h, as longer incubation times result in increased DNA fragmentation.

eSimilar kits for DNA clean-up (e.g. Qiagen) should also be suitable; however, these have not been tested in our laboratory.
fCommercial kits are available for the bisulfite treatment of DNA.

Protocol 2

Methylation-specific PCR

Equipment and Reagents
■ FastStart *Taq* DNA polymerase (Roche) and accompanying 10× *Taq* polymerase buffer
■ 50 mM MgCl₂ (Roche)
■ Methylation-specific oligonucleotide primers (100 µM)
■ 2 mM dNTPs (Invitrogen)
■ Sterile water
■ 2% Agarose gel containing 10 ng/ml ethidium bromide
■ 6× Orange loading dye solution (Fermentas)
■ Equipment and reagents for agarose gel electrophoresis including 1× TBE agarose gel running buffer (10.8 g/l Tris base; 5.5 g/l boric acid; 4 ml/l 0.5 M EDTA, pH 8.0, diluted from a 10× stock; Sigma)
■ DNA size marker (100 bp ladder; Invitrogen)
■ UV light source

Method
1. Design two sets of methylation-specific oligonucleotides, one for the unmethylated reaction and another for the methylated reaction, using METHYL PRIMER EXPRESS software (https://products.appliedbiosystems.com/ab/en/US/adirect/ab?cmd=catNavigate2&catID=602 121; Applied Biosystems) (see *Fig. 2*, also available in the color section).

2. Add 2 µl of DNA, treated with sodium bisulfite (see *Protocol 1*), into two different PCR tubes, one for the unmethylated reaction and another for the methylated reaction.

3. Prepare a PCR master mix containing (per reaction):
   ■ 2.5 µl of 10× of *Taq* polymerase buffer
   ■ 0.75 µl of 50 mM MgCl₂
   ■ 2.5 µl of 2 mM dNTPs
   ■ 0.0625 µl of each methylation-specific oligonucleotide (100 µM) for either the unmethylated or methylated reaction
   ■ 0.1 µl of FastStart *Taq* DNA polymerase
   ■ Sterile water up to a final volume of 23 µl

4. Mix briefly by vortexing or pipetting.

5. Add 23 µl of the master mix to each sample and amplify the DNA using the following cycling conditions:
   ■ 1 cycle of 5 min at 95°C
   ■ 35 cycles of 95°C for 30 s, 30 s at a primer-specific annealing temperature, and 72°C for 30 s
   ■ 1 cycle of 7 min at 72°C

6. Analyze the PCR products by mixing 5 µl of the reaction mix with 1 µl of 6× orange loading dye solution and resolving the sample by agarose gel electrophoresis alongside a DNA size marker (see *Fig. 3*).

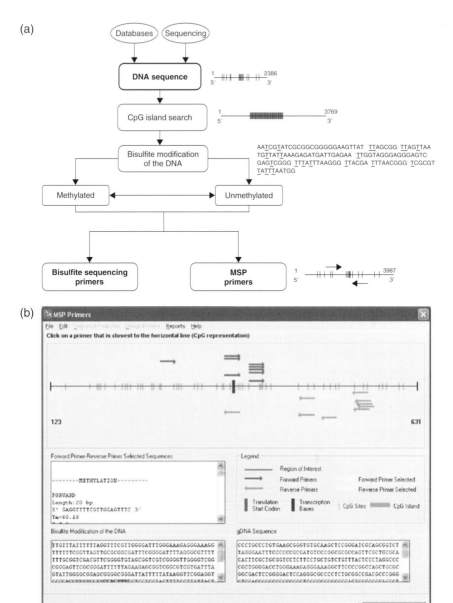

Figure 2. Designing experiments to study locus-specific DNA methylation status (see page xxv for color version).
(*a*) The first step is to identify a CpG island surrounding the transcription start point. Next, simulate the resulting DNA sequence after bisulfite modification of the DNA. Lastly, design oligonucleotide primers for MSP or BS. All of these steps can be achieved with the help of software such as METHYL PRIMER EXPRESS. (*b*) Representative output window in the design of MSP primers within the GSTP1 promoter obtained using METHYL PRIMER EXPRESS. The oligonucleotides are represented as red and orange arrows. Pink bars indicated CpG sites. The original and bisulfite-modified DNA sequences are shown in the lower boxes. The upper box shows the primer sequences.

Figure 3. Ethidium bromide agarose gel from an MSP experiment showing methylated and unmethylated samples.
U, PCR with primers for unmethylated DNA; M, PCR with primers for methylated DNA.

Protocol 3

PCR for direct bisulfite sequencing

Equipment and Reagents
- FastStart *Taq* DNA Polymerase (Roche) and accompanying 10× *Taq* polymerase buffer
- 50 mM MgCl$_2$ (Roche)
- Methylation-specific oligonucleotides (100 µM)
- 2 mM dNTPs (Invitrogen)
- Sterile water
- 0.8% Agarose gel containing 10 ng/ml ethidium bromide
- 6× Orange loading dye solution (Fermentas)
- Equipment and reagents for agarose gel electrophoresis including 1× TBE agarose gel running buffer (10.8 g/l Tris base; 5.5 g/l boric acid; 4 ml/l 0.5 M EDTA, pH 8.0, diluted from a 10× stock; Sigma)
- DNA size marker (100 bp ladder; Invitrogen)
- UV light source
- QIAquick gel extraction kit (Qiagen)
- BigDye terminator kit (Applied Biosystems)

Method
1. Design BS oligonucleotides using METHYL PRIMER EXPRESS software (Applied Biosystems) and referring to the criteria outlined in 2.1.2 (see *Fig.* 2, also available in the color section).

2. Combine (per reaction):
   - 4 µl of DNA treated with sodium bisulfite
   - 5 µl of 10× of *Taq* polymerase buffer
   - 1.5 µl of 50 mM MgCl$_2$
   - 5 µl of 2 mM dNTPs
   - 0.125 µl of each BS oligonucleotide primer (100 µM)
   - 0.2 µl of FastStart *Taq* DNA polymerase
   - Sterile water up to a final volume of 46 µl

3. Mix briefly by vortexing or pipetting.

4. Amplify the DNA using the following PCR conditions:
   - 1 cycle of 5 min at 95°C
   - 40 cycles of 30 s at 95°C, 30 s at a primer-specific annealing temperature, and 45 s at 72°C
   - 1 cycle of 7 min at 72°C

5. Analyze the PCR products by mixing 5 µl of the reaction mix with 1 µl of 6× orange loading dye solution and resolving the sample by agarose gel electrophoresis alongside a DNA size marker.

6. Extract and purify the PCR bands using a QIAquick gel extraction kit following the instructions of the supplier.

7. Using the PCR products as templates, perform a cycle sequencing reaction using the BigDye terminator kit version 3.1 following the instructions of the supplier[a].

8. Analyze the results using an ABI Prism 3130XL Applied Biosystems DNA sequencer or similar (see *Fig. 4*, also available in the color section).

Notes

[a]The quantity of PCR product used for sequencing varies depending on the size of the product. For 100–200 bp, use 1–3 ng of DNA; for 200–500 bp, use 3–10 ng of DNA; for 500–1000 bp, use 5–20 ng of DNA; for 1000–2000 bp, use 10–40 ng of DNA; for >2000 bp, use 20–50 ng of DNA (www.appliedbiosystems.com).

2.2 Summary

In contrast to genetic mutations, DNA methylation alters gene expression without DNA base changes. Currently, DNA methylation can be studied using a wide variety of experimental techniques. DNA methylation analyses can be global or locus specific. It is our opinion that the best way to analyze the DNA methylation status of a particular DNA sequence is the genomic sequencing of bisulfite-modified DNA. Once DNA methylation has been thoroughly characterized for a particular locus, MSP can be a useful tool for screening. Steps towards developing automation and multi-assay arrays will allow large numbers of samples to be checked simultaneously in the future.

3. TROUBLESHOOTING

- **No PCR bands for either MSP or BS**
  - Incorrect DNA modification: modify new DNA and use control modified DNA. Perform control PCR amplification with wild-type PCR primers to monitor the success of the bisufite treatment.
  - Poor-quality starting DNA: use new genomic DNA.
  - Deteriorated PCR reagents: change all of the PCR reagents.
  - Oligonucleotides not optimum for amplification: redesign the oligonucleotides.
  - Use a more sensitive technique such as MethyLight

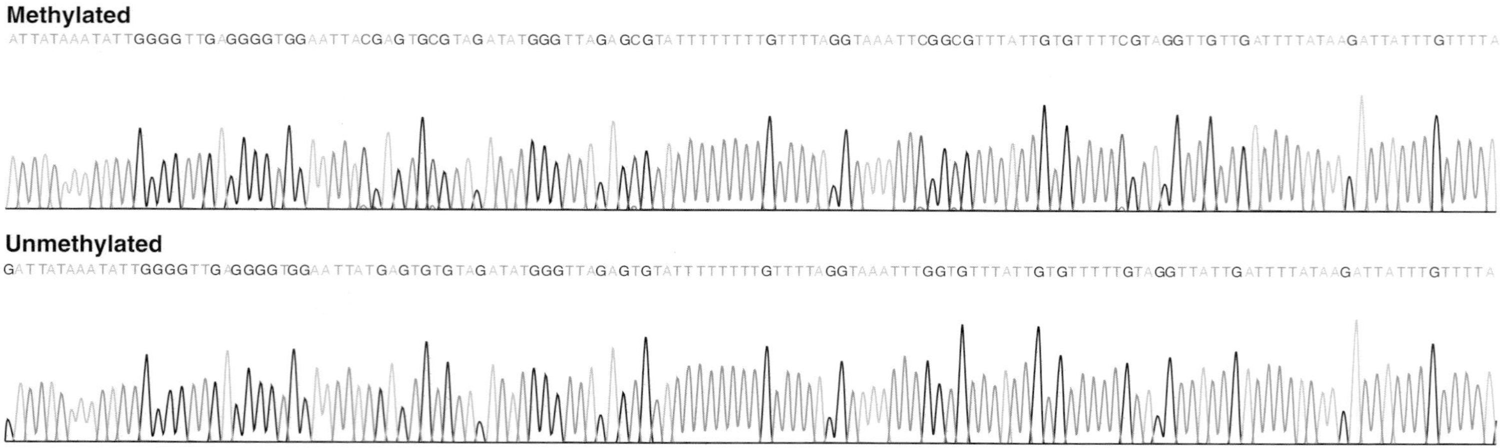

Methylated

ATTATAAATATTGGGGTTG.GGGGTGGAATTACG.GTGCGTAGATATGGGTTAG.GCGTATTTTTTTGTTTA.GGTAAATTCGGCGTTTATTGTTTTCGTAGGTGTTGATTTATAAGATTATTTGTTTA

Unmethylated

GATTATAAATATTGGGGTTG.GGGGTGG.ATTATG.GTGTGTAGATATGGGTTAG.GTGTATTTTTTTGTTTAGGTAAATTGGGTTTATTGTTTTTGT.GGTTATTGATTTATAAGATTATTTGTTTA

Figure 4. Bisulfite sequencing of methylated and unmethylated DNA (see page xxvi for color version).
Example electropherograms obtained after BS of the promoter region of a gene in a methylated (protected cytosines, blue traces) and unmethylated sample (no cytosines because they have been converted to thymines).

- Presence of bands in methylated and unmethylated reactions in MSP
  If this occurs in test samples, this might indicate a mixture of methylated and unmethylated clones in the sample, which is to be expected. However, if this happens in control samples and is thus not expected, it may indicate nonspecific oligonucleotides that will require redesigning with an increase in the number of CpGs per primer.
- Amplification of unmodified DNA in MSP and BS
  - Low modification rate: modify new DNA and use control modified DNA.
  - Low discriminatory strength primer: increase the number of cytosines not followed by guanines in the primer.
  - Difficult DNA region: pre-digest the DNA with restriction endonucleases.
- Bad sequencing traces in BS
  - Mixture of methylated and unmethylated clones in the sample: try cloning the PCR products and then sequence multiple clones.
  - Presence of unmodified DNA: modify new DNA and use control modified DNA.
  - Long amplicons: reduce the size of the PCR product.

4. REFERENCES

1. Bird AP (1986) *Nature*, **321**, 209–213.
2. Wolffe AP, Jones PL & Wade PA (1999) *Proc. Natl. Acad. Sci. U.S.A.* **96**, 5894–5896.
★★ 3. Fraga MF, Rodriguez R & Canal MJ (2000) *Electrophoresis*, **21**, 2990–2994. – *Paper describing methods for the rapid quantification of DNA methylation by HPCE.*
4. Shapiro R, DiFate V & Welcher M (1974) *J. Am. Chem. Soc.* **96**, 206–212.
5. Cedar H, Solage A, Glaser G & Razin A (1979) *Nucleic Acids Res.* **6**, 2125–2132.
6. Esteller M, Hamilton SR, Burger PC, Baylin SP & Herman JG (1999) *Cancer Res.* **59**, 793–797.
★★★ 7. Herman JG, Graff JR, Myohanen S, Nelkin BD & Baylin SB (1996) *Proc. Natl. Acad. Sci. U.S.A.* **93**, 9821–9826. – *Paper describing MSP for detecting the methylation status of CpG islands.*
8. Esteller M, Garcia-Foncillas J, Andion E, *et al.* (2000) *N. Engl. J. Med.* **343**, 1350–1354.
9. Goessl C, Muller M, Heicappell R, Krause H & Miller K (2001) *Ann. N.Y. Acad. Sci.* **945**, 51–58.
10. Baumer AU, Wiedemann M, Hergersberg & Schinzel A (2001) *Hum. Mutat.* **17**, 423–430.
11. Abrams ES, Murdaugh SE & Lerman LS (1990) *Genomics*, **7**, 463–475.
12. Melki JR, Vincent PC & Clark SJ (1999) *Cancer Res.* **59**, 3730–3740.
13. Radlinska M & Skowronek K (1998) *Acta Microbiol. Pol.* **47**, 327–334.
14. Xiong Z & Laird PW (1997) *Nucleic Acids Res.* **25**, 2532–2534.
15. Suzuki H, Itoh F, Toyota M, *et al.* (2000) *Electrophoresis*, **21**, 904–908.
16. Eads CA, Danenberg KD, Kawakami K, *et al.* (2000) *Nucleic Acids Res.* **28**, e32.

CHAPTER 17

PCR-based methods to determine DNA methylation status at specific CpG sites using methylation-sensitive restriction enzymes

Helmtrud I. Roach and Ko Hashimoto

1. INTRODUCTION

In recent years, there has been an explosion of interest in epigenetics, which refers to heritable changes in gene expression without alteration in the DNA sequence. The major epigenetic changes are histone acetylation and DNA methylation. Methylation takes place at cytosines that are adjacent to guanines, the so-called CpG sites. However, very little information is available on the methylation status of particular CpG sites in the promoter regions of specific genes from different cell types or cells from different developmental stages. This is due to several factors:

1. While DNA sequence analysis can be carried out on DNA extracted from any cell in the body, the methylation status for a particular gene is cell-type specific and thus needs to be analyzed separately for each tissue.
2. As PCR amplifications eliminate all CpG methylation, it is necessary to obtain sufficient genomic DNA directly from the relevant tissue/cells rather than by amplification of specific DNA segments.
3. Although the bisulfite modification method of detecting the presence or absence of methylation (see Chapter 16) gives valuable results, it is not best suited in all instances.

On the whole, conformationally relaxed chromatin (euchromatin) indicates transcriptionally active regions and is associated with hypomethylated DNA and acetylated histones, whereas compact chromatin (heterochromatin) is transcriptionally silent, hypermethylated, and bound to nonacetylated histones. DNA methylation is thus one of the principal mechanisms by which cells maintain

PCR: *Methods Express* (S. Hughes and A. Moody, eds.)
© Scion Publishing Limited, 2007

a stable chromatin configuration that regulates transcription. For further details, see (1–6).

2. METHODS AND APPROACHES

2.1 Methylation-sensitive restriction enzymes (MSREs)

The methylation sensitivity of certain restriction enzymes provides an elegant and straightforward method of assessing methylation status of specific CpG sites. There are over 50 such enzymes (see *Table 1*), all of which have at least one CpG site within their recognition sequence. However, the cleavage site is not necessarily within the CpG site or within the recognition sequence, as illustrated for *Bce*AI below, where the cleavage site is 12 bp downstream from the last base of the recognition sequence for the top strand and 14 bp upstream for the complementary strand:

$$5'-\text{ACGGCNNNNNNNNNNNNNN}\downarrow\text{NN}-3'$$
$$3'-\text{TGCCGNNNNNNNNNNNNNNN}\downarrow-5'$$

Table 1. Recognition sequences and sites of cleavage of commercially available MRSEs

In some cases, the cleavage site is some distance away from the recognition site. In this case, the numbers in parentheses indicate the distance between the last given nucleotide and the cleavage site. The first number is for the 5'→3' strand and the second for the complementary strand. W = A or T; Y = C or T; R = G or A. From: http://rebase.neb.com/rebase/rebase.html.

Enzyme	Site	Enzyme	Site	Enzyme	Site	Enzyme	Site
*Aat*II	GACGT↓C	*Bsm*BI	CGTCTC (1/5)	*Hae*II	RGCGC↓Y	*Nru*I	TCG↓CGA
*Aci*I	C↓CGC	*Bsp*DI	AT↓CGAT	*Hga*I	GACGC (5/10)	*Pae*R7I	C↓TCGAG
*Acl*I	AA↓CGTT	*Bsp*EI	T↓CCGGA	*Hha*I	GCG↓C	*Pml*I	CAC↓GTG
*Afe*I	AGC↓GCT	*Bsr*BI	CCGCTC (-3/-3)	*Hin*P1I	G↓CGC	*Pvu*I	CGAT⦰CG
*Age*I	A↓CCGGT	*Bsr*FI	R↓CCGGY	*Hpa*II	C↓CGG	*Rsr*II	CG↓GWCCG
*Asc*I	GG↓CGCGCC	*Bss*HII	G↓CGCGC	*Hpy*99I	CGWCG↓	*Sac*II	CCGC↓GG
*Asi*SI	GCGAT↓CGC	*Bst*BI	TT↓CGAA	*Hpy*CH4IV	A↓CGT	*Sal*I	G↓TCGAC
*Ava*I	C↓YCGRG	*Bst*UI	CG↓CG	*Kas*I	G↓GCGCC	*Sfo*I	GGC↓GCC
*Bce*AI	ACGGC (12/14)	*Cla*I	AT↓CGAT	*Mlu*I	A↓CGCGT	*Sgr*AI	CR↓CCGGYG
*Bmg*BI	CAC↓GTC	*Eag*I	C↓GGCCG	*Nae*I	GCC↓GGC	*Sma*I	CCC↓GGG
*Bsa*AI	YAC↓GTR	*Fau*I	CCCGC (4/6)	*Nar*I	GG↓CGCC	*Sna*BI	TAC↓GTA
*Bsa*HI	GR↓CGYC	*Fse*I	GGCCGG↓CC	*Ngo*MIV	G↓CCGGC	*Tli*I	C↓TCGAG
*Bsi*EI	CGRY↓CG	*Fsp*I	TGC↓GCA	*Not*I	GC↓GGCCGC	*Xho*I	C↓TCGAG
*Bsi*WI	C↓GTACG						

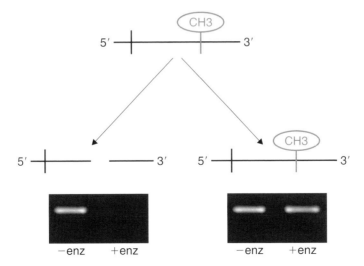

Figure 1. Principles of the MRSE assay for methylation status at specific CpG sites. In the absence of methylation (*left*), DNA is cleaved and cannot be amplified by PCR. In the presence of methylation (*right*), the DNA remains intact and can be amplified with suitable primer pairs.

2.2 Principle of the MSRE PCR method

The principle of this method is illustrated in *Fig. 1*. If the cytosine of a specific CpG site is not methylated, then the enzymes cleave as expected. However, if the cytosine of the CpG is methylated, then the enzymes cannot cleave and the DNA remains intact. By designing suitable primer pairs that bracket the region of interest, the presence or absence of methylation can be determined by the presence or absence of a PCR band following enzymatic digestion. The MSRE method is relatively easy and is particularly suitable for promoters with a limited number of CpG sites and if only small quantities of DNA are available. There are, however, two points for consideration:

- MSREs might not be available for a specific CpG site of interest.
- There may be several CpG sites cut by the same MSRE within the PCR-amplified region.

The latter is a particular problem for CpG island promoters, where the concentration of CpGs is very high and will require careful primer design (see section 2.3). However, if the aim is preliminary scanning of CpGs, for example to provide 'proof of concept' data that a change in methylation has taken place somewhere, then the MSRE method will provide useful results.

2.3 Identifying CpG sites and suitable MSREs

To find the promoter region sequence for the gene of interest, we recommend using the National Center for Biotechnology Information (NCBI) website (http://www.ncbi.nlm.nih.gov/).

- Select the 'Gene' option in the drop-down search box on the left. Type in the name of the gene of interest and then select the appropriate species from the list given. Scroll down to the section titled 'Related sequences', which, in most cases, will provide a list of accession numbers for genomic DNA. Clicking on the separate accession numbers should identify a file containing 1000–2000 bp of sequence upstream of the transcription start site, which corresponds to the promoter. Knowledge of the characteristics of the promoter will help to identify the most relevant area. If this information is not available, concentrate initially on the sequence ~1000 bp upstream of exon 1.
- Copy the sequence into Microsoft Word, then use the 'Find/replace' facility to mark all 'CG's by replacing them with **CG** (in bold, larger font and maybe change the color to red). Do the same for 'C G', then scan the right-hand edge for single Cs and determine whether the next line contains a 'G'. This will immediately give you useful information about whether the promoter contains a CpG island (many closely spaced CpG sites) or is a sparse CpG promoter.
- To find out which MSREs cut your sequence, go to http://www.restrictionmapper.org/ and select all 53 MSREs (listed in *Table 1*).
- Paste your sequence of interest into the box in the 'Sequence Info' section and click on 'Map sites'. This will generate a list of MRSEs that cut within your sequence and their cut positions. If you have more than one sequence to map, it is useful to do this straight away, as the program remembers the selected MSREs. Use the information to map the cut positions onto your sequence.
- Design PCR primers to bracket the region of interest. Primers should be located in regions with no CpG sites. An example of a sparse CpG promoter is the matrix metalloproteinase 13 (MMP-13) gene promoter, shown diagrammatically in *Fig. 2*. For this gene, the methylation status has been determined successfully for six out of the ten CpG sites in a 600 bp promoter region (7), using unique enzyme/primer combinations.

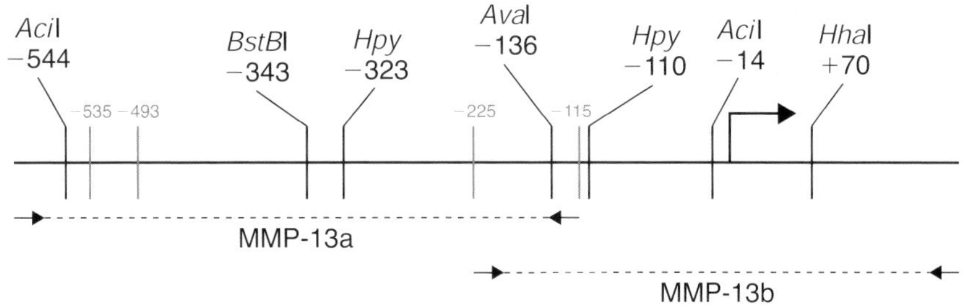

Figure 2. Example of a promoter suitable for the MSRE assay.
Ten CpG sites, cut by four different enzymes, are present within a 550 bp promoter region upstream from the transcription start site of MMP-13. Six of these sites (plus one in the coding region) can be assessed uniquely by various enzyme/primer combinations. However, four sites cannot be examined, as no MSREs are available to cut at these sites. The *Ava*I site at −136 was only evaluated with primers MMP-13b. Hpy, *Hpy*CH4IV.

2.4 Extraction of nucleic acids

In most studies, the aim is to link DNA methylation status directly to mRNA expression, for example, to determine whether loss of methylation at a particular CpG site is associated with induction of gene expression. In this case, simultaneous extraction of genomic DNA and RNA from the same specimen is highly desirable. However, in practice, this is not always possible. A particular problem arises if the tissue contains a lot of extracellular matrix with relatively few cells, as is the case for adult articular cartilage or muscle fibers. There are many commercially available kits (e.g. Qiagen) for the extraction of either RNA or DNA. However, combined kits for simultaneous RNA and DNA extraction are so far only available as microkits suitable for cells, not for whole tissues.

If DNA (or RNA) is to be extracted directly from whole tissues, a freezer mill is ideal to grind the tissue under liquid nitrogen into a fine powder (see *Protocol 1*). We have successfully used the Spex Certiprep 6750 for grinding human articular cartilage.

Protocol 1

Preparation of tissue for DNA or RNA extraction using a freezer mill

Equipment and Reagents
- Freezer mill (Spex Certiprep or similar)
- 10 ml Sterile tube
- Small spatula (to scoop out smashed specimen)
- Weighing scales

Method
1. Work in a well-ventilated room.

2. Pre-cool the freezer mill as instructed by the manufacturer.

3. Pour liquid nitrogen into the freezer mill[a]. Close the lid slowly to avoid spitting of the liquid nitrogen, as it boils vigorously the first time it is poured.

4. Weigh a 10 ml sterile tube.

5. Transfer samples into freezer mill cylinders with a metal rod, pre-cool, and run the milling cycles as instructed. Two samples can be pre-cooled while the first is being milled.

6. Scoop out the smashed samples and put them into the 10 ml tube using the spatula[b].

7. Reweigh the tube and calculate the milled weight of the sample (100–500 mg is ideal)[c].

8. Add lysis buffer as used in the DNA or RNA extraction kit and store the sample at –20°C or proceed directly to nucleic acid extraction (see *Protocol 2*).

9. Wash the cylinders and prepare the next samples.

Notes
[a]Wear a long-sleeved laboratory coat, cryo-resistant gloves, and goggles when handling liquid nitrogen.
[b]The yield of milled powder may be low as some sample inevitably remains inside the cylinder or attached to the metal rods. If complete recovery of the sample is critical, lysis buffer may be added directly to the mill cylinders.
[c]If both RNA and DNA are to be extracted from the same sample, it is best to split the milled powder into two and then use a maxi or midi kit for RNA or DNA, respectively. Alternatively, isolate the cells from the tissue prior to RNA/DNA extraction, as in *Protocol 2*.

2.4.1 Simultaneous extraction of RNA and DNA from cells

Depending on the tissue, it may be possible to isolate the cells from their matrix prior to DNA or RNA extraction. This has the advantage that no freezer mill is required and kits for simultaneous extraction of RNA and DNA can be used (see *Protocol 2*).

Protocol 2

Simultaneous extraction of RNA and DNA from cells

Equipment and Reagents
■ AllPrep DNA/RNA mini kit (Qiagen)
■ High-speed centrifuge
■ 2-Mercaptoethanol (Sigma)
■ 70% Ethanol
■ DNase I (Qiagen)

Method
1. The initial number of cells should not exceed 1×10^7 cells and it recommended to start with approximately 3×10^6 to 4×10^6 cells[a,b].

2. Add 2-mercaptoethanol or ethanol to the relevant buffers following the manufacturer's instructions.

3. Add 350–600 ml[c] of lysis buffer to 3×10^6 to 1×10^7 cells. Vortex and pass the lysate at least five times through a 20-gauge needle fitted to an RNase-free syringe, which homogenizes the cells.

4. Apply the lysate to an AllPrep DNA spin column placed in a 2 ml collection tube. Genomic DNA will bind to the DNA spin column.

5. Centrifuge for 1 min at ≥10 000 r.p.m. at room temperature.

6. Place the AllPrep DNA spin column in a new 2 ml collection tube and store at 4°C for later DNA purification. The flow-through will contain RNA.

7. RNA and DNA can be purified from the flow-through and spin column, respectively, following the manufacturer's instructions.

8. If possible, reverse transcribe the RNA immediately[d].

9. Quantify the RNA and DNA using a spectrophotometer[e].

Notes

[a]In human articular cartilage, cells only represent 3–5% of the tissue. Cells can be isolated by sequential treatment with:
• Trypsin (10% in PBS) for 30 min at 37°C
• Hyaluronidase (1 mg/ml in PBS) for 15 min at 37°C
• Collagenase B (10 mg/ml, Roche) for 12–15 h
The cells can then be washed in PBS and the cell suspension filtered through a sterile 70 μm sterile cell strainer, followed by centrifugation for 5 min at 1700 r.p.m. The main steps in this procedure are summarized in *Fig. 3*.
[b]In practice, a confluent T25 flask provides just enough cells, whilst a confluent T80 flask is ample.
[c]The volume of lysis buffer and wash buffer depends on the number of starting cells: for up to 3×10^6 cells use 350 ml of lysis buffer, and for up to 1×10^7 cells use 600 ml of lysis buffer.
[d]It is highly advisable to reverse transcribe RNA immediately into cDNA, which only takes a further 1.5 h. Alternatively, RNA can be frozen at –80°C.
[e]Quantification of RNA with the spectrophotometer often gives a low reading. This is because the RNA spin column preferentially binds mRNA, whereas small ribosomal RNA flows through. When total RNA is extracted by the Trizol method, the majority of RNA is ribosomal, which can easily be quantified. Using the column method, good PCR bands are often obtained even when the RNA released from the spin column cannot be quantified.

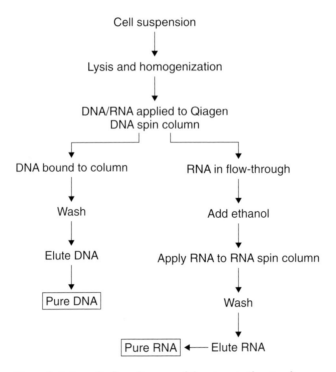

Figure 3. Schematic flow diagram of the steps in the simultaneous extraction of RNA and genomic DNA from the same specimen.

2.5 Detection of methylation status using MSREs

Although the bisulfite modification method has been used more extensively for determining methylation status, several groups have successfully used the MSRE PCR method (see *Protocols 3* and *4*) (7–9).

Protocol 3

Restriction enzyme digest

Equipment and Reagents
- Genomic DNA (5–10 ng/µl)
- Thermal cycler with heated lid[a] or water bath (37 or 65°C)
- Restriction enzyme(s)
- Relevant buffer and 100× bovine serum albumin (BSA) if required (New England Biolabs or similar)
- PCR-grade DNase/RNAse-free water
- PCR tubes, pipettes, and tips

Method
1. Check the concentration of supplied enzyme. For a 10 ml reaction, ~2–4 units are required[b].

2. For each reaction, set up[c]:
   - 9 µl of extracted genomic DNA (5 ng/µl)
   - 1 µl of 10× buffer
   - 0.1 µl of 100× BSA (if required)
   - 0.2–2 µl of restriction enzyme, containing 2–4 units of activity

3. For each specimen, set up a parallel sample, omitting the enzyme. This will be the no-enzyme control for the PCR.

4. Incubate at the recommended temperature for the enzyme (usually 37°C) for the recommended time (usually overnight).

5. To denature the enzyme, heat at 65°C (some enzymes need 80°C or cooling at 4°C) for 10–20 min.

6. Enzyme-treated DNA may be stored for up to 1 week at 4°C, but it is best to carry out the PCR immediately after enzyme digestion.

Notes
[a]The heated lid of a thermal cycler prevents condensation of the reaction mixture on the lid of the tube, which may be a problem in a water bath or incubator.
[b]Too much enzyme or incubation for too long increases nonspecific digestion, whilst digestion will not be complete when using too little enzyme or too short a time interval. Hence, it is essential to perform preliminary studies for each enzyme, using universally methylated DNA (Chemicon International, cat. no. S7821) or universally nonmethylated DNA. The latter can be obtained, using the GenomiPhi v2 DNA amplification kit (GE Healthcare Life Sciences).
[c]If several samples are to be treated with the same restriction enzyme, it may be useful to prepare a 'master mix' of enzyme, buffer, and BSA.

Protocol 4

PCR

Equipment and Reagents
- Digested DNA from *Protocol 3*
- Platinum PCR SuperMix (Invitrogen) containing PCR buffer, $MgCl_2$, and dNTPs
- Oligonucleotides (10 µM)
- Thermal cycler
- 2% Agarose gel containing 10 ng/ml ethidium bromide
- 6× Orange loading dye solution (Fermentas)
- Equipment and reagents for agarose gel electrophoresis including 1× TBE agarose gel running buffer (10.8 g/l Tris base; 5.5 g/l boric acid; 4 ml/l 0.5 M EDTA, pH 8.0, diluted from a 10× stock; Sigma)
- DNA size marker (100 bp ladder; Invitrogen)
- UV light source

Method
1. Combine per 25 µl reaction (it is not necessary to have a 50 µl reaction mixture, as recommended by the Invitrogen protocol):
   - 23 µl Platinum PCR SuperMix
   - 0.5 µl of each forward and reverse primer
   - 1 µl of DNA[a]

2. Mix briefly by vortexing or pipetting. Centrifuge at 12 000 ***g*** for 5–10 s to consolidate the sample.

3. Amplify the DNA using the following PCR profile:
   - 94°C for 2 min
   - 35 cycles of 94°C for 30 s, annealing using at a primer-dependent temperature for 30 s, and 72°C for 1 min
   - 72°C for 5 min

4. Analyze the PCR products by mixing 10 µl of the reaction mix with 1 µl of 6× orange loading dye solution and resolving the sample by agarose gel electrophoresis alongside a DNA size marker[b].

Notes
[a]It is important to PCR amplify the no-enzyme control DNA in order to assess the effect of enzymatic digestion on the PCR. A negative control is also essential to determine any possible PCR contamination.

[b]The assessment of methylation status depends on the difference in intensity between the band from enzyme-digested and that of the no-enzyme control. In an ideal specimen, a strong band is present in the no-enzyme control, whilst no band is detectable in the corresponding enzyme-digested sample.

2.6 Applications

In normal adult articular cartilage, the expression of proteases, such as MMP-13, is silenced. However, in osteoarthritis, many cartilage cells abnormally produce the enzyme, as has been shown by immunocytochemistry. Moreover, this abnormal expression is stably transmitted to daughter cells. These observations suggest that demethylation at some CpG sites in the promoter of the MMP-13 might have 'unsilenced' the gene and thus permitted abnormal expression of the protease. To test this hypothesis, genomic DNA isolated directly from human articular cartilage was subject to MRSE analysis for MMP-13. As can be seen in *Fig. 4*, all sites were fully methylated in control cartilage, but loss of DNA methylation was found at two CpG sites at −110 and −136 bp in the osteoarthritic specimen. When 16 osteoarthritis samples were compared with ten controls, demethylation at the *Hpy*CH4IV-cleavable site at −110 bp was present in the majority of osteoarthritis samples (7), which suggests that this particular CpG site may be important in the epigenetic 'unsilencing' of gene expression.

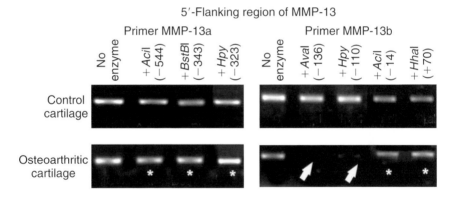

Figure 4. PCR of genomic DNA after treatment with MSREs.
Results are shown for the promoter region of MMP-13 in normal articular cartilage (top row) compared with osteoarthritic cartilage (bottom row). All of the CpG sites in normal articular cartilage are methylated, as indicated by the presence of PCR bands. Many of these CpG sites are also methylated in the osteoarthritic sample (asterisks). However, loss of methylation could be demonstrated in many osteoarthritic patients at the *Aval* site at −134 bp and the *Hpy*CH4IV site at −110 bp (arrows). Reprinted with permission from (7).

In parallel experiments, reverse transcriptase PCR demonstrated clear induction of MMP-13 in osteoarthritis samples with no expression in control (fracture neck of femur, #NOF) samples (see *Fig. 5*). MRSE analysis of the same samples using the *Hpy*CH4IV-cleavable site at −110 bp identified clear PCR bands for the controls (methylated, not susceptible to digestion; albeit reduced in intensity compared with undigested samples), whilst the corresponding bands were either very weak or absent in the osteoarthritis samples (unmethylated, susceptible to digestion). Although this example demonstrates the correlation between methylation and

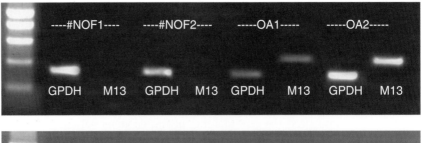

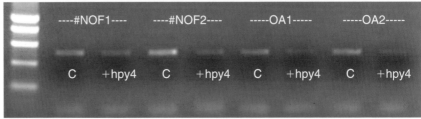

Figure 5. Combined determination of mRNA expression (top) and DNA methylation status (bottom).
Results are shown for two control samples, obtained from patients who had sustained a fracture of the neck of the femur (#NOF), and two osteoarthritic patients (OA). MMP-13 (M13) expression was absent in the control samples, but present in the OA samples. After digestion with *Hyp*CH4IV (hyp4) and PCR, a band was still present in the control samples, but was either very weak or absent in the OA samples. GPDH, glyceraldehyde 3-phosphate dehydrogenase; C, no-enzyme control for methylation status.

expression, it also pinpoints some of the problems of using conventional PCR for methylation detection (see section 3). The reduced intensity of the PCR bands for control samples could have been due to overdigestion or loss of methylation in some cells. In the latter case, this loss of methylation obviously was not yet sufficient to induce gene transcription. In the first osteoarthritis sample (OA1), a faint band was still present, but was very weak compared with the no-enzyme control. We would interpret this as absence of methylation.

3. TROUBLESHOOTING

- **A band is still present after enzyme digestion for the no-methylation control**
  As this negative control is universally unmethylated, one would expect an absence of bands for all enzyme/primer combinations. If a PCR band is still seen, then the conditions for enzyme digestion were not sufficient to produce complete digestion. Increase the enzyme concentration or the length of digestion (but see overdigestion below). As PCR is involved, a reduction in the number of cycles may also be beneficial.
- **Loss of band intensity is observed in the methylated control**
  As this control is universally methylated, one should not see any significant loss in band intensity after enzyme digestion. However, in practice we nearly

always observe some loss in intensity, presumably due to nonspecific DNA degradation. If the PCR band is considerably reduced in intensity compared with the no-enzyme control, then there is too much nonspecific degradation, possibly following overdigestion. Reduce the enzyme concentration or the length of digestion.

- **In a specimen where loss of methylation would be expected, a weak band is still visible**
  If conditions have been optimized using the positive and negative DNA controls, the presence of a weak band in enzyme-treated samples may be due in part to the nature of PCR and in part to the heterogeneity of the cells. If a strong band is found in the no-enzyme control together with a weak band in the enzyme-treated sample, a reduction in the number of PCR cycles may eliminate the weak band, whilst still showing the presence of a band in the no-enzyme control.

- **Heterogeneity of the cells with respect to the methylation status of a particular CpG site**
  This is a problem in both the MSRE PCR and the bisulfite modification methods. In any one cell, a particular CpG site is either methylated or not methylated, but this methylation status may vary within the cell population. For example, assume that a particular gene is induced in a specific situation so that 30% of the cells express this gene. Reverse transcriptase PCR will easily demonstrate expression where there was none prior to induction. Let us further assume that the induction was associated with loss of methylation at a specific CpG site, i.e. this CpG site had become demethylated in 30% of the cells. However, as this CpG site is still methylated in 70% of the cells, a strong PCR band would be present. Thus, it would be impossible to demonstrate changes in methylation status, even though these were present. To overcome these difficulties, we would recommend developing real-time PCR assays to measure the degree of methylation.

- **No PCR band is visible in the no-enzyme control**
  If the total amount of genomic DNA is too low, no PCR band may be visible in the no-enzyme control. In this case, either amplify for a further 5–10 cycles (recommended if a very faint band is visible) or use the PCR product for a further round of 30–35 cycles of PCR using the same primers. This is quite successful for most primers, but nonspecific amplification and additional random bands can also occur. In this case, design nested primers for the second round of PCR to amplify a region just inside the first primer product. This will usually eliminate the spurious additional bands.

4. REFERENCES

1. Roach HI & Aigner T (2006) *Osteoarthritis Cartilage*, **15**, 128–137.
★★ 2. Rodenhiser D & Mann M (2006) *CMAJ*. **174**, 341–348. – *A good review of the epigenetic literature in a clinical setting.*
3. Kress C, Thomassin H & Grange T (2001) *FEBS Lett*. **494**, 135–140.
4. Hendrich B & Tweedie S (2003) *Trends Genet*. **19**, 269–277.

5.   Fuks F (2005) *Curr. Opin. Genet. Dev.* **15**, 490–495.
6.   Davis CD & Uthus EO (2004) *Exp. Biol. Med. (Maywood)* **229**, 988–995.
★★★ 7.   Roach HI, Yamada N, Cheung KS, *et al.* (2005) *Arthritis Rheum.* **52**, 3110–3124. – *A good review of the methods described in this chapter.*
8.   Singer-Sam J, Goldstein L, Dai A, Gartler SM & Riggs AD (1992) *Proc. Natl. Acad. Sci. U.S.A.* **89**, 1413–1417.
9.   Pogribny IP, Pogribna M, Christman JK & James SJ (2000) *Cancer Res.* **60**, 588–594.

CHAPTER 18

PCR-based whole genome amplification

Nona Arneson, Simon Hughes, Richard Houlston, and Susan Done

1. INTRODUCTION

The use of PCR for whole genome amplification (WGA) is the opposite of the single-locus or limited-locus amplification function of PCR described elsewhere in this handbook. Specifically, PCR-based WGA has the goal of generating microgram quantities of genome-representative DNA from picogram or nanogram amounts of starting material. Furthermore, this amplification should introduce little, or ideally no, representational bias.

In both the research and clinical environment, the availability of genomic DNA in adequate quantities can be a major limiting factor with respect to the types of genomic analysis that can be used. An insufficient quantity naturally precludes those approaches that require large amounts DNA and as a result limits the total number of experiments that can be performed. Such experiments might include microsatellite analysis (1), single-nucleotide polymorphism (SNP) analysis (2) and comparative genomic hybridization (CGH) (3).

In researching human disease, tissues acquired for analysis (such as homogeneous cell populations generated by laser-capture microdissection, unique subpopulations of immune cells isolated by fluorescence-activated cell sorting, and buccal swabs) represent precious resources that often only consist of a few thousand cells. Unfortunately, this only provides sufficient DNA for performing basic genetic tests; however, WGA has the potential solution to this problem.

The choice of WGA methodology is often dependent on the source of the DNA. For instance, multiple-displacement amplification (4, 5), often viewed as the gold standard for WGA, has a requirement for high-quality, high-molecular-weight DNA, usually in excess of 2 kb, which can often only be obtained from fresh tissue or blood, and not from fixed tissue. In contrast, PCR-based methods are generally less affected by DNA quality and are more applicable to DNA extracted from various sources (fixed and fresh tissue).

PCR: *Methods Express* (S. Hughes and A. Moody, eds.)

The first PCR-based WGA approach, interspersed repetitive sequence PCR (6), utilized primers complementary to abundant genomic DNA repeats (Alu repeats) present in the genome. However, its success depends on these repeats being spread evenly throughout the genome and in close enough proximity to one another to generate a PCR product. Unfortunately, this is not the case and, as a result, interspersed repetitive sequence PCR preferentially amplifies regions of the genome rich in Alu repeats, thus introducing significant representational bias.

The next generation of PCR-based WGA methods used partially degenerate or random primers to initiate the PCR. These methods are known as degenerate-oligonucleotide-primed PCR (DOP–PCR) (7) and primer-extension pre-amplification PCR (PEP–PCR) (8), respectively, and have been widely used in research. Over time, these techniques have been adapted and improved (long products from low DNA quantities DOP–PCR (9) and improved PEP (I-PEP) (10)), which has led to more robust and less biased representations of the genome.

In more recent years, alternative approaches based on DNA fragmentation followed by the ligation of oligonucleotide adaptors have been developed (ligation-mediated PCR) (11–16). The DNA can be fragmented using sequence-specific restriction endonucleases (11), random endonucleolytic cleavage (12), or random shearing (16). The process begins by converting the complete genome into an *in vitro* molecular library of DNA fragments. This is followed by the addition of adaptor sequences, containing PCR priming sites, to both ends of every fragment. The fragments can then be amplified in a PCR using a single primer sequence (see *Fig. 1*).

The advantages of WGA are that it provides sufficient DNA for numerous assays to be performed immediately, as well as providing DNA that can be archived for future investigation. This chapter will provide an overview of the most commonly used PCR-based WGA techniques. The methods presented here demonstrate the use of WGA for generating large quantities of DNA for use with an array of biological assays.

2. METHODS AND APPROACHES

When implementing WGA in the laboratory, it is important to assess the entire experimental process closely, including sample collection, fixation, storage, and initial DNA extraction procedures, as all of these factors can affect DNA quality and thus have some bearing on the selection of the WGA technique. When using WGA, it is important first to validate the method selected and to become proficient in the technique before applying it to actual samples. Irrespective of the method selected, it is essential to establish that the results generated from the amplified DNA are indistinguishable from the results obtained from the original genomic DNA.

The DNA that can be amplified by PCR-based WGA includes DNA extracted from fixed, frozen, or archival tissue, whole blood, buccal swabs, single cells, sorted chromosomes, and laser-capture microdissected tissue (for microdissection of tissues embedded in paraffin, the sections must be deparaffinized prior to

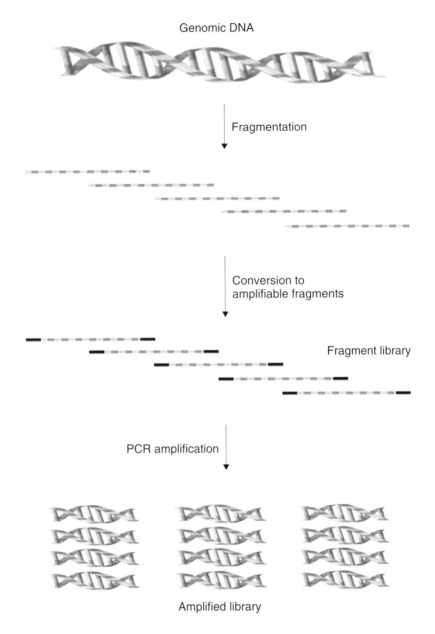

Genomic DNA

Fragmentation

Conversion to
amplifiable fragments

Fragment library

PCR amplification

Amplified library

Figure 1. Graphical representation of ligation-mediated PCR.
dsDNA is fragmented by enzymatic digestion (SCOMP, see section 2.3.1), random
shearing (PRSG, see section 2.3.2), or chemical cleavage (GenomePlex, see section
2.3.3) to generate a genomic fragment library. Adaptor complexes are then ligated onto
the fragment ends generating amplifiable fragments. Primers complementary to
sequences within the adaptors are then used to PCR amplify the fragments, generating an
amplified library. The PCR products can then be purified and assessed by agarose gel
electrophoresis.

microdissection). Genomic DNA may be extracted using a variety of commercially available methods such as the QIAmp DNA mini kit (Qiagen). Fixation of tissues can introduce sequence variations and reduce overall DNA quality. When studying such tissues, prior examination of the DNA by agarose gel electrophoresis will help determine the DNA quality.

The amplified DNA produced is suitable for a range of downstream genetic assays and thus has the potential for use not only in academic research, but also in commercial, forensic, and diagnostic laboratories.

Although we have named specific suppliers for the reagents and equipment used in this chapter, other manufacturers' products are likely to generate similar results (although it is clearly important for the user to verify this).

2.1 DOP–PCR

DOP–PCR was first described by Telenius *et al.* (7) and allows complete genome coverage in a single reaction. In contrast to the pairs of target-specific primer sequences used in traditional PCR, only a single primer, which has defined sequences at its 5′ end (containing an *Xho*l restriction site) and 3′ end and a random hexamer sequence between them is used for DOP–PCR. The primer sequence (with the *Xho*l restriction site highlighted in bold) is 5′-CCGA**CTCGAG**NNNNNNATGTGG-3′ where N represents A, G, C, or T.

DOP–PCR comprises two different cycling stages:

- Stage 1 – low stringency
- Stage 2 – high stringency

In stage 1, low-temperature annealing and extension in the first five to eight cycles occurs at many binding sites in the genome. At this stage, the 3′ end of the primer binds at sites in the genome complementary to the 6 bp well-defined sequence at the 3′ end of the primer (approximately 10^6 sites in the human genome). The adjacent random hexamer sequence (displaying all possible combinations of the nucleotides A, G, C, and T) can then anneal and tags these sequences with the DOP primer. In stage 2 (>25 cycles), the PCR annealing temperature is raised, which increases priming specificity during amplification of the tagged sequence. DOP–PCR as described in *Protocol 1* generates a smear of DNA fragments (200–1000 bp) that are visible on an agarose gel (see *Fig. 2a*).

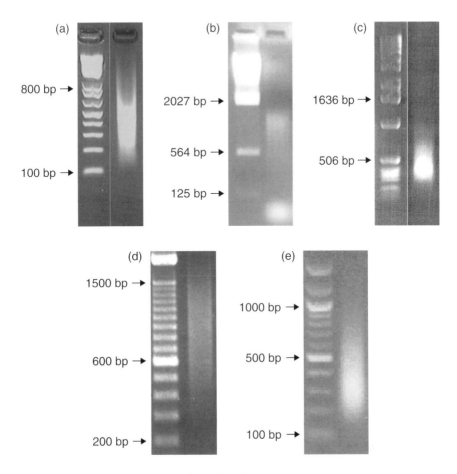

Figure 2. Amplification products of PCR-based WGA.
Examples of amplified DNA generated by DOP–PCR (*a*), I–PEP-PCR (*b*), single-cell comparative genomic hybridization (*c*), PCR of randomly sheared genomic DNA (*d*), and GenomePlex (*e*). The DNA used in all examples was obtained from fresh tissue or blood. The smears that would be obtained from formaldehyde-fixed, paraffin-embedded tissue are likely to be shorter.

Protocol 1

DOP–PCR

Equipment and Reagents
- Genomic DNA (5 ng/µl)
- 1 µM DOP–PCR primer (5′-CCGACTCGAGNNNNNNNATGTGG-3′)
- Thermosequenase (5 units/µl) and accompanying 10× Thermosequenase buffer (Amersham)
- *Taq* DNA polymerase (5 units/µl) and accompanying 10× PCR buffer (Invitrogen)
- 50 mM MgCl$_2$ (Invitrogen)
- 250 mM dNTP mix (Invitrogen)
- Nuclease-free water (Promega)
- 10 mM Tris/HCl (pH 8.0) containing 0.1 mM EDTA
- Thermal cycler (MJ Research)
- Agarose (Sigma)
- Ethidium bromide (10 mg/ml) (Sigma)
- 1× TBE buffer (10.8 g/l Tris base, 5.5 g/l boric acid, 4 ml/l 0.5 M EDTA, pH 8, diluted from a 10× stock; Sigma)
- DNA size marker (100 bp ladder; Invitrogen)
- 6× Orange loading dye (Fermentas)
- Electrophoresis apparatus

Method
Stage 1 – Low stringency[a]

1. Combine:
    - 2 µl of genomic DNA (DNA concentration 5 ng/µl)[b]
    - 1 µl of 10× Thermosequenase buffer
    - 1 µl of DOP–PCR primer
    - 0.1 µl of dNTP mix
    - 1 µl of Thermosequenase
    - Nuclease-free water up to a final volume of 10 µl[c,d]

2. PCR amplify the sample(s) in a thermal cycler using the following conditions:
    - 5 cycles of denaturation at 94°C for 60 s, annealing at 30°C for 60 s, ramping to 72°C over a 3 min period (3.5°C/15 s), and extension at 72°C for 2 min
    - Final extension at 72°C for 10 min

3. Store the reaction mixtures at −20°C for up to 3 days or continue immediately with step 2.

Stage 2 – High stringency[e]

1. Mix:
    - 4 µl of 10× PCR buffer
    - 2 µl of MgCl$_2$
    - 0.3 µl of DOP–PCR primer
    - 1.6 µl of dNTP mix
    - 0.4 µl of *Taq* polymerase
    - Nuclease-free water up to a final volume of 40 µl

2. Combine the step 1 and 2 reaction mixtures.

3. PCR amplify the sample(s) in a thermal cycler using the following conditions:
   - 30 cycles of denaturation at 94°C for 60 s, annealing at 55°C for 60 s, and extension at 72°C for 2 min
   - Final extension at 72°C for 10 min

4. Determine the size of the products by mixing 5 µl of the reaction mix with 1 µl of 6× orange loading dye solution and resolving the aliquot by agarose gel electrophoresis (1% agarose gel containing 20 µg of ethidium bromide (10 mg/ml) per 100 ml of agarose) alongside a DNA size marker[f].

5. Purify the amplified DNA[g,h,i].

Notes

[a]Low-stringency conditions are low-temperature annealing and extension that occur at several binding sites across the genome.

[b]It is important to determine the DNA concentration (µg/µl) accurately. This can be determined by using either the RediPlate 96 PicoGreen dsDNA quantitation kit (or similar) in conjunction with a fluorescence-based microplate reader or a standard spectrophotometer by taking the absorbance reading at 260 nm and multiplying it by 50 and then by the dilution factor.

[c]It is advisable to set up at least three reactions per sample (if there is sufficient DNA). This will provide enough amplified DNA to perform the required experiments and sufficient DNA for archiving.

[d]It is important to include a negative control, which includes all of the reaction constituents with the exception of DNA. The negative-control lane should not show any amplification. If it is does, this suggests possible contamination and therefore reactions must be repeated.

[e]High-stringency conditions are an elevated annealing temperature, allowing more specific priming of the fragments tagged with the primer sequence.

[f]The size of the amplification product depends on the quality of the starting DNA. If it is high-molecular-weight DNA (extracted from fresh tissue or cells), the amplification product smear will range from 50 to 1500 bp. However, if the DNA is of low molecular weight (e.g. from fixed tissue), then the size of the amplification product smear will generally be below 500 bp.

[g]For some downstream applications, it may be necessary to both purify and quantify the amplified DNA before use. The choice of DNA purification procedure is up to the user.

[h]DNA yields of approximately 4–6 µg are typically obtained in a 100 µl reaction.

[i]Standard protocols for array CGH can be used, but when using amplified DNA, we recommend labeling 3–4 µg of DNA instead of the standard 1 µg.

2.2 PEP–PCR

PEP–PCR (see *Protocol 2*) was first reported by Zhang *et al.* (8) and in contrast to DOP–PCR uses totally degenerate 15 mer PCR primers (5′-NNNNNNNNNNNNNNN-3′, where N can be any one of the four nucleotides A, G, C, or T). An additional difference between DOP–PCR and PEP–PCR is that in the latter the number of potential priming sites is orders of magnitude larger.

The effectiveness of PEP–PCR has been increased by several alterations. The I-PEP-PCR (10) approach combines the methods of PEP–PCR and 'long PCR' (17) and uses a DNA polymerase cocktail including *Taq* DNA polymerase (to carry out the primer extension as in a traditional PCR) and a proofreading DNA polymerase (to provide 3'→5' exonuclease activity to excise misincorporated nucleotides that slow the progression of *Taq* DNA polymerase). The result is far more efficient WGA because of increased fidelity due to the removal of the misincorporated nucleotides. Similar to DOP–PCR, PEP–PCR generates a smear of DNA fragments that are visible on an agarose gel (see *Fig. 2b*).

Protocol 2

I-PEP–PCR

Equipment and Reagents
- Genomic DNA (5 ng/μl)
- 280 μM 15 mer random primer (5'-NNNNNNNNNNNNNNN-3')
- Expand HiFi polymerase (3.6 unit/μl) and accompanying 10× Expand HiFi buffer (Roche)
- 25 mM MgCl$_2$ (Roche)
- 10 mM dNTP mix (Invitrogen)
- 1 mg/ml Gelatine (Sigma)
- Nuclease-free water (Sigma)
- Thermal cycler (MJ Research)
- SpeedVac SC110 (Savant)
- Agarose (Sigma)
- Ethidium bromide (10 mg/ml; Sigma)
- 1× TBE buffer (10.8 g/l Tris base, 5.5 g/l boric acid, 4 ml/l 0.5 M EDTA, pH 8, diluted from a 10× stock; Sigma)
- 6× Orange loading dye (Fermentes)
- DNA size marker (100 bp ladder; Invitrogen)
- Electrophoresis apparatus

Method
1. Combine:
   - 10 μl of genomic DNA (DNA concentration 5 ng/μl)
   - 6 μl of 10× Expand HiFi buffer
   - 6 μl of MgCl$_2$
   - 0.6 μl of dNTP mix
   - 3.4 μl of random primer
   - 3 μl of gelatine
   - 1.4 μl Expand HiFi polymerase
   - Nuclease-free water up to a final volume of 60 μl[a,b]

2. PCR amplify the sample(s) in a thermal cycler using the following conditions:
   - Initial denaturation at 94°C for 2 min
   - 50 cycles of denaturation at 94°C for 60 s, annealing at 28°C ramping to 55°C over a 2 min period (0.1°C/s), 55°C for 4 min, and elongation at 68°C for 30 s
   - Final elongation step of 8 min at 68°C

3. Determine the size of the products by mixing 5 μl of the reaction mix with 1 μl of 6× orange loading dye solution and resolving the aliquot by agarose gel electrophoresis (1% agarose gel containing 20 μg of ethidium bromide (10 mg/ml) per 100 ml of agarose) alongside a DNA size marker[c].

4. Purify the amplified DNA[d,e,f,g].

Notes

[a]It is advisable to set up at least three reactions per sample (if there is sufficient DNA). This will provide enough amplified DNA to perform the required experiments and sufficient DNA for archiving.

[b]It is important to include a negative control that includes all of the reaction constituents with the exception of DNA. The negative-control lane should not show any amplification. If it is does, this suggests possible contamination and therefore reactions must be repeated.

[c]Typically, a DNA smear ranging from 150 to 1500 bp is generated (see *Fig. 2b*). However, similar to DOP–PCR, this smear will be shorter when using DNA obtained from fixed material.

[d]For some downstream applications, it may be necessary to both purify and quantify the amplified DNA before use. The choice of DNA purification is up to the user.

[e]DNA yields of approximately 3–4 μg are typically obtained in a 60 μl reaction.

[f]The amplification efficiency may be improved in some cases by adding dimethyl sulfoxide to a final concentration of 5% (Merck).

[g]It is advisable to use 25–50 ng of amplified DNA for subsequent gene/region-specific PCR.

2.3 Ligation-mediated PCR

This set of methods involves ligating an adaptor sequence onto a 'representation' of DNA molecules generated following enzymatic digestion, random shearing, or chemical cleavage.

2.3.1 Single-cell comparative genomic hybridization (SCOMP)

SCOMP (*Protocols 3–6*) was first reported by Klein *et al.* (1999) (11) for the genetic analysis of single cells. It is a form of ligation-mediated PCR that was specifically designed for WGA of extremely limited sources of genomic DNA. It has the advantage that the reaction volume is purposely kept to a minimum and all buffers are optimized to eliminate the need to purify the reaction between steps. In addition, the entire reaction is performed in a single tube. This avoids initial template loss and reduces the risk of PCR contamination occurring.

SCOMP begins by converting the genome to a high-complexity representation (13) with a fragment size of less than 2 kb by digesting with the restriction enzyme *Mse*I. This results in a smear in the range of 100–1500 bp (11). Following enzyme digestion, adaptors containing specific primer sequences (specific to the restriction enzyme used) are ligated onto the ends of the genomic DNA and subsequently amplified in a high-stringency PCR. This results in a smear of PCR products in the range of 100–1500 bp, which can be visualized by agarose gel electrophoresis (see *Fig. 2c*).

Protocol 3

Restriction enzyme digestion of template genomic DNA

Reagents
- Genomic DNA (5 ng/μl)
- One-Phor-All Buffer Plus (GE Healthcare)
- *Mse*I (20 units/μl; New England Biolabs)
- Nuclease-free water (Promega)

Method
1. Combine:
   - 0.2 μl One-Phor-All Buffer
   - 0.2 μl *Mse*I
   - 3 μl DNA template[a]
   - Nuclease-free water up to a final reaction volume of 5 μl

2. Mix well by pipetting up and down and incubate at 37°C for 3 h[b].

3. Inactivate the *Mse*I by incubating the reaction at 65°C for 5 min.

4. Consolidate the reaction mixture by centrifugation at 12 000 *g* for 10–20 s.

Notes

[a]The 3 μl of template can be added directly after proteinase K inactivation without any purification. This is useful if the DNA has been laser-capture microdissected or where the entire amount of template is expected to be extremely limited (<100 ng). If using purified genomic DNA, the concentration can be in the range of 10–100 ng/μl. A negative control using 3 μl of purified water should also be used to monitor for contamination. If desired, a positive control using 10 ng of good-quality genomic DNA may also be used to monitor successful PCR.

[b]If several SCOMP reactions are going to be generated for subsequent analysis (i.e. more than 2–3 μg is required), it is possible to perform a larger volume digestion and aliquot it prior to Protocol 4.

Protocol 4

SCOMP adaptor ligation

Equipment and Reagents
- One-Phor-All Buffer Plus (GE Healthcare)
- LIB1 oligonucleotide 5′-AGTGGGATTCCTGCTGTCAGT-3′ (100 μM)
- ddMse11 oligonucleotide 5′-TAACTGACAGCdd-3′ (100 μM)
- ATP (Roche)
- T4 DNA ligase (high concentration; Roche)
- Nuclease-free water (Promega)
- Thermal cycler

Method
1. Combine:
   - 0.5 μl One-Phor-All Plus buffer
   - 0.5 μl 100 μM LIB1
   - 0.5 μl 100 μM ddMse11
   - Nuclease-free water up to a final reaction volume of 3 μl[a]

2. To form the adaptor complexes, incubate the sample using a step-down program on a thermal cycler going from 65 to 15°C, ramping at 1°C/min.

3. Leaving the samples at 15°C, add to each reaction:
   - 1 μl of 10 mM ATP
   - 1 μl of T4 DNA ligase (5 units/μl)

4. Add the total volume (5 μl) of the *Mse*I-digested genomic DNA (from *Protocol 3*) and mix well by pipetting up and down.

5. Incubate overnight (~12–16 h) at 15°C.

6. Spin the reactions in a microcentrifuge to collect the droplets formed by evaporation.

Note
[a]Prepare an adaptor for each sample from *Protocol 3*, including the negative control.

Protocol 5

SCOMP PCR amplification

Equipment and Reagents
- Expand Long Template PCR system (3.5 units/μl; Roche)[a]
- 10 mM dNTP mix (2.5 mM each dATP, dCTP, dGTP, and dTTP) (Invitrogen)
- Nuclease-free water (Promega)
- Thermal cycler

Method
1. Prepare a PCR master mix with the following reagents for each reaction from *Protocol 4*, including the negative control:
   - 4 μl of Expand Long Template Buffer 1
   - 2 μl of dNTPs mix
   - 1 μl of Expand Long Template polymerase
   - Nuclease-free water up to a final reaction volume of 40 μl

2. Add 40 μl of the PCR master mix to the ligated genomic DNA fragments from *Protocol 4* and place in a thermal cycler with the following three-stage combined program[b]:
   - Stage 1: initial incubation at 68°C for 3 min[c], followed by 15 cycles of denaturation at 94°C for 40 s, annealing at 57°C for 30 s, and elongation 68°C for 1 min 30 s with an increment of 1 s/cycle
   - Stage 2: 8 cycles of denaturation at 94°C for 40 s, annealing at 57°C for 30 s with an increment of 1°C/cycle, and elongation at 68°C for 1 min 45 s with an increment of 1 s/cycle
   - Stage 3: 22 cycles of denaturation at 94°C for 40 s, annealing at 65°C for 30 s, and elongation at 68°C for 1 min 30 s with an increment of 1 s/cycle, followed by a final extension at 68°C for 3 min 40 s

Notes
[a]Other PCR systems designed for high-fidelity PCR may be used in this protocol but have not been tested.
[b]The thermal cycler program was provided by Dr C. Klein. Further optimization of these cycling parameters has not been tested.
[c]The initial incubation at 68°C is required to fill in the recessive 3′ end of the lower DNA strand to generate a complementary primer sequence.

Protocol 6

Assessment, cleaning, and quantification of PCR products

Equipment and Reagents

- Agarose (Sigma)
- Ethidium bromide (10 mg/ml; Sigma)
- 1× TBE buffer (10.8 g/l Tris base, 5.5 g/l boric acid, 4 ml/l 0.5 M EDTA, pH 8, diluted from a 10× stock; Sigma)
- DNA size marker (100 bp ladder; Invitrogen)
- Electrophoresis apparatus
- QIAquick PCR purification kit (Qiagen)
- 6× Orange loading dye (Fermentas)
- Spectrophotometer

Method

1. Determine the size of the products from *Protocol 5* by mixing 5 μl of the reaction mix with 1 μl of 6× orange loading dye solution and resolving the aliquot by agarose gel electrophoresis (1% agarose gel containing 20 μg of ethidium bromide (10 mg/ml) per 100 ml of agarose) alongside a DNA size marker[a] (see *Fig. 2c*).

2. Clean each sample using the QIAquick PCR purification kit (Qiagen) according to the manufacturer's instructions with the exception of eluting the PCR products in 50 μl water[b].

3. Quantify the products using spectrophotometry.

Notes

[a]The negative-control lane should be free of any product formation. Ideally, fragments generated from a good-quality genomic DNA sample should range from 100 to 1500 bp (*Fig. 2c*). The more degraded the original genomic DNA sample, the smaller the PCR fragments will be.

[b]It is recommended that samples be eluted in water. If higher concentrations are required for downstream applications, the elution volume can be decreased from 50 to 30 μl, as recommended by Qiagen.

2.3.2 Adaptor-ligation PCR of randomly sheared genomic DNA (PRSG)

PRSG, another method based on ligation-mediated PCR that was designed to improve genome coverage (*Protocols 7–10*) was first reported by Tanabe *et al.* (16). Rather than using enzymatically generated fragments (13), this method uses randomly fragmented DNA as the template. The process involves three steps:

1. Hydrodynamic shearing of genomic DNA to a 0.5–2 kb size range.
2. End filling and adaptor ligation.
3. High-stringency PCR for faithful replication of the resulting fragments.

An assessment of the genome reproducibility provided by PRSG showed a failure rate of PCR of less than 1% when PRSG products were generated from high-quality DNA. However, the PCR failure rate increased (ranged from 30–50%) and was sample dependent when poor-quality genomic DNA obtained from formalin-fixed, paraffin-embedded samples was tested (16).

Protocol 7

Shearing of genomic DNA[a]

Equipment and Reagents
- Genomic DNA (up to 1 µg)
- HydroShear (GeneMachines)
- Phenol
- Chloroform
- Glycogen (20 µg/ml; Invitrogen)
- 7.5 M Ammonium acetate
- Isopropanol
- Ethanol
- TE buffer (10 mM Tris/HCl, pH 7.5, 1 mM EDTA)

Method

1. Shear 1 µg (200 µl) of high-molecular-weight genomic DNA using an automated hydrodynamic shearing machine (such as the HydroShear machine), according to the supplier's instructions (18)[b,c].

2. Add an equal volume of phenol to the solution of randomly fragmented DNA and mix for 5 min on a rotating platform.

3. Centrifuge the samples for 10 min at 12 000 *g*.

4. Transfer the upper aqueous layer to a fresh tube and add an equal volume of chloroform. Mix for 5 min on a rotating platform.

5. Centrifuge for 10 min at 12 000 *g*.

6. Transfer the upper aqueous layer to a fresh tube and add 1 ml of glycogen, 0.5 vols of 7.5 M ammonium acetate and 2.5 vols of 100% isopropanol. Mix well and incubate the solution at room temperature for 20 min.

7. Centrifuge for 10 min at 12 000 *g* to pellet the DNA.

8. Remove the supernatant and wash the pellet with 1 ml of 70% ethanol. Allow the DNA pellet to air dry.

9. Dissolve the pellet in 10 µl of TE buffer.

Notes

[a]If the genomic DNA is already degraded, such as that obtained from laser-capture microdissected tissues, the procedure can be started at *Protocol 8* after purification of the genomic DNA.

[b]The Hydroshear machine uses a ruby with a 0.05 mm diameter hole to shear the DNA, an approach specific to this piece of equipment. Alternative machines or methods that generate DNA fragments within the desired size range (0.5–2 kb) will likely yield comparable results.

[c]When using the Hydroshear machine, the DNA solution (200 µl) was randomly fragmented at appropriate flow rates (speed codes (s.c.) 4 or 5) for 20 iterations.

Protocol 8

Bal31 treatment of DNA

Reagents
- BAL31 nuclease (4 units/μl) and accompanying 2× reaction buffer (Fermentas)
- TE buffer (10 mM Tris/HCl, pH 7.5, 1 mM EDTA)

Method
1. Mix 5 μl (approximately 500 ng) of DNA solution from *Protocol 7* with 50 μl of BAL31 reaction buffer and incubate at 70°C for 5 min, followed by 5 min at 30°C.

2. Add 1 μl of BAL31 nuclease and incubate at 30°C for 1 min.

3. Purify the DNA fragments by phenol extraction (*Protocol 7*, steps 4–8).

4. Dissolve the pellet in 7 μl of TE buffer.

Protocol 9

End filling

Reagents
- T4 DNA polymerase (5 units/μl) and the accompanying 5× reaction buffer (Invitrogen)
- TE buffer (10 mM Tris/HCl, pH 7.5, 1 mM EDTA)

Method
1. Add 2 μl of T4 DNA polymerase buffer to the DNA from *Protocol 8*. Incubate the solution at 70°C for 5 min and then at 30°C for 5 min.

2. Add 1 μl of T4 DNA polymerase and incubate at 37°C for 5 min.

3. Purify the DNA fragments by phenol extraction (*Protocol 7*, steps 4–8).

4. Dissolve the pellet in 25 μl of TE.

Protocol 10

PRSG ligation of adaptors and PCR amplification

Equipment and Reagents

- Adaptor oligonucleotide 1: 5′-AATTCGGCGGCCGCGGATCC-3′ (100 μM)
- Adaptor oligonucleotide 2: 5′-GCCGCCGGCGCCTAGG-3′ (100 μM)
- T4 DNA ligase (5 units/μl) and accompanying 5× ligase buffer (Invitrogen)
- DNase-free water (Promega)
- 10 mM ATP (Invitrogen)
- ER-1 PCR primer: 5′-GGAATTCGGCGGCCGCGGATCC-3′ (100 μM)
- Platinum *Taq* DNA polymerase (5 units/μl) and accompanying 10× PCR buffer (Invitrogen)
- Nuclease-free water (Promega)
- TE buffer (10 mM Tris/HCl, pH 7.5, 1 mM EDTA)
- Agarose (Sigma)
- Ethidium bromide (10 mg/ml) (Sigma)
- 1× TBE buffer (10.8 g/l Tris base, 5.5 g/l boric acid, 4 ml/l 0.5 M EDTA, pH 8, diluted from a 10× stock; Sigma)
- DNA size marker (100 bp ladder; Invitrogen)
- 6× Orange loading dye (Fermentas)
- Electrophoresis apparatus

Method

1. Combine the following and then incubate at 16°C for 12 h:
   - 1 μl of the DNA solution from *Protocol 9* (20 ng)
   - 4 μl of T4 DNA ligase reaction buffer
   - 1 μl of preformed adaptor[a]
   - 1 μl of 10 mM ATP
   - 12 μl of nuclease-free water
   - 1 μl of T4 DNA ligase

2. For PCR amplification combine:
   - 1 μl (1 ng) of the adaptor-ligated DNA mixture (Step 1)
   - 77 μl of nuclease-free water
   - 10 μl of 10× PCR buffer
   - 1 μl of ER-1 primer
   - 10 μl of 2.5 mM dNTP mix
   - 1 μl of Platinum *Taq* DNA polymerase

3. Carry out the following cycling program:
   - 15–20 cycles of denaturation at 94°C for 1 min and annealing/extension at 72°C for 3 min
   - Final extension at 72°C for 10 min[b]

4. Aliquot the sample into five separate tubes (20 μl each) and add:
   - 58 μl of nuclease-free water
   - 10 μl of 10× PCR buffer
   - 1 μl of ER-1 primer
   - 10 μl of 2.5 mM dNTP mix
   - 1 μl of Platinum *Taq* NA polymerase

5. PCR amplify each aliquot for an additional 5–10 cycles using the same PCR cycling conditions as above[c].

6. Purify the DNA fragments by phenol extraction followed by isopropanol precipitation as described in *Protocol 7*[d].

7. Dissolve the DNA pellet in 100 μl of TE buffer.

8. Determine the size of the products by mixing 5 μl of the reaction mix with 1 μl of 6× orange loading dye solution and resolving the aliquot by agarose gel electrophoresis (1% agarose gel containing 20 μg of ethidium bromide (10 mg/ml) per 100 ml of agarose) alongside a DNA size marker (see *Fig. 2d*)[e].

Notes

[a]The preformed adaptor was made as described in SCOMP *Protocol 4* with the exception that the primers from this protocol were replaced with adaptor oligonucleotides 1 and 2.

[b]Some DNA fragments have a high GC content and may form a stable secondary structure, which often prevents PCR amplification at a standard annealing temperature of 50–60°C. However, by using an adaptor in conjunction with a high annealing temperature (72°C), the amplification bias during PCR, due to differences in template sequence composition, can be minimized. This in turn allows better genome coverage.

[c]To obtain sufficient yields of DNA, using a high-cycle-number PCR is not recommended as this may introduce sequence bias caused by mispriming and preferential amplification of shorter fragments. It is suggested to split the first PCR five ways after the initial 15–20 cycles and then perform an additional 5–10 cycles on each aliquot.

[d]Other DNA purification kits may be a suitable substitute for phenol extraction.

[e]The expected yield of products is 5–10 μg ranging in size from 0.4 to 1.5 kb (see *Fig. 2d*).

2.3.3 GenomePlex

GenomePlex WGA (*Protocols 11–13*) is a proprietary amplification technology that is based on nonenzymatic random fragmentation of genomic DNA. The protocol involves conversion of the genome into an *in vitro* molecular library of DNA fragments, followed by incubation at various temperatures to add adaptor sequences with specific PCR priming sites to both ends of every fragment (15, 19). The fragment library can then be amplified several 1000-fold to generate milligram quantities of DNA starting with as little as 10–100 ng.

Protocol 11

GenomePlex fragmentation

Equipment and Reagents
- Genomic DNA (1–10 ng/μl)
- GenomePlex Whole Genome Amplification kit (Sigma)[a]
- Thermal cycler

Method
1. Combine:
   - 10 μl of DNA sample (final concentration 10–100 ng)
   - 1 μl of 10× Fragmentation Buffer (blue-capped tube) from the GenomePlex Whole Genome Amplification kit[b,c].

2. Mix the sample by pipetting or brief vortexing.

3. Consolidate the sample by centrifugation (5–10 s).

4. Incubate at 95°C for 4 min in a thermal cycler[d].

5. Following incubation, cool the sample on ice for 5 min.

Notes

[a]The GenomePlex kit was originally developed by Rubicon Genomics, but is now available from Sigma. The constituents of the GenomePlex WGA kit buffers are proprietary and are therefore unknown.

[b]When handling many samples (>20), it is best to use multi-well strips or 96-well PCR plates, as this will help decrease the set-up time. However, caution must be taken when removing the strip caps to avoid cross-contamination of tube contents. For 96-well plates, use adhesive metal or plastic films. The advantage of these over strip caps is that they do not need to be removed and can be pierced using a pipette tip to allow the addition of the Library Preparation solution. The plate can then be resealed by placing a second film over the top of the first film.

[c]Users should prepare at least two GenomePlex reactions for each sample, as this will provide a greater yield of DNA. In addition, better results have been obtained in downstream applications when the products from at least two reactions are combined.

[d]As stated by Sigma in the manual for the GenomePlex kit, adhering to this incubation time is essential as longer or shorter times can affect results. However, when using degraded DNA in our experience, decreasing the fragmentation time by 30 s at a time can improve results.

Protocol 12

Library preparation

Equipment and Reagents
- GenomePlex Whole Genome Amplification kit (Sigma)
- Thermal cycler

Method

1. Add 2 μl of 1× Library Preparation buffer (green-capped tube) and 1 μl of Library Stabilization solution (yellow-capped tube) to each sample.

2. Mix the sample by pipetting or brief vortexing.

3. Consolidate the sample by centrifugation (5–10 s).

4. Incubate at 95°C for 2 min in a thermal cycler.

5. Following incubation, cool the sample on ice for 5 min and consolidate by centrifugation (5–10 s).

6. Add 1 μl of Library Preparation enzyme (orange-capped tube), mix by pipetting or vortexing, and centrifuge briefly.

7. Incubate the samples in a thermal cycler using the following conditions:
   - 16°C for 20 min
   - 24°C for 20 min
   - 37°C for 20 min
   - 75°C for 5 min

8. Store the reaction mixtures at –20°C for up to 3 days or continue with PCR amplification[a].

Note

[a]Following library preparation, samples can be stored at –20°C for up to 3 days. The effect of long-term storage on WGA DNA and subsequent downstream applications has not been assessed.

Protocol 13

GenomePlex PCR amplification

Equipment and Reagents
- GenomePlex Whole Genome Amplification kit (Sigma)
- JumpStart *Taq* DNA polymerase (Sigma)
- Nuclease-free water (Promega)
- Agarose (Sigma)
- Ethidium bromide (10 mg/ml) (Sigma)
- 1× TBE buffer (10.8 g/l Tris base, 5.5 g/l boric acid, 4 ml/l 0.5 M EDTA, pH 8, diluted from a 10× stock; Sigma)
- 6× Orange loading dye (Fermentas)
- DNA size marker (100 bp ladder; Invitrogen)
- Electrophoresis apparatus
- 6× Orange loading dye

Methods
1. Per reaction, combine:
   - 7.5 µl of 10× Amplification Master Mix (red-capped tube)
   - 47.5 µl of sterile water
   - 5 µl of JumpStart *Taq* DNA polymerase
   - 15 µl of solution from the library preparation (*Protocol 12*)[a]

2. Mix the reaction constituents thoroughly by pipetting or vortexing and centrifuge briefly.

3. Use the following PCR profile for amplification:
   - Initial denaturation at 95°C for 3 min
   - 14 cycles of denaturation at 94°C for 15 s
   - Annealing/extension at 65°C for 5 min[b]

4. Determine the size of the products by mixing 5 µl of the reaction mix with 1 µl of 6× orange loading dye solution and resolving the aliquot by agarose gel electrophoresis (1% agarose gel containing 20 µg of ethidium bromide (10 mg/ml) per 100 ml of agarose) alongside a DNA size marker[c] (see *Fig. 2e*).

5. Detect the DNA smears under UV light.

6. Store the reaction mixtures at −20°C prior to purification.

7. Clean the GenomePlex WGA products to remove unincorporated primers and other reaction constituents that may interfere with downstream applications[d].

8. Quantitate the GenomePlex WGA products using spectrophotometry.

Notes
[a]It is strongly recommended to use the JumpStart or BD TITANIUM *Taq* DNA polymerase, as both have been optimized for use with the GenomePlex kit. Other sources of *Taq* DNA polymerase may not produce optimal results.

[b]If more than one reaction has been set up for each sample (which is recommended), combine reactions at this point.

[c]The size of the amplification product is dependent on the quality of the starting DNA. If this was high-molecular-weight DNA (extracted from fresh tissue or cells), the amplification product smear will range from 50 to 2000 bp (see *Fig. 2e*). However, if the DNA was of low molecular weight (from fixed tissue), then the size of the amplification product smear will generally be below 500 bp.

[d]Any PCR product clean-up method is acceptable including the MinElute 96 UF PCR Purification kit (Qiagen), QIAquick PCR Purification kit (Qiagen), DNA Clean & Concentrator 5 (Genetix), Microcon YM-30 Centrifugal Filter Unit (Millipore), or phenol/chloroform extraction.

2.4 Downstream applications

2.4.1 DOP–PCR

DOP–PCR-amplified DNA has been widely used, but precise measurement of the amount of starting template DNA is important. A shortage of genomic DNA template sometimes leads to a lower reliability of results in downstream applications due to allele drop-out and representational bias. Good results have been obtained for array CGH (20) for which DOP–PCR has been demonstrated to provide the most reliable results when using minute quantities of DNA (21).

SNP typing (22) and microsatellite genotyping (23) have demonstrated that a large proportion of the genome can be amplified by DOP–PCR. However, there is some debate as to whether there is preferential amplification of shorter alleles (24). It is important to take these points into consideration when using DOP–PCR-amplified DNA.

2.4.2 I-PEP–PCR

It has been reported (for review, see 25) that the use of random primers in PEP or I-PEP is likely to generate less-biased representations than DOP–PCR due to the greater number of potential priming sites. A mathematical model for two different PCR-based WGA reactions, PEP–PCR and tagged random primer PCR (not covered in this chapter), was developed by Sun *et al.* (26) to explore predictions of target yield and coverage. This study determined that the use of a DNA polymerase with high processivity, as with I-PEP–PCR, would lead to increased amplification efficiency and locus coverage.

Previous work by Dietmaier *et al.* (10) has demonstrated the efficacy of I-PEP for WGA. In addition, loss of heterozygosity and CGH analysis using I-PEP-amplified DNA, when studying flat urothelial hyperplasias and bladder cancer, have yielded good results (27).

2.4.3 SCOMP

The PCR products resulting from SCOMP have been used for a number of downstream applications including chromosomal CGH (11), array CGH (28), loss of heterozygosity (11) analysis, and direct sequencing (11). Both chromosomal CGH and array CGH analysis require the SCOMP products to be labeled in an additional reaction with appropriate dyes. When using SCOMP products for a comparative analysis (test and control), it is essential to treat both the DNA of interest and the control genomic DNA in the same manner. As SCOMP reduces the genome to only a representation, it is important that control DNA is also a representation.

Moreover, it is also important that the control DNA be of similar quality to the DNA of interest. For example, if the DNA of interest was extracted from formalin-fixed, paraffin-embedded material, then the control 'normal' DNA should also be from this source and preferably from the same patient. Pooling several replicate SCOMP reactions, from the same sample, may also help to reduce any bias that may be introduced in the PCR amplification step.

2.4.4 PRSG

PRSG has been used successfully for SNP analysis, microsatellite analysis, Southern blotting, and array CGH. In an analysis of 307 microsatellites distributed throughout the genome, 84% were reproducibly amplified in PRSG DNA and of these 99% showed a consistent pattern between the PRSG product and the original genomic DNA (16). Array CGH experiments using an esophageal cancer cell line, TE6, showed more than 90% concordance of the fluorescence ratios between the PRSG-amplified and matched nonamplified DNA (16).

2.4.5 GenomePlex

When using GenomePlex for region-specific PCR we have found that the ideal range of PCR size was 100–600 bp. If using GenomePlex with DNA from formalin-fixed, paraffin-embedded tissues, it is likely that this range will be smaller. Microsatellite analysis of GenomePlex products worked well in our hands for tri- and tetranucleotide repeats, but the amplification of dinucleotide repeats did not demonstrate concordance between the amplified and nonamplified samples. The use of GenomePlex to amplify DNA obtained from formalin-fixed, paraffin-embedded tissue has been proven to provide good array CGH results (12), which is also the case for the concordance (99.89% when compared with non-amplified DNA) determined using the Illumina SNP platform (29). However, SNP genotyping using the Affymetrix experimental protocol is not recommended for GenomePlex WGA products (data not shown) due to poor results. This is likely due to the loss of *Xba*I restriction sites required for this protocol, but has yet to be tested.

3. TROUBLESHOOTING

3.1 General troubleshooting

- **Amplification in the negative control**
  Due to the manipulation of PCR products in all PCR-based WGA methods, the reactions can easily be contaminated. If negative controls produce a DNA smear, a number of steps can be tried to eliminate this:
  ○ Repeat using fresh reagents.
  ○ Use filtered tips to avoid introduction of contaminants via aerosol from the pipette.

- ○ Physically separate the areas in the laboratory where reactions are set up. Prepare and pipette the PCR mixture at one bench and then add the DNA to the reaction in a different location in the laboratory. Also, use different pipettes for reaction preparation and pipetting of DNA.
- ○ It is strongly advisable to aliquot all reaction constituents; if an aliquot becomes contaminated, only that aliquot will be lost and not the entire stock.

- **Unpredictable amplification**
  - ○ A positive control (such as starting with 50 ng of good-quality genomic DNA) can be used to ensure that the reaction is working optimally.
  - ○ It is recommended to store small aliquots of dNTPs and oligonucleotides, as repeated freeze/thaw cycles of a single stock can affect the integrity of these reagents and thereby affect the efficiency of WGA.
  - ○ The starting concentration of DNA is crucial. Best results are obtained when starting with 10 ng (or greater) of DNA extracted from fresh tissue/blood or 100 ng of DNA extracted from fixed tissue. Lower amounts of DNA (<10 ng from fresh tissue or <100 ng from fixed tissue) will generate amplification products, but results obtained in downstream applications may not be faithful to the result that would have been obtained from nonamplified DNA.
  - ○ For ligation-mediated PCR techniques, lack of amplification or poor amplification may be due to problems with either the DNA or the adaptor ligation. For the former, always include a positive control of 10–100 ng of good-quality, high-molecular-weight DNA. If this sample produces good results, the assay problems are a product of the DNA sample under investigation. However, if the positive control does not amplify, it suggests a problem with the reaction constituents or the thermal cycling. In such cases, it is recommended to use fresh aliquots of reagents and repeat the experiment. If the problem still persists, it may be necessary to order fresh reagents.

- **Smaller DNA fragments than expected in the amplification smear**
  - ○ Tissue fixation causes degradation of DNA within the sample. Although PCR-based WGA will amplify DNA with an average length of approximately 200 bp, it is essential to use increased quantities of starting DNA (100 ng) to guarantee a satisfactory yield of final product.
  - ○ The efficiency of amplification is dependent on the quality of the starting DNA. Although PCR-based WGA is tolerant to mild or moderate DNA degradation, moderate to severe degradation will reduce WGA efficiency and subsequently decrease the quality of results obtained in downstream applications. Care must be taken to evaluate the success of downstream applications in any given sample set.
  - ○ The size of the amplification products will be template dependent. If the template is degraded, as when using DNA from formalin-fixed, paraffin-embedded material, the smear will be smaller in size. Unfortunately, there is no way of improving this, so care must be taken when using such products for downstream applications, as there may be more bias in these samples.

- **Insufficient DNA for downstream reactions**
  Combining at least two individual WGA amplifications for each DNA to be studied is recommended, as this has produced better results in downstream experiments.
- **Unpredictable results from downstream applications**
  Successful PCR-based WGA and successful downstream results depend on the quality of starting DNA. For instance, if there is ineffective PCR amplification of a specific control gene with nonamplified DNA, negative results after WGA are to be expected. If using laser-capture microdissected material, a standard control PCR and DNA from nonlaser-capture microdissected tissue should be performed prior to WGA. In some cases, it may be necessary to try different amounts of amplified DNA in downstream PCR amplification procedures, e.g. 12, 25, 50, and 100 ng.

3.2 I-PEP–PCR

- **Poor I-PEP–PCR results**
  If the I-PEP–PCR performance is poor, the starting amount of DNA should be modified. If possible, try adding 10-fold more and 10-fold less starting DNA than first used. This incremental modification will help prevent inhibition of amplification by overloading effects and reaction failure due to insufficient template.

3.3 PRSG

- **Inefficient shearing of DNA**
  When using automated hydrodynamic shearing, it is necessary to avoid using samples containing more than 5 μg of DNA, insoluble DNA, or other contaminants, as the small hole in the ruby of the HydroShear is easily blocked. When using an alternative approach for shearing DNA other than the HydroShear, it is suggested that the protocol first be performed using a DNA sample for which results are known.
- **Sample processing**
  Of all the processes involved in PRSG, the DNA shearing is the most time-consuming and hence the processing of multiple samples is best carried out using an automated system.

4. REFERENCES

1. Engel C, Forberg J, Holinski-Feder E, *et al.* (2006) *Int. J. Cancer,* **118**, 115–122.
2. Hu N, Wang C, Hu Y, *et al.* (2005) *Cancer Res.* **65**, 2542–2546.
3. Davies JJ, Wilson IM & Lam WL (2005) *Chromosome Res.* **13**, 237–248.
4. Dean FB, Hosono S, Fang L, *et al.* (2002) *Proc. Natl. Acad. Sci. U. S. A.* **99**, 5261–5266.
5. Lage JM, Leamon JH, Pejovic T, *et al.* (2003) *Genome Res.* **13**, 294–307.
6. Ledbetter SA, Nelson DL, Warren ST & Ledbetter DH (1990) *Genomics,* **6**, 475–481.

★★★ 7. Telenius H, Carter NP, Bebb CE, Nordenskjold M, Ponder BA & Tunnacliffe A (1992) *Genomics*, **13**, 718–725. – *First report of the DOP–PCR technique.*

★★★ 8. Zhang L, Cui X, Schmitt K, Hubert R, Navidi W & Arnheim N (1992) *Proc. Natl. Acad. Sci. U.S.A.* **89**, 5847–5851. – *First report of the PEP technique.*

9. Kittler R, Stoneking M & Kayser M (2002) *Anal. Biochem.* **300**, 237–244.

★★ 10. Dietmaier W, Hartmann A, Wallinger S, *et al.* (1999) *Am. J. Pathol.* **154**, 83–95. – *Description of the I–PEP technique.*

★★★ 11. Klein CA, Schmidt-Kittler O, Schardt JA, Pantel K, Speicher MR & Riethmüller G (1999) *Proc. Natl. Acad. Sci. U.S.A.* **96**, 4494–4499. – *First report of the SCOMP technique.*

★ 12. Little SE, Vuononvirta R, Reis-Filho JS, *et al.* (2006) *Genomics*, **87**, 298–306. – *Description of the use of GenomePlex for comparative genomic hybridization.*

13. Lucito R, Nakimura M, West JA, *et al.* (1998) *Proc. Natl. Acad. Sci. U.S.A.* **95**, 4487–4492.

14. Pirker C, Raidl M, Steiner E, *et al.* (2004) *Cytometry A*, **61**, 26–34.

15. Saunders RD, Glover DM, Ashburner M, *et al.* (1989) *Nucleic Acids Res.* **17**, 9027–9037.

★★★ 16. Tanabe C, Aoyagi K, Sakiyama T, *et al.* (2003) *Genes Chromosomes Cancer*, **38**, 168–176. – *First report of the PRSG technique.*

17. Barnes WM (1994) *Proc. Natl. Acad. Sci. U.S.A.* **91**, 2216–2220.

18. Thorstenson YR, Hunicke-Smith SP, Oefner PJ & Davis RW (1998) *Genome Res.* **8**, 848–855.

19. Ludecke HJ, Senger G, Claussen U & Horsthempke B (1989) *Nature*, **338**, 348–350.

20. Peng DF, Sugihara H, Mukaisho K, TsubosaY & Hattori T (2003) *J. Pathol.* **201**, 439–450.

21. Ng G, Roberts I, and Coleman N (2005) *Diagn. Mol. Pathol.* **14**, 203–212.

22. Bannai M, Higuchi K, Akesaka T, *et al.* (2004) *Anal. Biochem.* **327**, 215–221.

23. Cheung VG & Nelson SF (1996) *Proc. Natl. Acad. Sci. U.S.A.* **93**, 14676–14679.

24. Grant SF, Steinlicht S, Nentwich U, Kern R, Burwinkel B & Tolle R (2002) *Nucleic Acids Res.* **30**, e125.

25. Hughes S, Arneson N, Done S & Squire J (2005) *Prog. Biophys. Mol. Biol.* **88**, 173–189.

26. Sun F, Arnheim N & Waterman MS (1995) *Nucleic Acids Res.* **23**, 3034–3040.

27. Obermann EC, Junker K, Stoehr R, *et al.* (2003) *J. Pathol.* **199**, 50–57.

28. Ghazani AA, Arneson NC, Warren K & Done SJ (2006) *J. Clin. Pathol.* **59**, 311–315.

29. Barker DL, Hansen MST, Faruqi AF, *et al.* (2004) *Genome Res.* **14**, 901–907.

CHAPTER 19

PCR sequencing of human genes for the discovery of DNA sequence variants

Abizar Lakdawalla

1. INTRODUCTION

Resequencing or PCR sequencing is an approach that allows the comparison of DNA sequences obtained from one or multiple samples to a known reference sequence in order to determine genetic changes. Resequencing of genomic DNA is utilized for a number of applications including:

- Identifying underlying genetic changes linked to certain phenotypes
- Determining evolutionary origins (1)
- Identifying organisms by DNA sequence (2)
- Ascertaining the functional attributes of genes by comparing sequences from different species with associated gene areas with conserved functions (3)
- The discovery of single-nucleotide polymorphisms (SNPs)
- Identification of molecular predictors of drug response
- Predicting inherited pre-disposition to diseases (4)

In this chapter, a method for determining sequence variants in human genomic DNA (and human mitochondrial DNA) using Sanger dideoxy sequencing is described.

1.1 DNA sequence variations

DNA variations are subclassified as base substitutions, insertions, or deletions. Base substitutions – in which one base is replaced by another – can be either a *transition* (a pyrimidine (C or T) is replaced by a pyrimidine or a purine (A or G) with a purine) or a *transversion* (a pyrimidine is replaced with a purine, or a purine with a pyrimidine). If occurring within a coding sequence, base substitutions can have a direct impact on the protein product:

PCR: *Methods Express* (S. Hughes and A. Moody, eds.)
© Scion Publishing Limited, 2007

- In *missense* substitutions, an amino acid is replaced either with a chemically similar amino acid (conservative substitution) or with a chemically dissimilar amino acid (nonconservative substitution).
- *Nonsense* substitutions are characterized by the modification of an amino acid codon into a stop codon (UAG, UAA, or UGA) leading to the premature truncation of the translated protein.
- *Synonymous* or *silent* substitutions do not modify the amino acid sequence as the variant base is typically located at the third base position of the codon (the 'wobble' base).

DNA variation can also impact the expression of a gene by modifying:

- Splice recognition sites
- Interaction of regulatory microRNA
- Stability of the encoded RNA
- Binding of structural or regulatory proteins to the DNA or RNA

The terminology used to describe mutations may differ depending on the field of use. In the context of population genetics, the terms 'allelic variation' or 'polymorphism' are used synonymously, although polymorphisms are specifically defined as a sequence variation that occurs at a frequency greater than 1% in the general population. Polymorphisms are often assumed not to have a phenotypic effect. In contrast, the terms 'mutation' and 'sequence variation' are often used to describe a disease-causing change in the DNA sequence (5). Just as not all 'mutations' are functionally significant, not all 'polymorphisms' are functionally benign (i.e. there are many common polymorphisms that have a clear phenotype, for example the 32 nt deletion (Δ32) within the chemokine co-receptor (CCR5) gene). As it is not central to the theme of this chapter, and to avoid confusion, we will use the term 'mutation' to describe all DNA sequence changes.

In section 2, recommended protocols and critical issues that may need to be addressed for each of the steps are provided. For the purpose of the methods in this chapter, the term 'mutation' is defined as any sequence change in a test sample compared with the sequence of a reference sample or standard.

2. METHODS AND APPROACHES

2.1 Resequencing methods

Resequencing methods are based on three primary steps (see *Fig. 1*):

1. Extraction and preparation of the nucleic acid.
2. Amplification of a target region by PCR. This step involves two key decision points:
   - Selection of the target nucleic acid for resequencing
   - Selecting the region for resequencing
3. Sequencing of the amplified template and analysis.

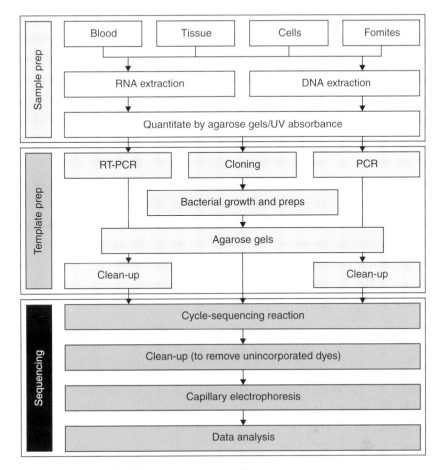

Figure 1. Steps involved in a resequencing project
A standard resequencing experiment is divided into three major steps: sample preparation (extraction of nucleic acid), template preparation (amplification of the region to be sequenced), and sequencing.

A resequencing project may involve PCR amplification and sequencing of tens to hundreds of genes with about 20 amplicons per gene for tens to hundreds of individual genomic DNA samples. This process is required to provide complete mutation detection, i.e. to reveal sequence variants in the sample compared with a reference sequence with high sensitivity (a very high percentage of true positives and a very low percentage of false negatives). To achieve the high level of sensitivity, the PCR fragments are sequenced bidirectionally.

2.1.1 Extraction of RNA or DNA

Suitable DNA and RNA extraction methods are dependent on the type of sample (for example blood, tissue culture cell lines, tissue homogenates, bacterial colonies, environmental samples, laser capture microscopy samples) and can range from simple boiling of a bacterial colony in Tris/EDTA buffer to methods based on capture of DNA on silica or ion-exchange resins. Although such methods are not covered in this chapter, it is essential that the selected method should purify the DNA (or RNA) away from inhibitors of PCR. The quality and quantity of the extracted DNA (or RNA) should be determined by UV spectroscopy with a micro-volume spectrophotometer (e.g. Nanodrop) or by agarose gel electrophoresis.

2.1.2 Selection of the target nucleic acid for resequencing

In some situations, it may be advantageous to sequence a cDNA copy of a mRNA instead of the genomic DNA. cDNA sequencing is useful for detecting mutations in the coding region of a gene, particularly those with large and/or many introns such as dystropin. By preparing the sequencing template by reverse transcriptase PCR (RT-PCR) on extracted mRNA, the target region of the gene can be sequenced more efficiently and without interference from pseudogenes. Sequencing of cDNA also allows the detection of gross structural changes in the mRNA (alternative splice forms, large insertions/deletions). However, sequencing cDNA can introduce additional errors as a result of the reverse transcription step, and some genes are known to undergo RNA editing (where the RNA sequence is not a faithful copy of the genomic DNA sequence) and therefore some caution in interpreting data is required.

Reporting a mutation in a cDNA sequence requires unambiguous gene annotations (a clear delineation of the transcription initiation site, alternative splice forms, polyadenylation site, and length variants). In general, resequencing of genomic DNA is preferred, as changes in the RNA sequence are an outcome of a genomic DNA mutation and not a mutation per se.

2.1.3 Selecting the region for resequencing

Selecting a region for resequencing may be dictated by:

- Annotated genomic features, i.e. coding exons, intron/exon boundaries, promoter regions, etc.
- The prevalence of currently identified DNA variations, i.e. data from SNP databases.
- The presence of predicted 'variation' hotspots.
- Known changes in the amino acid coding sequence.
- Results from genetic/association studies.
- Screening by heteroduplex analysis or by denaturing gradient gel electrophoresis (6).

The current paradigm for discovering DNA sequence changes associated with a specific gene requires that, at a minimum, a 1 kb region upstream of the promoter and all exons (preferably including exon/intron boundaries) should be sequenced. However, the exon-centric approach (focusing on the analysis of exons) to resequencing may result in missing alterations that are functionally significant to the phenotype under investigation. For example, in patients with the cardiac version of Fabry disease (a fat storage disorder caused by a deficiency of an enzyme involved in the biodegradation of lipids), a base substitution in an intron results in the insertion of a 'pseudo exon' of 57 bp between the normal exons 4 and 5 (7). An ideal solution would be to resequence the whole genome, which may become possible in a few years. However, it is often not possible, practically or financially, to resequence an entire gene in its genomic context. Therefore, for pragmatic reasons the exon-centric approach is widely used.

2.1.4 Primer design

PCR amplification requires robust selection of primer pairs:

- Primer sequences should amplify only the region of interest and not a similar region (gene duplication, pseudogene, or an additional member of the gene family).
- The primers should amplify all genotypes, i.e. a DNA variation in the target region should not affect the efficiency of PCR amplification (for example, a sequence change located directly at the 3′ end of a primer may result in no amplification).

Primer sequences that have been published in journals or publicly available primer sequence repositories may reduce the need for testing and validating PCR primers. The NCBI has created a Probe Database that lists resequencing primers, VariantSEQr™ primer sequences, for the exon coding regions and the 5′ promoter regions for about 16 000 human genes. The approximately 420 000 pre-designed PCR primer pair sequences are designed for uniform PCR and sequencing conditions and include an M13 sequence tail to simplify the setting up of sequencing reactions. The primer sequences can be downloaded from www.ncbi.nih.gov/genome/probe. The VariantSEQr primers are mapped to gene annotations that display known SNPs to allow the selection of specific primer pairs for a region of interest (see *Fig. 2*, also available in the color section).

2.1.5 Preparation of the sequencing template by PCR

Amplification of a region of genomic DNA is performed according to *Protocol 1*. For initial experiments, it is recommended that the PCR product be checked on an agarose gel for the presence of spurious bands or primer dimers. If extra bands are found, it is recommended that an alternative set of PCR primers be used to amplify this region to avoid interference from the spurious bands. In some instances, it may be necessary to excise and purify the band of interest from an agarose gel before proceeding with resequencing.

Current capillary electrophoresis-based methods for resequencing provide approximately 500–1000 bp of high-quality sequence per read. Thus, for greatest efficiency, PCR amplification should produce a product in this size range. Unfortunately, as an average exon is only about 170 bp (9), when amplifying from genomic DNA a substantial proportion of the non-coding region will also be sequenced in a 500-1000 bp amplicon. A method proposed by Wallace *et al.* (1999) (10) can potentially increase the efficiency by sequencing just coding regions. The method is based on concatenating exonic regions into a single template for sequencing. An overview of the 'meta-PCR' approach is summarized in *Fig. 3*.

1. Length of the genomic segment to be resequenced (does not include M13 primer sequences).
2. Notes
   V = Validated experimentally (29 probes)
   L = Low-complexity sequences within amplicon (28 probes)
   S = SNPs within primer binding sites (0 probes)
   M = Probes in minimal tiling path set (45 probes)

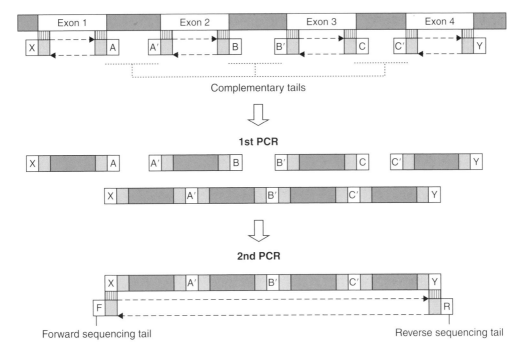

Figure 3. Sequencing multiple exons by meta-PCR.

Meta-PCR creates a concatemer of exonic regions only. PCR primers are designed that are specific to each exon. The meta-PCR is carried out as two sequential PCRs. The first PCR is carried out with primers against multiple exons at a limiting concentration of 0.04 μM each. The reverse primer for exon 1 contains a tail that is complementary to the tail on the forward primer of exon 2. Similarly, the reverse primer of exon 2 contains a tail that is complementary to the forward primer of exon 3, and so on. During the first few cycles of the first PCR, each exon region is amplified with the tail sequences becoming incorporated into the amplicon. During the latter cycles, a concatenated product is produced that includes all of the individual amplicons. For the second PCR, 0.1 vols of the inactivated first PCR is used as template to amplify the full-length concatenated product with primers that are complementary to the most upstream and downstream tails. A sequencing tail is added to the primers of the second PCR to simplify the downstream sequencing reaction set-up. Multiple complementary sequences can be used to create a larger concatemer in a single PCR.

Figure 2. VariantSEQr resequencing primers for the breast cancer 1, early onset gene (BRCA1) (see page xxvii for color version).

Resequencing amplicons (RSAs) are mapped to all known transcripts of BRCA1. Known SNPs are indicated by red slashes. Amplicons for a specific transcript can be selected by clicking * on the right. Individual amplicons can be selected by clicking on any individual amplicon in the map or in the table. The primer sequences (with or without an M13 sequencing tail) can be downloaded for the selected amplicons by clicking the 'Download' button.

2.1.6 Resequencing of mitochondrial DNA (mtDNA)

The mitochondrial gene is prone to a higher incidence of phenotypic effects because of the higher mutation rate. Due to the maternal inheritance of mtDNA, the sequencing and identification of variants is of utility in understanding human evolutionary origins (1). Design of primers for PCR amplification of mtDNA is challenging due to the presence of homopolymeric regions (regions with a continuous run of a particular base that result in premature termination of sequencing) and the increased number of SNPs and multiple-nucleotide polymorphisms. A robust set of resequencing primers for human mtDNA, the mitoSEQr™ set, are available from Applied Biosystems.

2.1.7 Purification of the PCR product

The PCR product is purified to remove excess primers and nucleotides that would compromise the sequencing reaction (see *Protocol 2*). Although multiple options are available for purification of PCR products (e.g. binding to silica, alcohol precipitation), we recommend an enzymatic clean-up method based on enzymatic digestion using ExoSAP-IT reagent (USB). The Exonuclease 1 enzyme (Exo) in the mixture removes any remaining single-stranded PCR primers as well as single-stranded extensions of the amplicon. The shrimp alkaline phosphatase (SAP) in the mixture removes the residual dNTPs in the PCR mixture.

Protocol 1

PCR amplification of a genomic region for sequencing

Equipment and Reagents
- Genomic DNA (5–10 ng/μl)
- PCR-grade water
- AmpliTaq Gold PCR Master Mix (2×)[a]
- Forward PCR primer with M13 tail (0.6 μM/μl)
- Reverse PCR primer with M13 tail (0.6 μM/μl)
- Ultrapure glycerol
- Thermal cycler with heated lid
- 1% Agarose gel containing 10 ng/ml ethidium bromide
- 6× Orange loading dye solution (Fermentas)
- Equipment and reagents for agarose gel electrophoresis including 1× TBE agarose gel running buffer (10.8 g/l Tris base; 5.5 g/l boric acid; 4 ml/l 0.5 M EDTA, pH 8.0, diluted from a 10× stock; Sigma)
- DNA size marker (100 bp ladder; Invitrogen)
- UV light source

Method
1. Combine the following to prepare a PCR master mix for the number of reactions required:
   - 5.0 μl of AmpliTaq Gold Master Mix
   - 1.6 μl of 50% UltraPure glycerol (Invitrogen)
   - 1.0 μl of forward primer (0.6 μM/μl)
   - 1.0 μl of reverse primer (0.6 μM/μl)
   - 0.4 μl of PCR-grade water

2. Add 9.0 μl of the prepared master mix to separate 0.2 ml PCR tubes or to wells of a 96-well plate.

3. Add 1.0 μl of genomic DNA (10 ng/μl) to the respective tube or well[b].

4. Seal the tubes/microplate and place in an ABI 9700 thermal cycler or equivalent.

5. Amplify using the following program:
   - 1 cycle of heat activation at 96°C for 5 min
   - 40 cycles at 94°C for 30 s, 55°C for 45 s, and 72°C for 45 s
   - 1 cycle of a final extension at 72°C for 10 min
   - Hold at 4°C

6. Resolve 2 μl of the PCR products on a 1% agarose gel alongside a DNA size marker to estimate the amount of amplified DNA and to determine that multiple bands and primer dimers are not present.

Notes
[a]Different PCR reagents and primers may require different thermocycling conditions.
[b]For amplification of human mtDNA, reduce the DNA concentration to 0.5–1.0 ng/μl.

Protocol 2

Post-PCR clean up with ExoSAP-IT

Equipment and Reagents
- ExoSAP-IT (USB)[a]
- Thermal cycler with heated lid

Method
1. Remove the sealing film (or caps) from the microplate or tubes that contain the PCR product (from *Protocol 1*).

2. Add 2 μl of ExoSAP-IT to the PCR tubes. Seal the tubes/plates and place back in the thermal cycler. Set the thermal cycler to incubate the reaction at 37°C for 30 min. Heat-inactivate the exonuclease and alkaline phosphatase at 80°C for 15 min.

Note

[a]The enzymes are heat labile and therefore need to be stored at –20°C and should be handled on ice. Running an aliquot of a PCR on an agarose gel after the clean-up should show no primer bands.

2.1.8 Performing the sequencing reaction

The DNA sequencing method (see *Protocol 3*) is based on the chain-termination method developed by Sanger *et al.* (1977) (8). DNA synthesis is performed in the presence of a mixture of deoxy- and dideoxynucleotides. As the dideoxynucleotides (ddNTP) do not contain a 3′-OH, the extending DNA chain is terminated on incorporation of a ddNTP. The ratio of dNTP to ddNTP is adjusted so that a proportion of the growing DNA chains are terminated at each nucleotide position resulting in a population of truncated fragments of different lengths. Cycle sequencing reactions are robust and easy to perform. The high cycle temperatures reduce secondary structure and allow precise priming, template annealing, and more thorough extension. The same protocol can be used for double- and single-stranded DNA, for PCR products, and for difficult templates, such as bacterial artificial chromosomes.

In this method, DNA is synthesized by DNA polymerases from a specific starting point defined by the primer-binding site on a DNA template (see *Fig. 4*):

- *Using the PCR primer as a sequencing primer.* This decreases the need for synthesizing specific sequencing primers as the PCR primers can also be used for setting up the sequencing reaction. The disadvantage is that separate sequencing reactions have to be set up for each sequencing direction and for each amplicon, increasing the amount of pipetting steps and the possibility of error.

- *Designing PCR primers with a sequencing tail.* For most resequencing projects, it has become customary to include a standard primer tail on the PCR primers, as this substantially simplifies the sequencing set-up. The most common tail

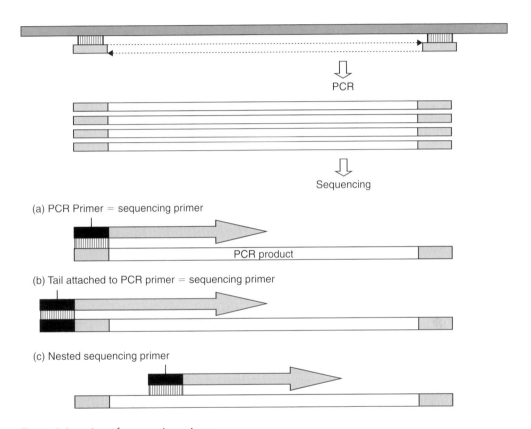

Figure 4. Location of sequencing primers.
(*a*) The sequencing primer can be the same as the PCR primer used to amplify the template, offering the advantage of reducing the number of primers needed. (*b*) The sequencing primer can bind to a universal tail, such as the M13 sequence, that is attached to all of the PCR primers. This significantly simplifies the setting up of sequencing reactions for different amplicons. In addition, sequence data closer to the amplicon ends can be obtained. (*c*) To sequence long amplicons, it may be necessary to use multiple nested sequencing primers that are spaced approximately 500 bp apart. Internal sequencing primers also confer greater specificity for sequencing PCR products with the potential for multiple amplicons from pseudogenes, related genes, etc. Careful selection of the location of nested sequencing primers would reduce interference from related amplicons.

Protocol 3

Sequencing reaction

Equipment and Reagents
- BigDye Terminator v3.1 Cycle Sequencing kit (Applied Biosystems)
- Forward sequencing primer (M13 for VariantSEQr amplicons) (3.2 pmol/µl)
- Reverse sequencing primer (M13 for VariantSEQr amplicons) (3.2 pmol/µl)
- Deionized water
- Thermal cycler with heated lid

Method
1. Prepare a forward sequencing master mix by adding the following to a tube[a,b]:
   - 4 µl of BigDye(R) Terminator Ready Reaction Mix v3.1
   - 1 µl of the forward sequencing primer
   - 3 µl of deionized water

2. Mix the reagents.

3. Dispense 8 µl of the prepared forward sequencing master mix into tubes or the wells of a microplate.

4. Add 2 µl (approximately 15 ng per 500 bp fragment) of the purified PCR product to the tube or well.

5. Seal the tubes or microplates.

6. Prepare a reverse sequencing master mix by adding the following to a tube:
   - 4 µl of BigDye Terminator Ready Reaction Mix v3.1
   - 1 µl of the reverse sequencing primer
   - 3 µl of deionized water

7. Mix the reagents.

8. Dispense 8 µl of the prepared reverse sequencing master mix into tubes or the wells of a microplate.

9. Add 2 µl (approximately 15 ng per 500 bp fragment) of the purified PCR product to the tube or well.

10. Seal the tubes or microplates

11. Perform the sequencing reaction using the following program:
    - 1 cycle of heat activation at 96°C for 1 min
    - 25 cycles at 96°C for 10 s, 50°C for 5 s, and 60°C for 4 min
    - Hold at 4°C

12. Store the reaction products at −20°C or purify immediately.

Notes
[a]Volumes are for one reaction. Scale up the volume based on the number of reactions required.
[b]Prepare the reactions on ice to prevent inadvertent extensions by the sequencing enzyme at room temperature. Protect the reagents and the reaction products from strong light to minimize photobleaching of the fluorescent dyes.

that is included is the M13 sequence, which has been used historically for sequencing of clones constructed in the single-stranded bacteriophage M13. Applied Biosystems has developed the VariantSEQr resequencing primer set for human genes, which include M13 tails.

- *Using nested (internal) sequencing primers.* Designing PCR primers for amplification of closely related genes/pseudogenes can be challenging. Even though a PCR may produce a mixture of PCR fragments, these can be resolved by using an internal sequencing primer that is specific to only one of the PCR fragments.

Typically, each template is sequenced in both directions to ensure that sequencing errors are minimized.

It should be noted that service providers and core facilities offer reliable DNA sequencing services and can perform the sequencing reactions, sequencing clean-up, capillary electrophoresis, and data collection steps.

2.1.9 Purification of the sequencing reaction products

This step is essential in order to remove the excess of dye-labeled ddNTPs and buffer ions that would interfere with capillary electrophoresis. The BigDye fluorescent molecules require stringent clean-up methods to prevent the appearance of 'dye blobs' (high fluorescence signal seen early in the sequence trace due to unincorporated ddNTPs). Two sequencing reaction clean-up methods are provided below (see *Protocols 4* and *5*).

- *Protocol 4* describes an alcohol precipitation method that uses centrifugation to precipitate the dye-labeled sequencing fragments. This method may result in the loss of sequence data close to the sequencing primer due to the relative inefficiency in precipitating low-molecular-weight DNA fragments.
- *Protocol 5*, based on Centri·Sep gel plates, provides robust purification using a pre-packed, hydrated, cross-linked gel that is highly effective at removing excess terminators and nucleotides from the Dye Terminator sequencing reaction mixture.

Protocol 4

Sequencing reaction clean-up using ethanol precipitation[a]

Equipment and Reagents
- 125 mM Na-EDTA
- 100% Ethanol
- 70% Ethanol
- Hi-Dye formamide (Applied Biosystems)
- Centrifuge

Method

1. Add 2.5 μl of 125 mM Na-EDTA to the sequencing reaction product. Mix by pipetting.

2. Add 30 μl of 100% ethanol and mix by pipetting.

3. Incubate the solution at room temperature for 15 min.

4. Centrifuge at 2000 **g** for 45 min.

5. Remove the supernatant.

6. Add 30 μl of 70% ethanol and centrifuge at 2000 **g** for 15 min. Remove the supernatant. Air dry for approximately 10–15 min. Resuspend in 10 μl of Hi-Dye formamide.

Note

[a]This method may result in the loss of small molecular fragments and a decrease in sequence quality close to the primer.

2.1.10 Capillary electrophoresis of sequencing reaction products

Separation of DNA fragments on polyacrylamide gels poured between two glass plates has been largely replaced by capillary electrophoresis. Capillary electrophoresis uses a denaturing flowable polymer to separate the fluorescently labeled DNA fragments according to molecular weight (see *Protocol 6*). During capillary electrophoresis, the products of the cycle sequencing reaction are

Protocol 5

Sequencing reaction clean-up using Centri·Sep spin columns

Equipment and Reagents
- Centri·Sep 96 plates (Applied Biosystems)
- 96-Well PCR sample collection plates (Applied Biosystems)
- Centrifuge

Method
1. Remove the Centri·Sep plates from storage at 2–8°C before starting the sequencing reaction to allow them to equilibrate to room temperature (approximately 2 h)[a].

2. Remove the foil sealing film from the top and bottom of the Centri·Sep 96 plate.

3. Place the Centri·Sep plate on top of a deep-well plate and centrifuge at 1500 **g** for 2 min. Discard the eluant.

4. Pipette the sequencing reaction products (10 µl) into the center of the gel bed of the individual wells in the Centri·Sep plate[b].

5. Stack the Centri·Sep plate on top of a supported 96-well PCR plate.

6. Centrifuge at 1500 **g** for 2 min.

7. Remove the 96-well PCR plate and seal with a sealing film.

8. Store at –20°C for up to 48 h or run the samples directly on a DNA sequencer[c]

Notes
[a]It is critical that the Centri·Sep plates have reached room temperature before removing the foil seals.
[b]It is important to make sure that samples are pipetted slowly into the center of the well and not against the sides of the well.
[c]Centri·Sep column-cleaned samples do not require resuspension in Hi-Dye formamide if run within 6 h or stored at –20°C. If resuspension is desired, then the samples should be dried under vacuum (Speed-Vac) and resuspended in 10 µl of Hi-Dye formamide.

---

injected into capillaries filled with polymer. A high voltage is applied to move the DNA fragments through the polymer, and shortly before reaching the positive electrode, the fluorescently labeled DNA fragments are illuminated by a laser beam. A charge-coupled device camera detects the fluorescence and data collection software converts the fluorescence signal to digital data for all four bases.

Protocol 6

Capillary electrophoresis of sequencing products

Equipment and Reagents
- Genetic Analyzer running buffer with EDTA (Applied Biosystems)
- POP-6 or POP-7 electrophoresis polymer (Applied Biosystems)
- Applied Biosystems Genetic Analyzer 3100 (16 capillaries), Avant (four capillaries), 3130 (four capillaries), 3130xl (16 capillaries), 3730 (48 capillaries), or 3730xl (96 capillaries)

Method
1. First turn on the computer and then the sequencer.

2. Start the data collection software.

3. Create a new plate record and enter in the required information (file name, sample names, dye set, run module, mobility file, and analysis module to be used)[a,b].

4. Refill buffers, water, and the polymer containers if required.

5. Replace the caps or plate seals from the sequencing samples with the provided septa. Place the tubes or plates in the black carrier/tray and snap on the white cover.

6. Place the sample trays into the system.

7. Check that the running conditions and sample names are appropriate and click on the green arrow in the software to initiate the run. Run conditions can be monitored in real time. After a run, the extracted files are sent to the Extracted Run Folder.

8. The *.ab1 files created by the sequencer can be opened remotely with the free software[c] SEQUENCE SCANNER v1.0, available for download from www.appliedbiosystems.com/sequencescanner.

Notes

[a]For the 3130(xl), the recommend conditions are: POP-6 polymer; Genetic Analyzer running buffer with EDTA, and a 36 cm array. The run file is RapidSeq36_POP6_1; the mobility file is KB_3130_POP6_BDTv3.mob, and the analysis module is KB.bcp.

[b]For the 3730(xl), the recommended conditions are: POP-7, 3730/3730xl running buffer with EDTA, a 36 cm array, the RapidSeq36_POP7_1 run file, the mobility file KB_3730_POP7_BDTv3.mob, and the KB.bcp analysis module.

[c]The software names the extracted files with the well position, the capillary number, and the sample name.

2.2 Analysis of results

Data analysis consists of three steps

- Base calling
- Comparing sequences
- Confirming sequence variants

2.2.1 Base calling

Base calling for resequencing differs from *de novo* sequencing where a single base per location is assumed and instead has the possibility of multiple bases per location due to heterozygosity, mixed DNA samples, or mosaicism (not all cells carry the mutation). Base calling is typically derived from the software integral to the sequencer (the KB base caller in the AB sequencers) or may be derived from the raw trace files by an independent software package such as the free Applied Biosystems SEQUENCE SCANNER v1.0 software or by software designed specifically for the discovery of variants such as SEQUENCE PILOT (JSI Medical Systems GmbH), SEQSCAPE (Applied Biosystems), MUTATION SURVEYOR (Soft Genetics), or POLYPHRED (University of Washington). A flow diagram for the SEQSCAPE software is shown in *Fig. 5.*

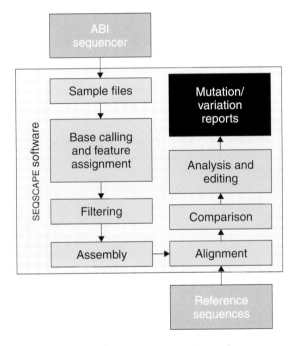

Figure 5. Flow chart for the use of SEQSCAPE resequencing software.
SEQSCAPE software requires two sets of inputs: the sequencing data files from the DNA sequencer – **sample files** (.ab1 files) – and a reference sequence or a VariantSEQr project template file for a specific gene (containing gene annotation, location of amplicons, and known variants). The software then performs **base calling** and **feature assignment** (converting the raw trace data to sequence data), which assigns a quality value to the called bases and then identifies mixed bases. In the **filtering** step, low-quality bases at the end of the sequences are trimmed. Traces that show poor quality overall are also flagged and removed from further analysis. In the **assembly** step, the remaining traces (forward and reverse and overlapping amplicons) are aligned (**alignment** step) to the provided reference sequence to generate a consensus specimen sequence. The base-calling quality for the assembled sequence is reviewed again to confirm the consensus sequence (**comparison** step). Variants such as SNPs, multiple-nucleotide polymorphisms, and indels (insertions or deletions) are identified by alignment with the reference (**analysis** and **editing**). The variants are presented as a table with links to the raw trace data for each identified mutation for editing and approval by a user. At the completion of analysis, a project file is created that contains the sample files, a consensus sequence for each specimen, and multiple variation reports (**mutation/variation reports**).

Resequencing software such as SEQUENCE PILOT and POLYPHRED compare each peak at a location in a sample with a statistically derived peak for the same location from multiple samples. Statistical variance from the normal indicates the presence of a mutation. SEQSCAPE utilizes an algorithm that looks at the multiple colors (bases) under each peak in isolation to derive the number of bases per peak. Resequencing software automatically flags variants or questionable bases for manual review. Ambiguous base calls are more common in the first 50 bases of the trace for the larger fragments due to lower electrophoretic resolution. The manual review process is based on a visual examination of the traces and on either approving the call made by the software or manually changing the base call.

2.2.2 Comparing sequences

The final sequence files are aligned and mapped to a template for a gene region. Overlapping sequences from forward and reverse sequencing runs and from overlapping amplicons are matched and a consensus sequence created. This newly derived 'consensus' sequence is then aligned with a reference sequence and a variation or mutation report is generated. Each mutation is typically linked to the traces so that a manual review can still be performed (see *Fig. 6*). Variants for all specimens are listed in a table in SEQSCAPE software showing the base change (in blue letters, hyperlinked to the raw trace), position, type of mutation

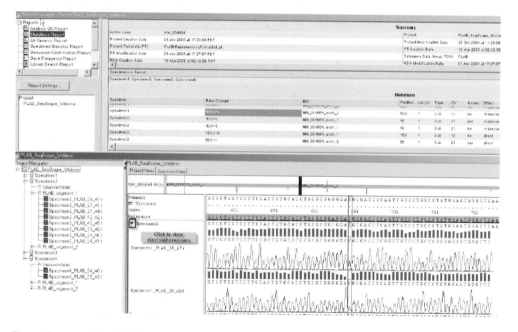

Figure 6. SEQSCAPE **Report Manager**
Sequence variants with the base change and position for every specimen are listed in the table (middle panel). Each specimen and variation is hyperlinked to the sequence trace data to allow manual review. Sequence traces for both strands are shown (the forward strand is indicated with letters in italics). Quality scores for each base are shown as vertical bars above the peak.

(substitution, etc.), whether the mutation is reported, and the impact on the protein sequence. The raw traces for the mutations in the specimens are shown in a lower panel for manual review and approval (see *Fig. 6*). The sequences of multiple specimens can be compared to highlight the presence of mutations. *Fig. 7* (also available in the color section) shows an alternative project view in SEQSCAPE software, with mutations being reported on the basis of amino acid codons (lower table). Changes in the sequence compared with a reference are indicated with pink highlighting (middle panel) and a summary of the region sequenced is shown in the topmost panel (*Fig. 7*). In *Fig. 8* (also available in the color section), data generated from DNA with a heterozygous insertion/deletion mutation are shown. Heterozygous insertions/deletions (indels) are listed in a table with the position and size of the indel. The raw traces (lower panel in *Fig. 8*) show that the 2 bp deletions have shifted the sequence for one of the two chromosomes by 2 nt in the three samples.

2.2.3 Confirming sequence variants

In addition to checking the trace files (chromatograms) for all discovered variants, a mutation should be confirmed by using a different set of PCR primers to amplify and sequence the same region of the genomic DNA.

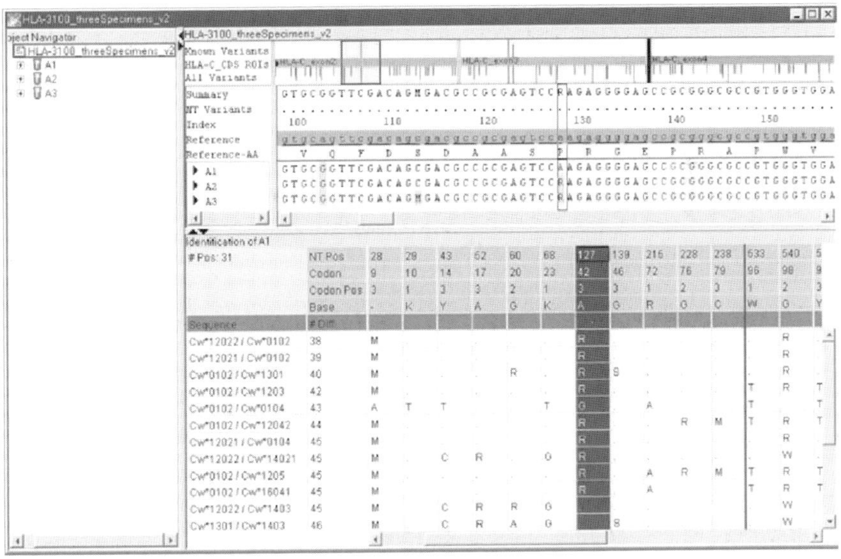

Figure 7. SEQSCAPE **Project Navigator (see page xxviii for color version).**
In the project view, a summary of all of the specimen sequences is shown. The location of variants is displayed in the top panel for the regions (amplicons) that have been sequenced. In the second panel, a subsection of the sequenced region (blue square in the top panel) is shown in more detail. The assembled sequence is aligned with the reference sequence and variations are highlighted for the different specimens. The table in the lower panel summarizes the impact that the mutations would have on the encoded protein.

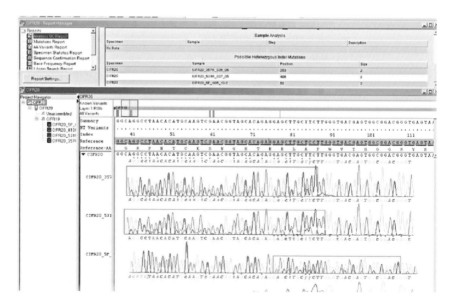

Figure 8. Indel mutations in the CIFR20 gene (see page xxviii for color version).
Heterozygous insertions or deletions are detected and shown by the position and length of the indel (top panel) in the mutation report, which contains hyperlinks to the trace files (shown in the bottom panel). Three samples are shown with 2 bp deletions in the CIFR20 gene in one of the two chromosomes (region boxed in red).

Detection of mutations can be complicated by mosaicism, mitochondrial heteroplasmy (where a mixture of mitochondria is present in a single cell), or somatic and oncogenic mosaicism. Mosaicism may be specific to a tissue type and may not be detectable in blood or oral epithelial cells. In mixed samples with low proportions of one sample, the genotype of the sample in low proportion may be completely masked by the dominant sample. For example, the detection of mutations in cells with mitochondrial heteroplasmy requires cloning of the mtDNA and sequencing a large number of clones to delineate mutations in the rare DNA.

It is also important to consider that mutations may originate from the amplification process. It is possible that a reverse transcription and/or a PCR may produce a mutation at an earlier amplification cycle. A repeat PCR or PCR with a different set of primers may not exhibit the same mutation. The presence of a SNP at the primer-binding site can suppress the amplification of one allele resulting in pseudo-homozygosity. A similar effect would be seen if one allele is deleted and cannot be amplified: the hemizygous allele will appear as homozygous. In addition, the cloning of a target region into a vector can result in the introduction of a mutation, or the selection of an insignificant species in the original sample. To attain a high level of confidence, multiple clones should be sequenced to confirm a mutation.

2.3 Mutation nomenclature

A mutation is compared with a reference sequence, either a genomic or a cDNA reference. Comparison with a genomic reference tends to be less confusing, although comparison with a cDNA reference may be useful to understand the effect of the mutation on a change in the protein.

2.3.1 Single base change (>)

- A mutation referenced to a cDNA uses the nomenclature Transcript No.:c.1234A>C (c = cDNA), i.e. the base A at position 1234 has been changed to a base C.
- The mutation could also be indicated at the RNA level with the notation, r.1234a>c (r = RNA) (small letters for the base change compared with capital letters for DNA).
- A mutation at the amino acid level would be indicated as p.T123Y (p = protein) (or p.Thr123Tyr), that is threonine-123 has been changed to tyrosine.
- A mutation referenced to genomic DNA would be indicated by GenBank Accession No.:g.12345678A>C (g = genomic DNA).
- The presence of two sequence variants in an allele are listed between square brackets separated by a semi-colon, e.g. [c.1234A>C; c.1344C>G].
- Two sequence variants in different alleles are listed between square brackets, separated by a '+', e.g. [c.76A>C]+[c.87delG].
- Two sequence changes with alleles unknown are listed between square brackets, separated by a '(+)', e.g. [c.76A>C (+) c.83G>C].
- Descriptions of sequence changes in different genes (e.g. for recessive diseases) are listed between brackets, separated by a '+' character, and include an identifier to the gene changed, e.g. [GJB2:c.76A>C]+[GJB:c.87delG].

2.3.2 Deletion (del)

A single base deletion is indicated by g.12345678delC. Multiple base deletions are indicated by a '_' underscore character that separates the first and last bases deleted, e.g., g.1234561_5delAGCT.

2.3.3 Insertions (ins)

A single base insertion is indicated by g.1234567_8insT, denoting that a T was inserted between nucleotides 1234567 and 1234568. Duplicating insertions (duplication or dup) are described as duplications, not as insertions; ACTTTGTGCC to ACTTTGTGGCC is described as c.8dupG and not as c.8_9insG. Alternatively, c.77_79dup (or c.77_79dupCTG) denotes that nt 77–79 were duplicated.

3. TROUBLESHOOTING

Sequence quality may be compromised by the quality of the templates, by the PCR and sequencing primer design, or by contaminants in the reagents. The list below summarizes the common artifacts seen in resequencing and their potential causes and solutions.

- **Cannot obtain any sequence data**
  - *Sequencing template may be absent or at low concentration.* Check the PCR product on a gel. No amplification may have occurred or the DNA may have been degraded. Larger amplicons or large vector DNA may require a substantially higher DNA concentration in the sequencing reaction.
  - *Priming site may not be present on the template.* Check the sequence of your sequencing primer with that of the PCR amplicon to ensure a perfect match.
  - *Template may contain inhibitors.* Remnants of phenol, alcohol, or ion-exchange resins used for purification of the template may inhibit the sequencing reaction. Excess RNA or polysaccharides are inhibitory (common in some plasmid preps). $CsCl_2$, EDTA, and KCl also cause inhibition.
  - *Sequencing reagents may have lost activity.* Inappropriate storage may have decreased the activity of the sequencing mixes. Use fresh lots.
- **High background noise in the sequencing results (many 'Ns' in the sequence)**
  - *Signal intensity may be too low due to reduced reagent concentration.* The data collection software 'corrects' for low signal by amplifying the signals, resulting in more noise and wrong base calls. Use more template and/or BigDye. Also check the sequencing primer concentration.
  - *Template may contain inhibitors.* Make sure that appropriate post-PCR purification methods are used to ensure removal of dNTPs, salts, etc.
  - *DNA may be degraded.* Degradation may be caused by nucleases, elimination of the heat-inactivation step after the ExoSAP-IT step, UV scission during agarose gel purification, or repeated freeze/thaw cycles of the PCR/plasmid template.
  - *Sequencing primer annealing temperature may be inappropriate for the sequencing reaction conditions.* The annealing temperature may be too high (nonspecific binding) or too low (inefficient binding).
  - *'Dye blobs' are seen early in the sequence trace.* Unincorporated dye-labeled ddNTPs have not been removed during the post-sequencing reaction clean-up.
  - *High spikes are seen at certain positions.* High-amplitude peaks that mask one to two bases usually indicate the presence of a bubble in the capillary. The capillary needs to be flushed with polymer before use.
  - *Peaks broaden progressively until the sequence is unreadable.* This is referred to as 'loss of resolution' and may be due to contaminants in the sample or in the capillary. Running sequencing standards should help indicate whether the effect is from a capillary. It may require extensive

flushing of the capillary with polymer or replacing capillaries that are at the end of their life.

- **Multiple peaks are seen**
  - ○ *Sequencing primer may be binding at multiple sites on the template.* If the sequencing primer matches an additional site perfectly, strong double peaks will be seen. If the match is weaker, then weaker secondary peaks will be seen. Check the primer and amplicon sequences for partial hybridization sites.
  - ○ *PCR primers may have amplified multiple regions of the genomic DNA.* Multiple bands may easily be visualized on agarose gels and will require gel purification of the required band. Co-amplification of pseudogenes or highly conserved regions may result in similar-sized bands that may not resolve on a gel. Redesign of PCR primers or an approach using nested PCR or sequencing primers may be required.
  - ○ *Presence of a 25–80 bp subsequence close to the start.* PCR primers can partially overlap with each other and be extended by PCR to form primer dimers. These appear as a <70 bp product on agarose gels. This requires optimization of PCR conditions or redesign of primers.
  - ○ *Carry-over of PCR primers or dNTPs into the sequencing reaction.* Incomplete removal of PCR primers during the post-PCR clean-up step will result in the PCR primers behaving as sequencing primers.
  - ○ *A secondary sequence is seen with the peaks displaced by n – 1. The secondary sequence may be very weak or may be up to 30% of the primary sequence.* A common problem due to primers containing $n – 1$ products. Inefficient oligonucleotide synthesis results in a population of primers with a variable length resulting in different 5′ ends but a common 3′ end. Use good-quality primers or purify primers by high-performance liquid chromatography.
  - ○ *Sequence quality deteriorates at or after a homopolymeric region.* The sequencing enzyme may not pass through homopolymeric regions. This may require design of a primer that is closer to the homopolymer region and/or may require the use of different sequencing chemistry specifically designed for sequencing regions with high levels of G or C. Similarly, for long stretches of A or T, an alternative chemistry (dRhodamine) may need to be tried.
  - ○ *Peaks appear to be compressed together.* This occurs as a result of the formation of strong secondary structures in some regions of the template. It may require additives like betaine, dimethyl sulfoxide, or glycerol in the sequencing mix to reduce the formation of secondary structures.
  - ○ *Strong overlapping peaks due to a frame-shift mutation.* Sequence quality deteriorates from the start of a heterozygous frame-shift mutation. Resequencing software specifically designed to detect frame shifts will produce correct base calls.
- **Sequence truncation**
  - ○ *Sequence may terminate at GC-rich regions.* Relocation of the sequencing primer to or beyond the GC-rich region may allow the remaining sequence

to be read. Addition of dimethyl sulfoxide or betaine to the sequencing mix may resolve the sequence.

○ *Template concentration may be too high.* Excess template will result in the depletion of dye-terminators in the first 100 bases or so (a top-heavy sequence trace). Reduce the concentration of the DNA template.

○ *Salt concentration is too high.* Processivity of the sequencing enzyme is impacted by salt concentration. Repurify the template to remove excess salts.

4. REFERENCES

1. Jobling MA, Hurles ME & Tyler-Smith C (2004) *Human Evolutionary Genetics: Origins, Peoples and Disease.* Garland Science, New York.
2. Hall L, Doerr KA, Wohlfiel SL & Roberts GD (2003) *J. Clin. Microbiol.* **41**, 1447–1453.
3. Wels M, Francke C, Kerkhoven R, Kleerebezem M & Siezen RJ (2006) *Nucleic Acids Res.* **34**, 1947–1958.
4. Torok HP, Glas J, Lohse P & Folwaczny C (2006) *Expert Opin. Pharmacother.* **7**, 1591–602.
★★ 5. den Dunnen JT (2004) *Nomenclature for the description of sequence variations.* Human Genome Variation Society. Available at: http://www.HGVS.org/mutnomen/
6. Futreal PA, Coin L, Marshall M, *et al.* (2004) *Nat. Rev. Cancer,* **4**, 177–183.
7. Ishii S, Nakao S, Minamikawa-Tachino R, Desnick RJ & Fan JQ (2002) *Am. J. Hum. Genet.* **70**, 994–1002.
8. Sanger F, Nicklen S & Coulson AR (1977) *Proc. Natl. Acad. Sci. U. S. A.* **74**, 5463–5467.
9. Sakharkar MK, Chow VT & Kangueane P (2004) *In Silico Biol.* **4**, 387–393.
10. Wallace AJ, Wu C-L& Elles RG (1999) *Genet. Test.* **3**, 173–183.

APPENDIX 1
List of suppliers

ABgene – www.abgene.com
Alexis Corporation – www.alexis-corp.com
Amersham Biosciences – www.amershambiosciences.com
Anachem Ltd – www.anachem.co.uk
Appleton Woods Ltd – www.appletonwoods.co.uk
Applied Biosystems – www.appliedbiosystems.com
Art Robbins Instruments – www.artrobbinsinstruments.com
AutoGen, Inc. – www.autogen.com
Axon Instruments – www.axon.com

Beckman Coulter, Inc. – www.beckman.com
Becton, Dickinson and Company – www.bd.com
Bio-Rad Laboratories, Inc. – www.bio-rad.com
Biotage Pyrosequencing – www.biotage.com
Biotecx / Cinna Laboratories – www.biotecx.com
BOC Group – www.boc.com
Brosch direct Ltd – www.broschdirect.com
BSD Robotics – www.bsdrobotics.com

Calbiochem – www.calbiochemicom
Cambridge Scientific Products – www.cambridgescientific.com
Carl Zeiss – www.zeiss.com
Chemicon International, Inc. – www.chemicon.com
Clontech Laboratories, Inc. – www.clontech.com
Corning, Inc. – www.corning.com

DakoCytomation – www.dakocytomation.com
Difco Laboratories – www.difco.com
Dionex Corporation – www.dionex.com
DNA Technology A/S – www.dna-technology.dk
DuPont – www.dupont.com

Elliot Scientific Ltd – www.elliotscientific.com
Eppendorf – www.eppendorf.com
European Collection of Animal Cell Culture – www.ecacc.org.uk
Eyela – www.eyelausa.com

Fermentas – www.fermentas.com
Findel Education Ltd – www.fipd.co.uk
Fiskars – www.fiskars.com
Fluka – www.sigma-aldrich.com
Fluorochem – www.fluorochem.co.uk

Gene Bridges GmbH – www.genebridges.com
GeneCodes – www.genecodes.com
Gibco BRL – www.invitrogen.com
Goodfellow Cambridge Ltd – www.goodfellow.com
Greiner Bio-One – www.gbo.com

Hamilton – www.hamiltoncompany.com
Harlan – www.harlan.com
Hybaid – www.hybaid.com
HyClone Laboratories – www.hyclone.com

ICN Biomedicals, Inc. – www.icnbiomed.com
Insight Biotechnology – www.insightbio.com
Invitrogen Corporation – www.invitrogen.com

Jencons-PLS – www.jencons.co.uk

Kendro Laboratory Products – www.kendro.com
Kodak: Eastman Fine Chemicals – www.eastman.com

Lab-Plant Ltd – www.labplant.com
Lancaster – www.lancastersynthesis.com
Lasergene – www.dnastar.com
Leica – www.leica.com
Life Technologies Inc. – www.lifetech.com
LOT-Oriel – www.lot-oriel.com

Merck, Sharp and Dohme – www.msd.com
MetaChem – www.metachem.com
Millipore Corporation – www.millipore.com
Miltenyi Biotec – www.miltenyibiotec.com
MJ Tetrad – www.biorad.com
MWG Biotech – www.mwg-biotech.com

National Diagnostics – www.nationaldiagnostics.com
New England BioLabs, Inc. – www.neb.com
Nikon Corporation – www.nikon.com
Olympus Corporation – www.olympus-global.com
Optivision Ltd – optivision.co.uk

Perbio Science – www.perbio.com
PerkinElmer, Inc. – www.perkinelmer.com
Pharmacia Biotech Europe – www.biochrom.co.uk
Photonic Solutions plc – www.psplc.com
Phred – www.phrap.org/phredphrapconsed.html
Promega Corporation – www.promega.com

Qiagen N.V. – www.qiagen.com

R&D Systems – www.rndsystems.com
Roche Diagnostics Corporation – www.roche-applied-science.com

Sanyo Gallenkamp – www.sanyogallenkamp.com
Sarstedt – www.sarstedt.com
Schleicher and Schuell Bioscience, Inc. – www.schleicher-schuell.com
Scientifica – www.scientifica.uk.com
Serotec – www.serotec.com
Shandon Scientific Ltd – www.shandon.com
Sigma-Aldrich Company Ltd – www.sigma-aldrich.com
Sorvall – www.sorvall.com
Stratagene Corporation – www.stratagene.com

Thames Restek – www.thamesrestek.co.uk
Thermo Fisher Scientific – www.thermofisher.com
Thistle Scientific – www.thistlescientific.co.uk

Vector Laboratories – www.vectorlabs.com
VWR International Ltd – www.bdh.com

Whatman International Ltd – www.whatman.com
Wolf Laboratories – www.wolflabs.co.uk

York Glassware Services Ltd – www.ygs.net

Index